In Silico Bees

In Silico Bees

EDITED BY
James Devillers

CRC Press
Taylor & Francis Group
Boca Raton London New York

CRC Press is an imprint of the
Taylor & Francis Group, an **informa** business

CRC Press
Taylor & Francis Group
6000 Broken Sound Parkway NW, Suite 300
Boca Raton, FL 33487-2742

First issued in paperback 2018

ISBN-13: 978-1-4665-1787-5 (hbk)
ISBN-13: 978-1-138-37470-6 (pbk)

Library of Congress Cataloging-in-Publication Data

In silico bees / edited by James Devillers.
     pages cm
   Includes bibliographical references and index.
   ISBN 978-1-4665-1787-5
  . Honeybee--Behavior--Mathematical models. 2. Honeybee--Effect of chemicals on--Mathematical models.. I. Devillers, James, 1956-

QL568.A6I57 2014
595.79'9--dc23   2013030987

Visit the Taylor & Francis Web site at
http://www.taylorandfrancis.com

and the CRC Press Web site at
http://www.crcpress.com

To the memory of Professor Davide Calamari (1941–2004),
a visionary in applied and computational ecotoxicology,
a man of deep humanity, and a great friend.

Contents

Acknowledgments

I am extremely grateful to the authors of the chapters for preparing valuable contributions. To ensure the scientific quality and clarity of the book, each chapter was sent to two referees for review. I would like to thank all the referees for their useful comments. Finally, I would like to thank the staff at CRC Press for making the publication of this book possible.

Contributors

Pierrick Aupinel
Unité expérimentale d'entomologie Le Magneraud
Institut National de la Recherche Agronomique
Surgères, France

Morgane Barbet-Massin
Département Ecologie et Gestion de la Biodiversité
Muséum National d'Histoire Naturelle
Paris, France

Kristen A. Baum
Department of Zoology
Oklahoma State University
Stillwater, Oklahoma

Andrew G. Birt
Department of Entomology
Texas A&M University
College Station, Texas

Robert N. Coulson
Department of Entomology
Texas A&M University
College Station, Texas

Axel Decourtye
Association de Coordination Technique Agricole
and
Unité Mixte Technologique Protection de l'Abeille dans l'Environnement
Avignon, France

Hugo Devillers
CTIS
Rillieux La Pape, France

James Devillers
CTIS
Rillieux La Pape, France

Carsten F. Dormann
Department of Computational Landscape Ecology
Helmholtz Centre for Environmental Research
Leipzig, Germany
and
Faculty of Forest and Environmental Sciences
University of Freiburg
Freiburg, Germany

Hermann J. Eberl
Department of Mathematics and Statistics
University of Guelph
Guelph, Ontario, Canada

James R. Edwards
School of Mathematics and Statistics
The University of Sydney
Sydney, New South Wales, Australia

Jeroen Everaars
Department of Computational Landscape Ecology
and
Department of Ecological Modelling
Helmholtz Centre for Environmental Research
Leipzig, Germany

Dominique Fortini
Unité expérimentale d'entomologie Le
   Magneraud
Institut National de la Recherche
   Agronomique
Surgères, France

Julie Fourrier
Association de Coordination Technique
   Agricole
Institut Claude Bourgelat—VetAgroSup
Marcy l'Etoile, France

Mark Hoogendoorn
Department of Computer
   Science
VU University Amsterdam
Amsterdam, the Netherlands

Frédéric Jiguet
Département Ecologie et Gestion de la
   Biodiversité
Muséum National d'Histoire Naturelle
Paris, France

Brian R. Johnson
Department of Entomology
University of California, Davis
Davis, California

Peter G. Kevan
School of Environmental Sciences
University of Guelph
Guelph, Ontario, Canada

Franck Muller
Département Systématique et
   Evolution
Muséum National d'Histoire Naturelle
Paris, France

Mary R. Myerscough
School of Mathematics and Statistics
The University of Sydney
Sydney, New South Wales, Australia

Adrien Perrard
Département Systématique et
   Evolution
Muséum National d'Histoire Naturelle
Paris, France

Vardayani Ratti
Department of Mathematics and
   Statistics
University of Guelph
Guelph, Ontario, Canada

Quentin Rome
Département Systématique et
   Evolution
Muséum National d'Histoire Naturelle
Paris, France

Timothy M. Schaerf
School of Biological Sciences
and
School of Mathematics and Statistics
The University of Sydney
Sydney, New South Wales, Australia

Martijn C. Schut
Department of Computer
   Science
VU University Amsterdam
Amsterdam, the Netherlands

Maria D. Tchakerian
Department of Entomology
Texas A&M University
College Station, Texas

Paolo Tremolada
Department of Biosciences
University of Milan
Milan, Italy

Jan Treur
Department of Computer
   Science
VU University Amsterdam
Amsterdam, the Netherlands

Marco Vighi
Department of Environmental and
   Landscape Sciences
University of Milano–Bicocca
Milan, Italy

Claire Villemant
Département Systématique et
   Evolution
Muséum National d'Histoire Naturelle
Paris, France

1 Automatic Systems for Capturing the Normal and Abnormal Behaviors of Honey Bees

James Devillers and Hugo Devillers

CONTENTS

ABSTRACT

Technologies for automated detection and recording of insect behavior and movements have evolved dramatically in the past decade. Because the flight activity of honey bees provides crucial information on their health conditions and a rough estimate of the state of the hive, tracking and recording the displacements of bees with automatic devices are of primary interest. This also allows us to better understand their mechanisms of orientation, their capacity to exploit their environment, and to better estimate the adverse effects of chemicals voluntarily or accidentally spilled in the aquatic and terrestrial ecosystems. In this context, an attempt is made to review all automatic systems that can be used to track honey bees. The problems linked to the analysis of recorded data are also discussed.

1

KEYWORDS

Automatic monitoring, Bee counter, Video camera, RFID tag, Harmonic radar, Radio transmitter

1.1 INTRODUCTION

It is widely appreciated that honey bee activity plays a key role in the agricultural economy worldwide. In addition to the production of honey, pollen, wax, propolis, and royal jelly, the honey bee constitutes an important, sometimes unique, pollinator of numerous crops [1–5], even if this bee–crop dependence has to be related to the evolution of agricultural practices in different countries [6]. The honey bee also plays an important functional role in most terrestrial ecosystems for maintaining the biodiversity of wild-plant communities [7–9].

Unfortunately, the continual expansion of industrial production and the growing use of pesticides in agriculture have led to an increase in the number and quantities of xenobiotics released into the environment. Thus, the amount of pesticides used worldwide was approximately 5.2 billion pounds both in 2006 and 2007, with the United States being the leader in development, production, and sales. Indeed, the United States accounted for 22% of the quantity used worldwide and 25%, 10%, and 14% of the amount of herbicide, insecticide, and fungicide used worldwide, respectively [10]. Analysis of US Customs Service shipping records for 2001–2003 also reveals that about 1.7 billion pounds of pesticides were exported from US ports at a rate of more than 32 tons/h. This includes nearly 28 million pounds of pesticides whose use is banned in the United States [11]. Between 2001 and 2003, more than 125 million pounds of pesticides were exported, which have been identified as *highly toxic* to bees. This represents a yearly average of 42 million pounds, or 2 tons/h. Exports of dimethoate accounted for about 65% of the 2001–2003 total [11]. Europe is also a major consumer of pesticides. In 2003, 219,771 tons of pesticides were used in the 25 countries of the European Union, including 107,574 tons of fungicides, 83,934 tons of herbicides, 21,404 tons of insecticides, and 6,859 tons of growth regulators. Five countries together accounted for nearly 75% of the total used in EU-25. France alone accounted for 28%, Spain and Italy 14% each, Germany 11%, and the United Kingdom 7% [12]. The situation is not much different as regards industrial chemicals. Thus, for example, the European Inventory of Existing Commercial Chemical Substances (EINECS) includes more than 100,000 substances that were deemed to be on the European Community market between 1971 and 1981 [13]. The potential adverse effects of these substances are often unknown, whereas they can contaminate the environment [14].

Foraging bees are directly affected by all these pollutants spilled voluntarily or accidentally into the environment. Depending on the chemical nature of the pollutants and the dose absorbed, foragers can die in the field, on the flight back to the hive, at the hive entrance, or a little later. Foragers can also contaminate hive inhabitants with poisoned nectar, pollen, and water brought back to the hive. It is well known that an acute poisoning event by pesticides or other chemicals is often detected by the presence of a large number of dead or dying bees at the hive entrance.

Unfortunately, sublethal effects are not detected so easily. Consequently, there is a need to evaluate the activity of the foragers for detecting abnormal behaviors.

Insect marking for scientific studies dates back to the beginning of the twentieth century, and all the methods in use have been comprehensively reviewed by Hagler and Jackson [15]. In this context, the aim of this chapter is to review automatic devices that allow us to evaluate normal and abnormal behaviors of foraging bees. Broadly speaking, there are systems that capture the overall activity of bees and those able to track them individually during their flight. Instead of cataloging the studies using these different systems, an attempt is made to stress their importance in ecotoxicology and ecology as well as their requirements in terms of data management, cleaning, and analysis.

1.2 MONITORING THE OVERALL ACTIVITY OF ADULT BEES

1.2.1 PHOTOELECTRIC SYSTEMS

When no significant mortality is observed in close proximity to the hive after insecticide treatment, it may be appropriate to quantify the bee in-and-out activity at the hive entrance for a few weeks in order to detect potential losses due to sublethal effects. Quantifying bee loss is also useful as a regular beekeeping practice for following hive activity. Several attempts have been made to design automatic counter devices for outgoing and incoming honey bees. The first prototypes were based on the recording of electrical impulses [16,17] or on the use of photoelectric signals [18]. Another system was designed by Chauvin [19,20] based on a mechanical setup that distinguished between outgoing and incoming flights, which was easily adapted to Dadant hives. Different initiatives have progressively improved the functioning and capacities of the photoelectric cell systems [21–30]. With time, the number of bees detected increased, the direction of bee movements was better discriminated, and the capacity of storage increased. The devices were more robust and adapted to outdoor conditions. Therefore, they were adequate for carrying out semifield or field trials to follow the daily activity of the colonies and detect bee losses due to predators, pathogens, or poisoning. The most interesting system currently commercialized is the BeeScan™ counter supplied by Lowland Electronics BVBA (Belgium). A basic description of this counter was published by Struye et al. [31]. More recent technical information can be obtained in the instruction manual [32]. Briefly, bee movement detection at the hive entrance is made through 32 bidirectional channels. The detection of a bee inside a channel is performed via an infrared diode exciting two photoreceptors. The counter is able to detect a distance of 1 mm between bees, which reduces the chance of miscounting them as they move head to tail through a channel. The counting period can be of 1, 5, 10, 15, 30, or 60 min. For the selected period, the "in," "out," and "out-in" (*i.e.*, difference between out and in) data and the corresponding cumulative values (*i.e.*, "IN," "OUT," and "OUT-IN") are recorded and stored. The memory can store 61 files, each corresponding to 24 h of recording. A BeeScan counter can be used between 0°C and 55°C and with a relative humidity of 0%–100% [32]. It is suited to study the nycthemeral activity of adult bees during the recorded time. Thus, for example, Figure 1.1a shows bee activity at the hive

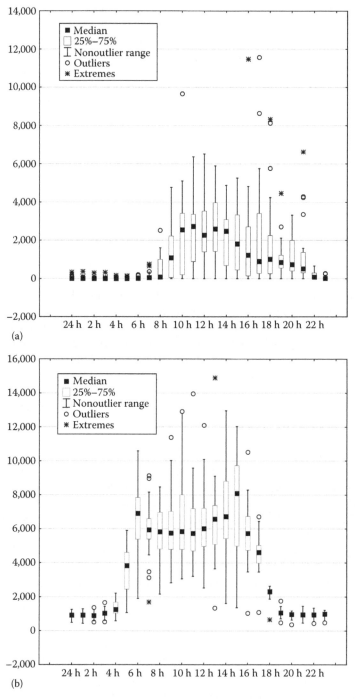

FIGURE 1.1 Nycthemeral activity of adult bees in (a) Montpellier and (b) Martinique recorded from a BeeScan counter.

entrance each hour of the day in July during an experiment performed in Montpellier (France), which has a Mediterranean climate. Bee activity began significantly around 8 a.m., then increased regularly until the beginning of the afternoon, followed by a decrease with a small peak at the end of the afternoon. In this hive, the activity during the evening and at night was near null. Figure 1.1b also shows the movements at the hive entrance during a cycle of 24 h, but the experiment was performed in Martinique Island (French Indies), which has a tropical climate. It was anticipated that the activity started and ended early, but more surprising was the discovery that the activity rapidly increased to reach a maximum and then a kind of plateau appeared till noon. Then an increase of activity was observed till a maximum around 3 p.m., followed by a quick decrease with very limited activity from 7 p.m. to 4 a.m. It is noteworthy that the nycthemeral activities shown in Figure 1.1 correspond to two hives, but during the two experiments, other hives, were equipped with BeeScan counters and showed similar patterns. From a practical point of view, knowledge of the nycthemeral cycle of bees in relation to local climatic characteristics allows us to select an optimal time to apply insecticide treatment in agriculture or in vector control. Obviously, this will be done when a minimum number of bees are outside the hive.

BeeScan counters are also particularly suited to study the influence of abiotic factors on bee activity. Thus, Figure 1.2 shows the daily activity of a hive from February 17 to March 11, 2012. There were strong rains on March 3, and as a

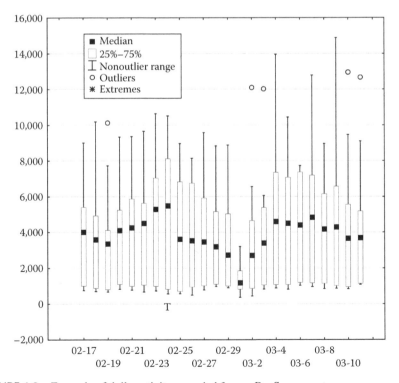

FIGURE 1.2   Example of daily activity recorded from a BeeScan counter.

result, bee activity was very limited. Obviously, this rather trivial example is only for illustration. To study the specific influence of the different abiotic factors on bee activity, it is necessary to take advantage of the numerous possibilities offered by the multivariate statistical analyses [33]. Devillers et al. [34] studied the influence of various abiotic factors on the daily activity of bees. Six hives equipped with BeeScan counters (BA–BF) and three hives with pollen traps (P1–P3) were installed near sunflower fields (*Helianthus annuus* L., cultivar *prodisol*) treated at the seed stage with Aventis Regent TS (500 g/L of fipronil) and located in the agricultural district of Birieux (France). Counter data were collected during 53 consecutive days from about mid-July to the beginning of September. Because the climatic data were only measured hourly, bee-counter data were converted to the same timescale. Queens in two hives equipped with counters were changed during the experiments. Consequently, these were excluded. This allowed us to design 53 d × 24 h data matrices of activity for three different hives and a 50 d × 24 h data matrix for the fourth hive because the data were not recorded for three consecutive days. Double paired data matrices with different climatic data were also designed. To pinpoint the relative importance of each abiotic factor on bee activity, co-inertia analysis (CIA) [35,36] was used. Briefly, CIA can be viewed as a generic multivariate approach used to find a costructure between two data tables (Figure 1.3). These two data matrices are first considered independently. They can be analyzed by means of different multivariate approaches such as principal components analysis (PCA) [37], correspondence factor analysis, or multiple correspondence factor analysis [38], depending on the nature of the data. These separate analyses emphasize the basic structure of the counter data matrix and the abiotic data matrices. Thus, for example, Figure 1.4 shows the PCA of the counter data obtained with the BB hive. The PC_1PC_2 map accounts for 79% of the total inertia of the system. Trajectories between objects (Figure 1.4a) and between variables (Figure 1.4b) were drawn to facilitate the interpretation of the two corresponding spaces. Figure 1.4a reveals three main clusters broadly related to the different months of the experiments. Data recorded in July are located in the right part of Figure 1.4a. The middle part is mostly occupied by the data recorded in August, while the left part deals with data collected at the end of August and at the beginning of September. These clusters correspond to a decreasing gradient of flying activity from right to left of Figure 1.4a. For example, the total number of outgoing bees on July 20 was 207,314 (point #9) while that recorded on September 4 was only 55,723 (point #53). Interestingly, some points in Figure 1.4a do not follow these rules. Thus, for example, point #4 (July 15) and point #7 (July 18) have an atypical location in PC_1PC_2. This was due to weather disturbances. Indeed, July 15 was cloudy and windy and July 18 was as rainy as August 19 (point #37). Figure 1.4b displays the classical nycthemeral activity of honey bees. However, PCA allows us to extract less obvious information on the flying behavior of bees. Indeed, the particular shape of the trajectory in Figure 1.4b is due to the fact that in the first part of July, the bees left the hive very early in the morning and stopped their gathering activity only in the evening. Conversely, in the beginning of September, they only worked significantly around 10 a.m. and, in addition, stopped fairly early in the afternoon. This pattern is graphically illustrated in Figure 1.4a for points 9 (right) and 53 (left), where the greater

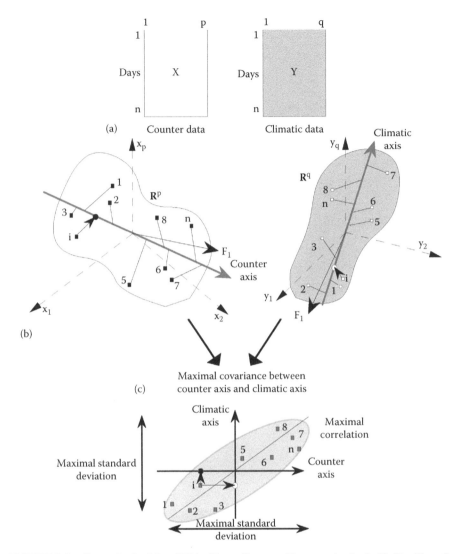

FIGURE 1.3 General principle of CIA. (From *Comput. Electron. Agric. J.*, 42, Devillers, J., Doré, J.C., Tisseur, M., Cluzeau, S., and Maurin, G., Modelling the flight activity of *Apis mellifera* at the hive entrance, 87–109, Copyright [2004], with permission from Elsevier.)

the length of a clockwise line, the higher the number of outgoing bees. The corresponding time is also represented. Similarly, a PCA was performed on each data matrix of abiotic factor (results not shown). In the second step of the CIA (Figure 1.3), a matching analysis is performed in order to detect a costructure between the two data matrices. This second analysis is based on the search of co-inertia axes, maximizing the covariance between the coordinates of the projections of the rows of each data table. Figure 1.5 summarizes the results obtained with CIA of the counter data of the BB matrix×temperature data matrix. A detailed analysis of this

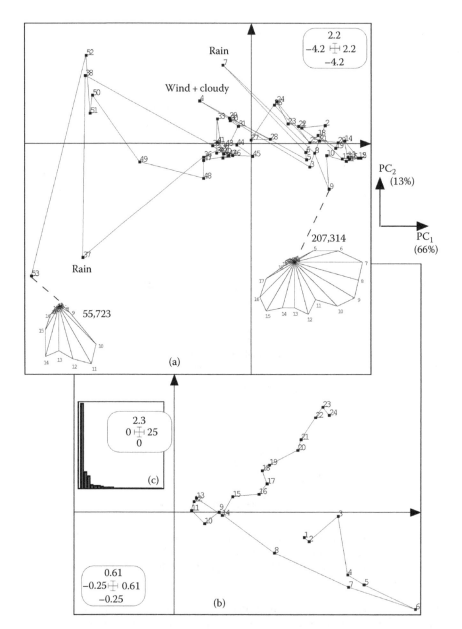

FIGURE 1.4 Pattern resulting from the PCA of the bee-counter data recorded at BB hive entrance. (a) Days, (b) hours, and (c) histogram of the eigenvalues. (From *Comput. Electron. Agric. J.*, 42, Devillers, J., Doré, J.C., Tisseur, M., Cluzeau, S., and Maurin, G., Modelling the flight activity of *Apis mellifera* at the hive entrance, 87–109, Copyright [2004], with permission from Elsevier.)

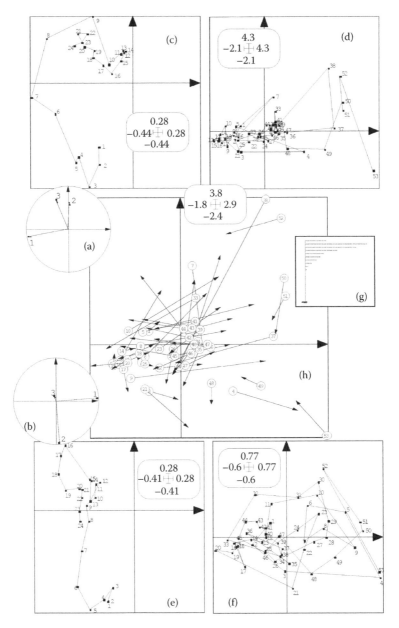

FIGURE 1.5 CIA between the bee counter of the BB hive and temperature data matrices. Projection of the axes resulting from each separate analysis on co-inertia axes for (a) the bee-counter data matrix and (b) the temperature data matrix. Projection of the 53 plots by 24 components array via CIA for bee-counter data matrix (c and d) and temperature data matrix (e and f). (g) Result of the permutation test. (h) Matching graph. (From *Comput. Electron. Agric. J.*, 42, Devillers, J., Doré, J.C., Tisseur, M., Cluzeau, S., and Maurin, G., Modelling the flight activity of *Apis mellifera* at the hive entrance, 87–109, Copyright [2004], with permission from Elsevier.)

figure is beyond the scope of this chapter and can be found in Devillers et al. [34]. Of prime importance is Figure 1.5g, which displays the results of a permutation test based on Monte Carlo simulations during which 1,000 CIAs using matrices with random permutations were performed. It shows that there exists a costructure between the two data matrices. Figure 1.5h shows the simultaneous graphical display obtained at the end of the matching process of CIA (Figure 1.3, bottom). The identification numbers 1–53, in small circles, stand for the different days of experiment and the length of the arrows visualizes the degree of costructure between the two matrices under study (*i.e.*, bee activity/temperature). The lower the length of the arrows, the higher the costructure between the two matrices. The lengthy arrows in Figure 1.5h can be easily explained. Thus, for example, when it rained, especially in July and for a short period of time, the temperature did not significantly decrease but the bee activity outside the hive ceased. Nevertheless, Figure 1.5h clearly shows that a strong costructure exists between the two data tables. It is noteworthy that Devillers et al. [34] used BeeScan counter data to derive quantitative models from the abiotic data by using partial least squares (PLS) analysis [39] and a three-layer perceptron [40].

Decourtye et al. [41] compared the sublethal effects of imidacloprid (24 μg/kg) and deltamethrin (500 μg/kg) on bees in both semifield and laboratory conditions. The main goal of the study was to compare the results obtained with restrained worker bees in the conditioned proboscis extension response (PER) laboratory assay [42] with those recorded by using free-flying foragers visiting an artificial flower feeder in an outdoor flight cage and undergoing an olfactory conditioning procedure. A BeeScan counter was used to measure the activity at the hive entrance. After feeding with sucrose solution contaminated with imidacloprid at the concentration of 24 μg/kg, negative effects on the learning performances were observed in foraging bees subjected to both olfactory conditioning procedures. In addition, the neonicotinoid insecticide induced a reduction of activity at the hive entrance. Conversely, deltamethrin 500 μg/kg did not impact the learning performance of bees. In addition, although a clear reduction of forager visits on the artificial flower feeder was observed, no changes in activity at the hive entrance were noted during the delivery of a food solution added with deltamethrin [41]. Recently, BeeScan counters were also used to assess the potential impact of vector control treatments with deltamethrin on honey bee colonies in Martinique [43].

1.2.2   VIDEO MONITORING

The use of video to study bee activity within and outside the hive is relatively recent. This relies on the availability, at affordable cost, of digital cameras and high-performance computing devices. Video has found applications in the analysis of bee movements within the hive, especially dances [44]. Bees perform two types of dances for communicating the location of food sites. When the site is within a radius of about 50 m from the colony, the bee performs a round dance, which consists of a series of alternating left-hand and right-hand loops. When the site is considerably far from the nest (>100 m), the bee performs the so-called waggle dance. In this dance, the transition between one loop and the next is punctuated by a phase during which the bee waggles her abdomen from side to side while moving in a more-or-less straight line. Thus, the bee executes a left-hand loop, performs a waggle, executes a right-hand loop, performs a waggle, executes a left-hand loop, and so on. During the waggle phase, the abdomen moves from side to side at an approximate frequency of 12–15 Hz. The longer the duration of the waggle phase, the greater the distance of the food source. Information about the direction of the food source and its attractiveness is also conveyed in the waggle dance [45–48].

Experimenters can manually study the videos of bee dances and annotate the various sequences of motions. Although the task is simple and does not require specific expertise, it is time consuming, tiring, and prone to errors. Recently, efforts into automating such tasks became evident with the advances made in vision-based tracking systems. Feldman and Balch [49] proposed a system to automatically analyze bee movements, and label it, on the basis of examples. First, a video camera records bees in the observation hive. This video is passed to a tracker, which extracts coordinate information to be used by the human labeler (creating the training set) and then by the k-nearest neighbor (KNN) classifier. KNN is a good clustering method, but it cannot account for time series information. To overcome this problem, a hidden Markov model (HMM) was used. Once the HMM was specified, the Viterbi algorithm was used to determine the most likely state sequence for a given observation sequence. Their system was able to label movements with 81.5% accuracy in a fraction of the time it would take a human [49]. Veeraraghavan et al. [48] proposed general principles that could help in simultaneous automatic tracking and behavior analysis with applications in tracking bees and recognizing specific behaviors exhibited by them. They suggested a joint tracking and behavior analysis instead of the traditional *track and then recognize* approach for activity analysis. They proposed the use of hierarchical Markov models to simulate various behaviors and presented methods for detecting and characterizing abnormal behaviors. They showed how it was possible to simultaneously track and analyze bee dances in their hive [48]. Campbell et al. [50] used a video camera positioned over the unmodified entrance of a standard hive for measuring the flight activity of bees. They distinguished four different behaviors, namely, loitering, crawling, flying out, and flying in, which were modeled. The task of tracking bees was treated from frame-to-frame as an assignment problem and solved using maximum weighted bipartite graph matching. The training dataset for the motion models consisted of 600 frames annotated with the motion type and frame-to-frame change in position and orientation of every visible bee. They annotated 1,800 additional frames with ground truth of arrivals, departures, and number of visible bees in each frame. The counter was able to detect bees with a precision of 94%. It overcounted arrival by 2% and undercounted departures by 7% on the annotated data set [50]. Chen et al. [51] fixed a charge-coupled device (CCD) camera capable of sensing infrared light at the entrance of a hive for monitoring the in-and-out activity of honey bees. Through the stable light provided by an infrared LED, the camera was able to acquire images with consistent light intensity and send them to the computer via a USB interface. The computer was also used to identify the tags on the backs of the bees and record the conditions of the entries to and exits from the hive. Tag identification was made from a model derived from a support vector machine [52]. The system was tested on

100 tagged bees during 15 days. The size of the colony was about 12,000 honey bees. On the first day, 82 honey bees were detected. On the second day, 59 honey bees were detected, which is a 31% reduction. Similarly, 40 honey bees were detected on the third day, which represents a 32% reduction [51]. Colin et al. [53] used a video camera to quantify the foraging activity of small colonies of bees confined in two semicylindrical insect-proof tunnels after a contamination by the neonicotinoid insecticide imidacloprid at 6 µg/kg or by the phenylpyrazole insecticide fipronil at 2 µg/kg. Contaminations were made via sucrose solutions filling a feeder. A control sugar solution was offered to all the colonies on the first day (day 0). During the four consecutive days, a contaminated sugar solution was given to three nuclei, while the other remained with a noncontaminated solution. The number of bees on the feeder was counted every 3 min using data from the video film. Bees sucking the sucrose solution were considered as active and the others as inactive. The first criterion selected by the authors [53] was the number of active bees relative to time and was called *attendance*. The second criterion was the ratio of number of inactive (I) to active (A) bees. Colin et al. [53] showed that imidacloprid at 6 µg/kg did not affect attendance, but a significant increase in the I/A ratio on day 4 was observed. Fipronil at 2 µg/kg showed a strong effect on bee attendance and also induced a drastic decrease in the number of foragers coupled with an increase in the number of inactive bees at the feeder [53].

1.3 MONITORING THE ACTIVITY OF INDIVIDUAL BEES

1.3.1 USE OF RADIOFREQUENCY IDENTIFICATION

Radiofrequency identification (RFID) systems are noncontact identification devices now commonly used in numerous domains, including the study of insect behaviors [54]. Indeed, tracking the movements of insects in their natural habitat is essential to understand their biology, ethology, and behavior when faced with pollutants. RFID systems consist of two components, which are the transponder or tag, attached to the object that needs to be identified, and the reader, used to detect the information stored on the transponder. Figure 1.6 shows a bee with a transponder glued to the back side of her thorax in such a way that the movements of the wings cannot be hampered. Decourtye et al. [55] developed a method under tunnel to automatically record the displacements of foragers individualized with RFID tags and to detect the alteration of the flight pattern between an artificial feeder and the hive after contamination with fipronil. RFID tags (mic3®-TAG 64-bit RO, iID2000, 13.56 MHz system, 1.0 mm × 1.6 mm × 0.5 mm; Microsensys GmbH, Erfurt, Germany) weighing about 3 mg were used. This represents about 3% of the weight of a bee and is considerably low, especially if we consider that she is able to carry up to 70 mg of nectar [56] and 10 mg of pollen [57]. Five readers were placed at the entrance of the hive and artificial feeder. A software called BeeReader (Tag Tracing Solution, Valence, France) recorded the identification codes and the exact time of the events in files having a .txt format. Data cleaning and analysis were made with TimeBee® (CTIS, Rillieux-La-Pape, France). Technical grade fipronil (98% pure) was tested at sublethal doses of 0.06 and 0.3 ng/bee. Stock solutions were prepared in ethanol.

FIGURE 1.6 **(See color insert.)** Bee with an RFID tag glued to the back side of her thorax. (Reproduced with kind permission of Axel Decourtye, ACTA.)

Fipronil was added to a sucrose solution (50% w/w). The final concentration of ethanol in the sucrose solution was 1% (v/v). After a 3 h starvation period, each group of tagged foragers received a volume of contaminated or control sugar solution. Decourtye et al. [55] showed that the percentage of disappearing bees between the feeder and the hive was similar for all treatments. For the overall period, the percentage of bees not returning to the hive was between 24% and 44%. Conversely, fipronil at the dose of 0.3 ng/bee induced an increase in the flight time between the artificial feeder and the hive in comparison to the control group and the group treated with 0.06 ng/bee. Bees treated with 0.3 ng/bee showed the lowest number of foraging trips within the first 24 h, but after this period, the number of trips was not different from that of the other groups. The next step was to use the RFID systems under field conditions. This was done by Henry et al. [58] in a study on thiamethoxam, a recently marketed neonicotinoid systemic insecticide currently used to protect oilseed rape, maize, and other blooming crops foraged by honey bees. These authors proceeded in two steps. First, they assessed mortality induced by homing failure (m_{hf}) in exposed foragers. The study was conducted in an intensive cereal farming system of western France and in a suburban area in Avignon, southern France. To simulate daily intoxication events, foragers received a sublethal dose of 1.34 ng of thiamethoxam in a 20 mL sucrose solution and were released away from their colony with an RFID tag glued on each bee's thorax. In a second step, they assessed the extent to which m_{hf}, in combination with natural forager mortality, might upset colony dynamics. For that purpose, m_{hf} was introduced into a model of honey bee population dynamics [59]. Henry et al. [58] revealed that exposure of foragers to nonlethal but commonly found doses of thiamethoxam could affect forager survival, with potential contributions to collapse risk. Furthermore, the extent to which exposures affected forager survival seemed to be dependent on the landscape context and the prior knowledge of foragers about this landscape. Higher risks were observed when the homing task was more challenging [59]. Schneider et al. [60] also used RFID tags to study the potential effects of two other neonicotinoid insecticides, namely, imidacloprid and clothianidin, on the foraging activity of *Apis mellifera*. In the trials conducted with imidacloprid, all bees of the control groups and the groups treated with doses of 0.15 and 1.5 ng/bee returned to the hive after posttreatment release at the feeder, 95% returned when treated at 3 ng/bee, but only 25% of the bees returned after administration of 6 ng of imidacloprid/bee. Among the bees treated with 3 and 6 ng of imidacloprid/bee that were not directly flying to the hive, the authors [60] observed reduced mobility, followed by a phase of motionlessness with occasional trembling and cleaning movements. The number of feeder visits for bees treated with 1.5 and 3 ng of imidacloprid compared to control bees during the 3 h observation period after administration was reduced by 47% and 98%, respectively. The administration of 1.5 and 3 ng of imidacloprid significantly prolonged the median duration needed for a single foraging trip by 50% and 130%, respectively, during the first 3 h observation period. The median flight time to the feeder was also prolonged by 64.7% and 241.1%, the time spent at the feeder by 27.5% and 45.6%, and the median flight time from the feeder to the hive by 20% and 210%, respectively. In the trials conducted with clothianidin, all bees treated with 0 and 0.05 ng/bee returned to the hive. At 0.5 ng/bee, 94.4% of them returned to the hive during a 3 h observation period immediately after treatment. Among the bees treated with 1 ng, only 73.8% returned to the hive and only 20.6% returned after the uptake of 2 ng/bee. Abnormal reactions were observed after the release at the feeder site following the administration of 1 and 2 ng/bee of clothianidin. Bees were moving around with an awkwardly arched abdomen, sometimes followed by a phase of turning upside down and lying on the back with paddling leg movements [60]. During the 3 h observation period after administration of 0.5 and 1 ng of clothianidin, the number of feeder visits per bee compared to the control group was reduced by 31% and 71%, respectively. The administration of 0.5, 1, and 2 ng of clothianidin/bee significantly prolonged the median duration of a foraging trip by 20%, 32.2%, and 109.3%, the time spent at the feeder by 14.1%, 39.9%, and 101.8%, the median flight time back to the hive by 30%, 40%, and 90%, and the median time interval of an in-hive stay between foraging trips by 15.8%, 36.7%, and 95.9%, respectively. For both insecticides, the adverse effects could persist after 24 h only for the highest doses.

Parasites are known to induce changes in the normal behavior of their hosts. Thus, the microsporidia *Nosema apis* and *Nosema ceranae* infect the epithelium of the ventriculum (midgut) of honey bees, inducing changes in their physiology and behavior. The former type of modifications relies mainly on the alteration of digestive functions [61]. Reduction in the development of the hypopharyngeal glands was also observed, leading to a more rapid switch from nursing to other tasks in infected bees [62,63]. A lifespan reduction was observed in *Nosema*-infected bees [64] as well as alterations of flight, orientation, and homing behavior [65]. Interestingly,

Schneider [66] used RFID tags in a 2-year experiment to monitor the potential behavioral changes of honey bees experimentally infected by *Nosema* spores. A reduction of 14% and then of about 62% in the lifespan was observed in contaminated bees. Effects on the time spent outside the hive were also studied. When comparing the time duration that infected and noninfected bees spent outside their hive during the different weeks of their life, a significantly longer time of absence was found for infected bees during weeks three to five in the first year. The duration was 1.6, 1.74, and 1.54 times longer compared to noninfected bees, respectively. Similar effects were not detected in the second year of the experiments [66]. During the sixth and eighth week of life, longer foraging trip durations were recorded for inoculated bees. A prolongation factor of 3.3 and 2.3, respectively, was observed compared to noninfected bees. However, this difference was statistically not significant. A non-significant dependence was observed between increasing foraging activity and longevity in contaminated bees [66].

Obviously, RFID tags have also found application in the study of physiology, ecology, and ethology of *Apis mellifera*. Schneider [66] studied the behavioral changes in adult honey bees being incubated at different temperatures during their pupal development stage. The RFID method was used to monitor groups of bees reared at 32°C, 35°C, or 36°C over their lifetime. Immediately after hatching, these bees were labeled with RFID transponders and were introduced into a foster registration colony for lifelong monitoring of activity, flight durations, and lifespan. Schneider [66] observed that bees reared at above-optimal temperature of 36°C were overly active and spent the most time outside of the hive compared to the two other tested groups of bees (*i.e.*, 32°C and 35°C). Bees reared at an optimal temperature of 35°C lived longer than those reared at 36°C and the 32°C [66]. Pahl et al. [67] used RFID tags to estimate the homing capability of honey bees. They showed that most of the tested bees found their way back home even after blind displacement to unfamiliar areas, some of them from up to 11 km.

To conclude, RFID tags are very interesting devices that allow us to obtain a lot of information on bees during their normal activity or when they are faced with environmental pollutants. Figures 1.7 and 1.8 illustrate what kind of information can be easily extracted from the analysis of recorded data. Obviously, to be of interest and representative, the number of tagged bees has to be at a minimum of some hundreds.

It is worthy to note that RFID methodology has also found application in the study of ecology and ethology of bumblebees. Ohashi et al. [68] developed an experimental system for producing computer records of multiple foragers harvesting from simple artificial flowers with known rates of nectar secretion, using RFID tags to identify individual animals. However, the study is of limited interest because it was only performed with two tagged *Bombus impatiens*. Much more interesting is the work by Stelzer and Chittka [69], who studied the foraging rhythms of bumblebee colonies in northern Finland during the summer when the sun stays above the horizon for weeks. They showed that *Bombus terrestris* workers did not use the available 24 h foraging period but exhibited robust diurnal rhythms instead. Foraging activity took place mainly between 8 and 23 h, with reduced or no activity during the rest of the day. Activity levels increased steadily during the morning, reached a maximum around midday, and decreased again during late afternoon and early evening. It was

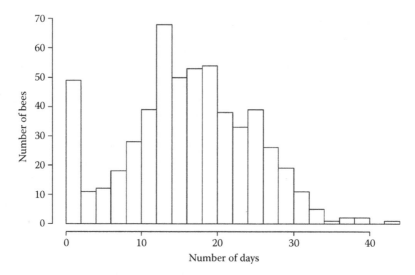

FIGURE 1.7   Distribution of survival times of tagged bees during the experiment.

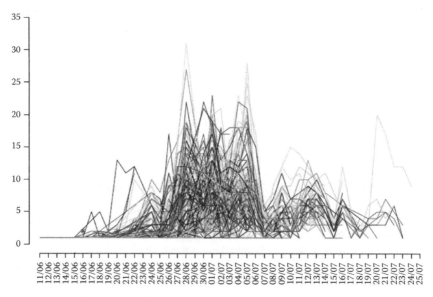

FIGURE 1.8   Number of outgoing trips per tagged bee and per day of experiment. Stars indicate minor problems in data recording.

previously shown that under controlled laboratory conditions, foragers showed free-running circadian rhythms both in constant light and darkness [70]. Foraging patterns at the nest entrance of native *B. pascuorum*, the native bumblebee, followed the same temporal organization, with the foraging activity being restricted to the period between 6 and 22 h [69].

1.3.2 OTHER TECHNIQUES

Attempts have been made to use bar codes to detect the individual behavior of honey bees [71]. However, although the method is cheap, it is not widely used because the reading of the codes is impossible when the bee is not perpendicular to the laser scanning reader or when the bar code is not clean enough [44]. Conversely, while the harmonic radars are expensive, they are increasingly used due to their ability to track in real time the flying activity of insects [44]. Riley et al. [72] were the first to test such a technology on honey bees. Figure 1.9 shows a bee equipped with a transponder for radar detection. Capaldi et al. [73] used harmonic radar to study the ontogeny of orientation flight of honey bees in a flat region of farmland (700×900 m) planted with cereal crops and surrounded by flowering crops, hedges, wooded areas, and buildings. Transponders weighing 0.8 and 12 mg were attached to bees as they departed from the hive, and orientation flight tracks were recorded by the radar. Of the 53 radio-tagged bees, the authors analyzed 29 full orientation flight sequences taken by bees of different ages (3–27, median = 6 days), with various degrees of experience (1–17 orientation flights). For comparison purposes, they also tracked flights (n = 15) taken by experienced foragers of unknown age from the same hive [73]. Capaldi et al. [73] demonstrated that with increased experience, honey bees hold trip duration constant but fly faster. Consequently, later trips covered a larger area than the earlier ones. Each flight was typically restricted to a narrow sector around the hive. Orientation flights provide bees with

FIGURE 1.9 A honey bee carrying a transponder for radar detection. (From *Comput. Electron. Agric.*, 35, Riley, J.R. and Smith, A.D., Design considerations for an harmonic radar to investigate the flight of insects at low altitude, 151–169, Copyright [2002], with permission from Elsevier.)

repeated opportunities to view the hive and its environment from different view-points, suggesting that they progressively learn the local landscape. The authors showed that changes in orientation flight were related to previous flights rather than chronological age, suggesting a learning process adapted to changes in the abiotic and biotic factors [73].

Harmonic radars were also used for capturing the activity of other Hymenoptera. Thus, Osborne et al. [74] used a harmonic radar (vertical dipole aerial 16 mm long, 12 mg) to study the foraging behavior of bumblebees (*Bombus terrestris*). The authors showed that outward tracks had a mean length of about 275 m with a range of 70–631 m and were often to forage destinations beyond the nearest available forage. Most bumblebees were constant to compass bearing and destination over successive trips. The bees' ground speeds ranged from 3 to 15.7 m/s in various wind conditions [74]. Lihoreau et al. [75] investigated the acquisition of long-distance traplines by foraging bumblebees (*B. terrestris*) with known foraging experience by first training them to forage on five artificial flowers arranged in a regular pentagon (50 m side length) within a field, then by tracking their complete flight paths with harmonic radar, and finally recording all their flower visits with motion-sensitive cameras. Stable traplines that linked together all the flowers in an optimal sequence were typically established after 26 foraging bouts, during which time only about 20 of the 120 possible routes were tried by the bumblebee. Radar tracking of selected flights revealed a strong decrease by 80% (ca. 1,500 m) of the total travel distance between the first and the last foraging bout. When the spatial configuration of flowers was modified, the bees engaged in localized search flights to find new flowers. Lihoreau et al. [75,76] suggested that bumblebees developed their traplines through trial and error by combining exploration, learning, and sequential optimization. It is notewor-thy that information on the use of harmonic radars to study the flying behavior of insects, especially honey bees and bumblebees, can be found in Riley and Smith [77] and Riley et al. [78].

The carpenter bee (*Xylocopa flavorufa*) is one of the main pollinators of cow-pea (*Vigna unguiculata*) in Kenya. This large solitary bee flies very quickly, ruling out most of the conventional techniques used to study its foraging flights [79]. Its nests are burrowed in dead branches located at the top of trees found in wooded landscapes. This excluded the use of harmonic radars, which only allow us to fol-low individuals in open habitat where radar beams are not blocked by vegetation and where bees do not fly outside the reference areas. To overcome this problem, Pasquet et al. [79] used small radio transmitters (0.35 g = a third of their average weight, antenna = 14 cm) to assess both the distance between the bee's nest and one of its foraging locations and how many places were visited during a single foraging trip. The range of the transmitters (around 200 m at soil level) and the large potential target area (near 300 km^2) requested the use of an aircraft for locating the transmit-ters. Experiments were made during the Kaskazi and Kusi seasons showing different climatic conditions. Pasquet et al. [79] found that carpenter bees were able to forage up to 6 km from their nest. Bees flew significantly shorter distances during poor weather conditions, even with roughly similar flower densities. Long flights were correlated with the presence of wind. The results obtained by Pasquet et al. [79] suggested *X. flavorufa* had the potential to move pollen over several kilometers and, hence, between domesticated and wild cowpea populations, leading to problems if the former population was genetically modified.

Neotropical orchid bees (Euglossini) can travel very large distances to collect fragrances, food, or nesting material. Wikelski et al. [80] used radio transmitters (Sparrow Systems, Fisher, IL, 300 mg, 2 radio pulses/s, 378 MHz, antenna length 42 mm) to track the movements of males of *Exaerete frontalis* in a complex and forested environment. This cleptoparasitic euglossine species was selected due to its abundance, ecology, and also large size (about 600 mg), which allowed the fixation of a relatively heavy transmitter. Of the 16 radio tagged bees, four were lost. The remaining bees were followed by ground surveys and helicopter tracking overflights. The results showed that bees habitually used large rainforest areas (at least 42–115 ha) on a daily basis. Aerial telemetry located individuals up to 5 km away from their core areas, and bees were often stationary, for variable periods, between flights to successive localities [80].

Recently, Hagen et al. [81] used micro radio telemetry (Advanced Telemetry Systems, Isanti, MN, Series A2405, 200 mg, antenna length shortened to ca. 3 cm) to track flight distances and space use of bumblebees. Using ground surveys and Cessna overflights in a Central European rural landscape mosaic, they obtained maximum flight distances of 2.5, 1.9, and 1.3 km for *Bombus terrestris* (workers), *Bombus ruderatus* (workers), and *Bombus hortorum* (young queens), respectively. Bumblebee individuals used large areas (0.25–43.53 ha) within one or a few days. Habitat analyses of one *B. hortorum* queen at the landscape scale revealed that gardens within villages were used more often than expected from habitat availability. Detailed movement trajectories of this individual revealed that prominent landscape structures (*e.g.*, trees) and flower patches were repeatedly visited. However, Hagen et al. [81] also observed long (*i.e.*, >45 min) resting periods between flights (*B. hortorum*) and differences in flower handling between bumblebees with and without transmitters (*B. terrestris*), suggesting that the weight of transmitters was too high, leading to a significant energetic cost.

1.4 CONCLUDING REMARKS

Different technologies are available to tag and track honey bees and to automatically record their activity. Although a fundamental difference has to be made between theoretical studies that attempt to show the applicability potential of a new technology and a field study that aims to respond to a specific problem, such as the potential adverse effects of an insecticide, choosing the best technique for marking and tracking bees relies on the goal, duration, and constraints of the study, cost-effectiveness ratio, and training experience, to only cite the most influential factors. If we consider the other hymenopterans discussed in this chapter, deciding which marking/tracking technique is best also includes considering their size, life stage, sex, ecology, and ethology.

Independent of these crucial aspects, tracking bees with these new devices implies that data are sequentially or continuously recorded and, hence, have to be analyzed. Data analysis often starts by a phase of cleaning, especially when numerous bees have been tagged and tracked. In practice, different situations can occur. If bee activity is recorded over a long period of time, some bees can lose their tag, or a problem could occur in recording data (*e.g.*, flattened battery). A cleaning phase also exists with systems recording the whole behavior of bees at the hive entrance. The most common approach to deal with missing data is to delete cases containing missing observations. Unfortunately, this always leads to a loss of information that can impact the interpretation of the results, especially if the number of tagged bees is limited. There exist methods to handle missing data (see, *e.g.*, [82–86]), and it would be highly beneficial to use them. On the contrary, the goal of the study may require that the recorded data be censored. Thus, for example, the use of RFID tags to study the flying behavior of honey bees requires deleting the recorded data corresponding to the time when the bee stays at the hive entrance and on the artificial feeder, if any.

Regression analyses are commonly used for discovering relationships between the recorded data and various predictors judged influential. The goal seems to be to obtain the highest possible correlation coefficient in order to orientate the ecological interpretation of the results. Unfortunately, very often, this is done without verifying whether the approach used was appropriate or not [87–89]. On the other hand, exploratory data analysis allows us to deeply investigate the data and optimally formulate hypotheses, leading to a better interpretation of results. Techniques exist to extract step by step the different influential variables [90] in order to explain how they are related to all the activity data automatically recorded during the experiments. These techniques also allow us to better interpret the outliers.

ACKNOWLEDGMENTS

JD would like to extend a warm thanks to his colleagues Pierrick Aupinel (INRA), Axel Decourtye (ACTA), Dominique Fortini (INRA), and Julie Fourrier (ACTA) for sharing interest and resources to work on these new methodologies.

REFERENCES

1. I.H. Williams, *The dependence of crop production within the European Union on pollination by honey bees*, Agric. Sci. Rev. 6 (1994), pp. 229–257.
2. I.H. Williams, *Insect pollination and crop production: A European perspective*, in *Pollinating Bees. The Conservation Link Between Agriculture and Nature*, P. Kevan and V.L. Imperatriz Fonseca, eds., Ministry of Environment of Brasilia, Brasilia, Brazil, 2002, pp. 59–65.
3. Anonymous, *Colony collapse disorder, action plan*, CCD Steering Committee, (2007) Available at http://www.ars.usda.gov/is/br/ccd/ccd_actionplan.pdf. Accessed on January 3, 2013.
4. Anonymous, *USDA's response to colony collapse disorder*, Audit Report 50099-0084-HY, United States Department of Agriculture, Office of Inspector General, Washington, DC, 2012.
5. A.M. Klein, B.E. Vaissière, J.H. Cane, I. Steffan-Dewenter, S.A. Cunningham, C. Kremen, and T. Tscharntke, *Importance of pollinators in changing landscapes for world crops*, Proc. R. Soc. B 274 (2007), pp. 303–313.
6. L.A. Garibaldi, M.A. Aizen, A.M. Klein, S.A. Cunningham, and L.D. Harder, *Global growth and stability of agricultural yield decrease with pollinator dependence*, Proc. Nat. Acad. Sci. 108 (2011), pp. 5909–5914.

7. C.A. Kearns and D.W. Inouye, *Pollinators, flowering plants, and conservation biology*, BioScience 47 (1997), pp. 297–307.

8. L. Comba, S.A. Corbet, L. Hunt, and B. Warren, *Flowers, nectar and insect visits: Evaluating British plant species for pollinator-friendly gardens*, Ann. Bot. 83 (1999), pp. 369–383.

9. J.P. Dijkstra and M.M. Kwak, *A meta-analysis on the pollination service of the honey bee* (Apis mellifera L.) *for the Dutch flora*, Proc. Neth. Entomol. Soc. Meet. 18 (2007), pp. 79–87.

10. A. Grube, D. Donaldson, T. Kiely, and L. Wu, *Pesticides Industry Sales and Usage. 2006 and 2007 Market Estimates*, Biological and Economic Analysis Division Office of Pesticide Programs, Office of Chemical Safety and Pollution Prevention, U.S. Environmental Protection Agency, Washington, DC, 2011.

11. C. Smith, K. Kerr, and A. Sadripour, *Pesticide exports from U.S. ports, 2001–2003*, Int. J. Occup. Environ. Health 14 (2008), pp. 167–177.

12. Eurostat, *The Use of Plant Protection Products in the European Union. Data 1992–2003*, Office for Official Publications of the European Communities, Luxembourg, 2007.

13. Anonymous, *European inventory of existing commercial chemical substances (EINECS)*, Available at http://ec.europa.eu/environment/chemicals/exist_subst/einecs.htm.

14. P. McCutcheon, O. Nørager, W. Karcher, and J. Devillers, Sources of data for risk evaluation and QSAR studies, in *Practical Applications of Quantitative Structure-Activity Relationships (QSAR) in Environmental Chemistry and Toxicology*, W. Karcher and J. Devillers, eds., Kluwer Academic Publishers, Dordrecht, the Netherlands, 1990, pp. 13–24.

15. J.R. Hagler and C.G. Jackson, *Methods for marking insects: Current techniques and future prospects*, Annu. Rev. Entomol. 46 (2001), pp. 511–543.

16. A.E. Lundie, *The Flight Activities of the Honey Bee*, US Dept. Agric. Bull. No. 1328, 1925.

17. A.C. Fabergé, *An apparatus for recording the bees leaving and entering a hive*, J. Sci. Instrum. 20 (1943), p. 28.

18. W.H. Brittain, *Apple pollination studies in the Annapolis Valley, NS, Canada, 1928–1932*, Can. Dept. Agric. Bull. 162 (1933), pp. 119–125.

19. R. Chauvin, *Nouvelle technique d'enregistrement de l'activité de la ruche*, Apiculteur 96 (1952), pp. 9–14.

20. R. Chauvin, *Sur la mesure de l'activité des abeilles au trou de vol d'une ruche à 10 cadres*, Ins. Soc. 35 (1976), pp. 75–82.

21. H.G. Spangler, *Photoelectrical counting of outgoing and incoming honey bees*, J. Econ. Entomol. 62 (1969), pp. 1183–1184.

22. R.M. Burril and A. Dietz, *An automatic honey bee counting and recording device for possible systems analysis of a standard colony*, Am. Bee J. 113 (1973), pp. 216–218.

23. E.H. Erickson, H.H. Miller, and D.J. Sikkema, *A method for separating and monitoring honeybee flight activity at the hive entrance*, J. Apic. Res. 14 (1975), pp. 119–125.

24. R.M. Burril, A. Dietz, and C.F. Kossac, *Flight activity analysis of honey bees using Apicard generated data*, Apiacta 12 (1977), pp. 110–112, 116.

25. G.A. Buckley, L.G. Davies, and D.T. Spindley, *A bee counter for monitoring bee activity and bee behaviour*, Br. J. Pharmacol. 64 (1978), pp. 475P.

26. R.S. Pickard and D. Hepworth, *A method for electronically monitoring the ambulatory activity of honeybees under dark conditions*, Behav. Res. Meth. Instrum. 11 (1979), pp. 433–436.

27. R.M. Burril and A. Dietz, *The response of honey bees to variations in solar radiation and temperature*, Apidologie 12 (1981), pp. 318–328.

28. F. Marletto and P. Piton, *Conta-api elettronico per la verifica dell'attività degli alveari*, Apic. Mod. 74 (1983), pp. 137–141.

29. M. Rickly, G. Buhlmann, L. Gerig, H. Herren, H.J. Schürch, W. Zeier, and A. Imdorf, *Zur Anwendung eines elektronischen Bienenzählgerätes am Flugloch eines Bienenvolkes*, Apidologie 20 (1989), pp. 305–315.

30. C. Liu, J.J. Leonard, and J.J. Feddes, *Automated monitoring of flight activity at a beehive entrance using infrared light sensors*, J. Apic. Res. 29 (1990), pp. 170–173.

31. M.H. Struye, H.J. Mortier, G. Arnold, C. Miniggio, and R. Borneck, *Microprocessor-controlled monitoring of honeybee flight activity at the hive entrance*, Apidologie 25 (1994), pp. 384–395.

32. Anonymous, *BeeSCAN Instruction Manual*, Lowland Electronics BVBA, Belgium.

33. J. Devillers and W. Karcher, *Applied Multivariate Analysis in SAR and Environmental Studies*, Kluwer Academic Publishers, Dordrecht, the Netherlands, 1991.

34. J. Devillers, J.C. Doré, M. Tisseur, S. Cluzeau, and G. Maurin, *Modelling the flight activity of* Apis mellifera *at the hive entrance*, Comput. Electron. Agric. 42 (2004), pp. 87–109.

35. S. Dolédec and D. Chessel, *Co-inertia analysis: An alternative method for studying species–environment relationships*, Freshwater Biol. 31 (1994), pp. 277–294.

36. J. Devillers and D. Chessel, *Comparison of* in vivo *and* in vitro *toxicity tests from co-inertia analysis*, in *Computer-Aided Molecular Design. Applications in Agrochemicals, Materials, and Pharmaceuticals*, C.H. Reynolds, M.K. Holloway, and H.K. Cox, eds., ACS Symposium Series No. 589. American Chemical Society, Washington, DC, 1995, pp. 250–266.

37. I.T. Jolliffe, *Principal Component Analysis*, Springer-Verlag, New York, 2010.

38. J. Devillers and W.Karcher, W. (1990), *Correspondence factor analysis as a tool in environmental SAR and QSAR studies*, in *Practical Applications of Quantitative Structure-Activity Relationships (QSAR) in Environmental Chemistry and Toxicology*, W. Karcher and J. Devillers, eds., Kluwer Academic Publishers, Dordrecht, the Netherlands, 1990, pp. 181–195.

39. J. Devillers, A. Chezeau, and E. Thybaud, *PLS-QSAR of the adult and developmental toxicity of chemicals to* Hydra attenuata, SAR QSAR Environ. Res. 13 (2002), pp. 705–712.

40. J. Devillers, *Neural Networks in QSAR and Drug Design*, Academic Press, London, UK, 1996.

41. A. Decourtye, J. Devillers, S. Cluzeau, M. Charreton, and M.H. Pham-Delègue, *Effects of imidacloprid and deltamethrin on associative learning in honeybees under semi-field and laboratory conditions*, Ecotoxicol. Environ. Safety 57 (2004), pp. 410–419.

42. A. Decourtye and M.H. Pham-Delègue, *The proboscis extension response: Assessing the sublethal effects of pesticides on the honey bee*, in *Honey Bees: Estimating the Environmental Impact of Chemicals*, J. Devillers and M.H. Pham-Delègue, eds., Taylor and Francis, London, UK, 2002, pp. 67–84.

43. J. Devillers, A. Decourtye, and J. Fourrier, *MAPA 32/10: Mission Scientifique d'Evaluation des Effets non Intentionnels des Larvicides et Adulticides sur les Pollinisateurs pour le Centre de Demoustication du Conseil Général de la Martinique*, LIFE 08 ENV/F/000488, Report phase 3, 2012.

44. S. Lefort, M. Tisseur, and A. Decourtye, *De la traçabilité même chez les butineuses!*, Bull. Tech. Apic. 32 (2005), pp. 153–164.

45. V. Frisch, *The Dance Language and Orientation of Bees*, Harvard University Press, Cambridge, MA, 1993.

46. T. Seeley, *The Wisdom of the Hive: The Social Physiology of Honey Bee Colonies*, Harvard University Press, Cambridge, MA, 1995.

47. M.H. Pham-Delègue, *Les Abeilles*, Editions de la Martinière, 1998.

48. A. Veeraraghavan, R. Chellappa, and M. Srinivasan, *Shape and behavior encoded tracking of bee dances*, IEEE Trans. Patt. Anal. Mach. Intell. 30 (2008), pp. 463–476.

49. A. Feldman and T. Balch, *Automatic identification of bee movement*, in *Proceedings 2nd International Workshop on the Mathematics and Algorithms of Social Insects*, C. Anderson and T. Balch, eds., Georgia Institute of Technology, Atlanta, GA, December 15–17, 2003, pp. 53–59.

50. J. Campbell, L. Mummert, and R. Sukthankar, *Video Monitoring of Honey Bee Colonies at the Hive Entrance*, Intel Research Pittsburgh, Pittsburgh, PA, pp. 1–4.

51. C. Chen, E.C. Yang, J.A Jiang, and T.T. Lin, *An imaging system for monitoring the in-and-out activity of honey bees*, Comput. Electron. Agric. 89 (2012), pp. 100–109.

52. N. Cristianini and J. Shawe-Taylor, *An Introduction to Support Vector Machines and Other Kernel-Based Learning Methods*, Cambridge University Press, New York, 2000.

53. M.E. Colin, J.M. Bonmatin, I. Moineau, C. Gaimon, S. Brun, and J.P. Vermandere, *A method to quantify and analyze the foraging activity of honey bees: Relevance to the sublethal effects induced by systemic insecticides*, Arch. Environ. Contam. Toxicol. 47 (2004), pp. 387–395.

54. S. Streit, F. Bock, C.W.W. Pirk, and J. Tautz, *Automatic life-long monitoring of individual insect behaviour now possible*, Zoology 106 (2003), pp. 169–171.

55. A. Decourtye, J. Devillers, P. Aupinel, F. Brun, C. Bagnis, J. Fourrier, and M. Gauthier, *Honeybee tracking with microchips: A new methodology to measure the effects of pesticides*, Ecotoxicology 20 (2011), pp. 429–437.

56. C.R. Ribbands, *The Behaviour and Social Life of Honeybees*, London Bee Research Association, London, UK, 1953.

57. D. Hodges, *The Pollen Load of the Honeybee*, London Bee Research Association, London, UK, 1952.

58. M. Henry, M. Béguin, F. Requier, O. Rollin, J.F. Odoux, P. Aupinel, J. Aptel, S. Tchamitchian, and A. Decourtye, *A common pesticide decreases foraging success and survival in honey bees*, Science 336 (2012), pp. 348–350.

59. D.S. Khoury, M.R. Myerscough, and A.B. Barron, *A quantitative model of honey bee colony population dynamics*, PLoS ONE 6, 2011, e18491. doi:10.1371/journal.pone.0018491.

60. C.W. Schneider, J. Tautz, B. Grünewald, and S. Fuchs, *RFID tracking of sublethal effects of two neonicotinoid insecticides on the foraging behavior of* Apis mellifera, PLoS ONE 7 (2012), e30023. doi:10.1371/journal.pone.0030023.

61. L. Malone and H.S Gatehouse, *Effects of* Nosema apis *infection on honey bee* (Apis mellifera) *digestive proteolytic enzyme activity*, J. Invert. Pathol. 71 (1998), pp. 169–174.

62. D.I. Wang and F.E. Moeller, *Histological comparisons of the development of hypopharyngeal glands in healthy and -infected worker honey bees*, J. Invert. Pathol. 14 (1969), pp. 135–142.

63. D.I. Wang and F.E. Moeller, *The division of labor and queen attendance behavior of Nosema-infected worker honeybees*, J. Econ. Entomol. 63 (1970), pp. 1539–1641.

64. R. Beutler, E. Opfinger, and O. Wahl, *Pollenernährung und Nosemabefall der Honigbiene* (Apis mellifica), Z. Vergl. Physiol. 32 (1949), pp. 338–421.

65. J. Kralj and S. Fuchs, Nosema sp. *influences flight behavior of infected honey bee* (Apis mellifera) *foragers*, Apidologie 41 (2009), pp. 21–28.

66. C. Schneider, *Detecting the Influence of Different Potential Stress Factors on the Behavior of the Honeybee* Apis mellifera *using Radiofrequency Identification (RFID)*, Thesis, Julius-Maximilians-Universität Würzburg, Würzburg, Germany, 2011.

67. M. Pahl, H. Zhu, J. Tautz, and S. Zhang, *Large scale homing in honeybees*, PLoS ONE 6 (2011), e19669. doi:10.1371/journal.pone.0019669.

68. K. Ohashi, D. D'Souza, and J.D. Thomson, *An automated system for tracking and identifying individual nectar foragers at multiple feeders*, Behav. Ecol. Sociobiol. 64 (2010), pp. 891–897.

69. R.J. Stelzer and L. Chittka, *Bumblebee foraging rhythms under the midnight sun measured with radiofrequency identification*, BMC Biol. 8 (2010), http://www.biomedcentral.com/1741-7007/8/93.

70. R.J. Stelzer, R. Stanewsky, and L. Chittka, *Circadian rhythms of complete forager castes of bumblebee colonies monitored by radio-frequency identification*, J. Biol. Rhythms 25 (2010), pp. 257–267.

71. M. Sasaki, *A trial of the micro-bar-code system for monitoring honeybee behavior*, Honeybee Sci. 10 (1989), pp. 182–183.

72. J.R. Riley, A.D. Smith, D.R. Reynolds, A.S. Edwards, J.L. Osborne, I.H. Williams, N.L. Carreck, and G.M. Poppy, *Tracking bees with harmonic radar*, Nature 379 (1996), pp. 29–30.

73. E.A. Capaldi, A.D. Smith, J.L. Osborne, S.E. Fahrbach, S.M. Farris, D.R. Reynolds, A.S. Edwards, A. Martin, G.E. Robinson, G.M. Poppy, and J.R. Riley, *Ontogeny of orientation flight in the honeybee revealed by harmonic radar*, Nature 403 (2000), pp. 537–540.

74. J.L. Osborne, S.J. Clark, R.J. Morris, I.H. Williams, J.R. Riley, A.D. Smith, D.R. Reynolds, and A.S. Edwards, *A landscape-scale study of bumble bee foraging and constancy, using harmonic radar*, J. Appl. Ecol. 36 (1999), pp. 519–533.

75. M. Lihoreau, N.E. Raine, A.M. Reynolds, R.J. Stelzer, K.S. Lim, A.D. Smith, J.L. Osborne, and L. Chittka, *Radar tracking and motion-sensitive cameras on flowers reveal the development of pollinator multi-destination routes over large spatial scales*, PLoS Biol. 10 (2012), e1001392. doi:10.1371/journal.pbio.1001392.

76. M. Lihoreau, N.E. Raine, A.M. Reynolds, R.J. Stelzer, K.S. Lim, A.D. Smith, J.L. Osborne, and L. Chittka, *Unravelling the mechanisms of trapline foraging in bees*, Commun. Integ. Biol. 6 (2013), pp. e22701–e22701.

77. J.R. Riley and A.D. Smith, *Design considerations for an harmonic radar to investigate the flight of insects at low altitude*, Comput. Electron. Agric. 35 (2002), pp. 151–169.

78. J.R. Riley, J.W. Chapman, D.R. Reynolds, and A.D. Smith, *Recent applications of radar to entomology*, Outlooks Pest Manag. Feb. (2007), pp. 1–7.

79. R.S. Pasquet, A. Peltier, M.B. Hufford, E. Oudin, J. Saulnier, L. Paul, J.T. Knudsen, H.R. Herren, and P. Gepts, *Long distance pollen flow assessment through evaluation of pollinator foraging range suggests transgene escape distances*, Proc. Natl. Acad. Sci. USA 105 (2008), pp. 13456–13461.

80. M. Wikelski, J. Moxley, A. Eaton-Mordas, M.M. López-Uribe, R. Holland, D. Moskowitz, D.W. Roubik, and R. Kays, *Large-range movements of neotropical orchid bees observed via radio telemetry*, PLoS ONE 5 (2010), P. e10738. doi:10.1371/journal.pone.0010738.

81. M. Hagen, M. Wikelski, and W.D. Kissling, *Space use of bumblebees* (Bombus spp.) *revealed by radio-tracking*, PLoS ONE 6 (2011), P. e19997. doi:10.1371/journal.pone.0019997.

82. N.J. Aebischer, P.A. Robertson, and R.E. Kenward, *Compositional analysis of habitat use from animal radio-tracking data*, Ecology 74 (1993), pp. 1313–1325.

83. R.J.A. Little and D.B. Rubin, *Statistical Analysis with Missing Data*, John Wiley & Sons, Hoboken, NJ, 2002.

84. J.L. Schafer and J.W. Graham, *Missing data: Our view of the state of the art*, Psychol. Methods 7 (2002), pp. 147–177.

85. S. Nakagawa and R.P. Freckleton, *Missing inaction: The dangers of ignoring missing data*, Trends Ecol. Evol. 23 (2008), pp. 592–596.

86. J.W. Graham, *Missing Data: Analysis and Design*, Springer-Verlag, New York, 2012.

87. N.G. Yocco, *Use, overuse, and misuse of significance tests in evolutionary biology and ecology*, Bull. Ecol. Soc. Am. 72 (1991), pp. 106–111.

88. A.J. Onwuegbuzie and L.G. Daniel, *Uses and misuses of the correlation coefficient*, Annual Meeting of the Mid-South Educational Research Association, Point Clear, AL, November 17–19, 1999.
89. J.F. Meilof, *The use and abuse of correlation coefficients*, Arch. Neurol. 58 (2001), pp. 833–834.
90. J. Devillers, J. Thioulouse, and W. Karcher, *Chemometrical evaluation of multispecies-multichemical data by means of graphical techniques combined with multivariate analyses*, Ecotoxicol. Environ. Safety 26 (1993), pp. 333–345.

2 Computational Modeling of Organization in Honey Bee Societies Based on Adaptive Role Allocation

Mark Hoogendoorn, Martijn C. Schut, and Jan Treur

CONTENTS

ABSTRACT

One of the unique features of the organization in honey bee societies is the ability to adapt to environmental circumstances in a highly decentralized way. This adaptation takes the form of changes in the allocation of bees to roles, whereby bees individually decide to take up certain roles that require attention. Within the domain of computational modeling of multiagent systems, a trend has developed to model such systems from an organizational perspective. Usually this is done by describing multiagent systems as organizations using structural elements such as roles and groups and behavioral properties of these elements, for example, expressions for role behavior, specifying how agents should behave once they fulfill a certain role. In dynamic environments, changes in environmental circumstances may require changes in such organizations as well. In this chapter, the change process as seen in honey bee colonies has been modeled by means of techniques from the domain of multiagent organizations. This results in an adaptive multiagent organizational model that is able to cope with changing environmental circumstances.

KEYWORDS

Organizational modeling, Multiagent organizations, Agent systems, Adaptive role allocation, Honey bee colonies

2.1 INTRODUCTION

In the literature, it has been shown that honey bees (*Apis mellifera*) are very effective in adapting their organization based on the environmental circumstances that the colony is experiencing [1,2]. Within the colony, several specialized roles are present, such as *brood carers* that take care of feeding larvae, *patrollers* that guard the hive against enemies, *foragers* who harvest food outside of the hive, *undertakers* who remove corpses from the hive, and *resting workers* who do not engage in any activities. Bees switch between these roles triggered by changes that they observe in the environment. Such observations differ per bee. Each role has a specific trigger for which a bee has a certain threshold that determines whether this is the role it should play. The bee always plays the role for which it is most triggered. For example, bees are triggered to start playing the *brood carer* role when they observe the larvae emitting a very high level of hunger pheromones. Once they are allocated to the role, they start getting food from the combs and feed the larvae that are emitting the pheromones. A trigger for the *patroller* role is the number of enemies observed around the hive. *Foragers* that have returned from their hunt for food communicate the location where they found the food by means of the honey bee dance (see [3]). For other bees currently not playing the *forager* role, such a dance is a trigger to start playing the *forager* role. The more corpses there are, the more bees are being triggered to switch from their current role to being *undertaker*. Bees perform the *resting worker* role in case they are not sufficiently triggered for any other role.

The aforementioned provides an excellent mechanism to change an organization effectively. Within artificial intelligence, modeling of organizations [4] has recently received attention in the discipline of multiagent systems [5–7]. Within multiagent systems, there is a tendency to analyze and design more and more complex systems consisting of larger numbers of agents (*e.g.*, in nature, society, software). Due to this increasing complexity, the need arises for a concept of higher abstraction than the concept agent in order to still oversee the functioning of such a system. To this end, organizational modeling is becoming a practiced stage in the analysis and design of multiagent systems. Hereby, the environment in which the multiagent organization participates has to be taken into consideration. An environment can have a high degree of variability, which might require organizations that change to adapt to the environment's dynamics to ensure a continuous proper functioning of the organization. Hence, such change processes are a crucial function of the organization and should be part of the organizational model. In order to model these change processes, a more central perspective can be taken, which is, for instance, presented in Hoogendoorn et al. [8]. Hereby, there is a central authority in place that performs changes in case it is necessary to do so, given the goals of the organization and the environmental circumstances. An alternative is to take a decentralized perspective (for instance, in case there is no such central authority) whereby each individual member of the organization decides upon a change by himself or herself. In order to model such decentralized change, the mechanism as explained for honey bees earlier can be used.

This chapter presents a model for decentralized organization change that incorporates the change mechanism as seen in honey bee colonies and extends the work presented by the authors [9]. Such a model can aid developers of multiagent systems in creating and analyzing such an organization with change characteristics and also allows the studying of these kinds of organizations (*e.g.*, studying changes in honey bee colonies). The description of the model abstracts from the actual tasks being performed by the organization. The scope of the model is broader than simply being able to model social insects: the mechanisms incorporated in the model facilitating decentralized organizational change may work in other types of organizations as well. In Bonebeau and Theraulaz [10], for example, a comparable approach is used for finding an optimal allocation of cars to paint booths. Note that the model is illustrated by means of examples from the domain of honey bees. These illustrations are simplifications of the actual processes encountered in honey bee colonies and are not meant to model the real process, but as an illustration of the application and expressiveness of the proposed modeling approach.

Section 2.2 presents the modeling approach used. The model describing organizational change using the approach as seen in honey bee colonies is described in Sections 2.3 and 2.4. Results of a simulation of the organizational model for a bee colony are shown in Section 2.5, and finally Section 2.6 concludes the chapter.

2.2   MODELING ORGANIZATIONAL DYNAMICS

In order to model the phenomena that occur within honey bee organizations, two elements are needed. First, a way to represent the *structure* of the organization and second, a representation of the *behavior* of the organization is required. For the representation of the structure of the organization, AGR (agent/group/role) as introduced by Ferber and Gutknecht [11,12] is used, which is presented in Section 2.2.1. Section 2.2.2 introduces the approach used to express the behavior of the multiagent organization.

2.2.1   Representing Organizational Structure

For the structural description of multiagent organizations, the AGR model has been adopted [7,11,12]. The three primitive definitions are as follows.

The *agents*: The model places no constraints on the internal architecture of agents. An agent is only specified as an active communicating entity that plays roles

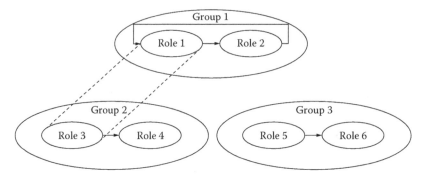

FIGURE 2.1   Example organization modeled within AGR.

within groups. This agent definition is intentionally general to allow agent designers to adopt the most accurate definition of agenthood relative to their application.

A *group* is defined as an atomic set of roles. Each agent plays a role in one or more groups. In its most basic form, the group is only a way to tag a set of roles. An agent can contribute to multiple groups at the same time. A major point of these groups is that they can freely overlap.

A *role* is an abstract representation of an agent function, service, or identification within a group. Each agent can handle multiple roles, and each role handled by an agent is local to a group.

AGR distinguishes three aggregation levels: the organization as a whole, groups, and roles, as illustrated in Figure 2.1. The large ovals denote groups whereas the smaller ovals denote the roles within the organizations. Furthermore, the solid arrows denote intragroup interactions between roles within a given group, and the dashed lines represent intergroup interactions. Agents realizing the roles are not depicted. However, the specification of the aggregation levels can place additional constraints on the agents that are to realize the roles. For example, the dashed lines between role 1 and role 3 could indicate that those roles will have to be fulfilled by the same agent.

2.2.2   REPRESENTING ORGANIZATIONAL BEHAVIOR

Describing the structure of an organization is not enough; the behavior has to be described as well. For example, the intragroup interactions in Figure 2.1 describe that role 5 can communicate to role 6, but it does not describe when this should occur nor what content is to be communicated.

The specification of behavior follows the same aggregation levels as identified in AGR, namely, the level of roles, groups, and the organization as a whole. The importance of such aggregation levels and the relation between these aggregation levels are emphasized by Lomi and Larsen [13]. In the introduction to their book, they describe the following as main challenges in the field:

- "Given a set of assumptions about (different forms of) individual behavior, how can the aggregate properties of a system be determined (or predicted) that are generated by the repeated interaction among those individual units?"

- "Given observable regularities in the behavior of a composite system, which rules and procedures—if adopted by the individual units—induce and sustain these regularities?"

Both views and problems require means to express relationships between dynamics of different elements and different levels of aggregation within an organization. The different aggregation levels of the behavioral specification are shown in Figure 2.2 in the form of an AND tree.

As can be seen, on the highest level of the tree, organizational properties are shown, which are properties the organization as a whole needs to achieve. At the level below the organizational level, group properties and intergroup interaction properties are specified, which together entail the organizational properties. The group properties are entailed by the lowest level, namely, role properties and transfer properties that specify interactions between roles within the same group.

The language TTL (Temporal Trace Language), described in Bosse et al. [14] and Sharpanskykh and Treur [15], has been adopted for the specification of behavior in organizational models; for more formal details on TTL, see Box 2.1. To model direct causal or temporal dependencies between two state properties, not the expressive language TTL, but the simpler *leads to* format is used. This is an executable format that can be used to obtain a specification of a simulation model in terms of local dynamic properties (the leaves of the tree in Figure 2.2). The format is defined as follows. Let α and β be state properties of the form "conjunction of literals" (where a literal is an atom or the negation of an atom) and e, f, g, h be nonnegative real numbers. In the *leads to* language, $\alpha \twoheadrightarrow_{e,f,g,h} \beta$, which means:

If state property α holds for a certain time interval with duration g, then after some delay (between e and f) state property β will hold for a certain time interval of length h.

A specification of dynamic properties in *leads to* format has the advantages that it is executable and that it can be depicted graphically in a causal graph-like style. TTL comes with a dedicated software environment that has been implemented in SWI-Prolog. This software environment allows for the specification of properties (both in TTL and the executable subset *leads to*), a simulation engine that can execute *leads to* specifications, and an automated property checking tool that verifies properties against traces (for instance, resulting from the simulation engine). The software is available on a variety of platforms, including Windows and Linux.

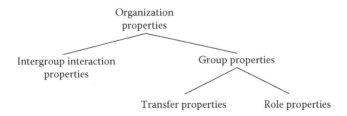

FIGURE 2.2   AND tree of behavioral properties.

BOX 2.1   FORMAL CONCEPTS USED IN TTL

In TTL [14], ontologies for states are formalized as sets of symbols in sorted predicate logic. For any ontology Ont, the ground atoms form the set of *basic state properties* BSTATPROP(Ont). Basic state properties can be defined by nullary predicates (or proposition symbols), such as incident, or by using n-ary predicates (with $n > 0$) like observes (number_of_casualties, 7). The *state properties* based on a certain ontology Ont are formalized by the propositions (using conjunction, negation, disjunction, implication) made from the basic state properties and constitute the set STATPROP(Ont).

Important concepts to express dynamics in TTL are *states*, *time points*, and *traces*. A *state* S is an indication of which basic state properties are true and which are false, that is, a mapping S: BSTATPROP(Ont) → {true, false}. The set of all possible states for ontology Ont is denoted by STATES(Ont). Moreover, a fixed *time frame* T is assumed, which is linearly ordered. Then, a *trace* γ over a state ontology Ont and time frame T is a mapping γ: T → STATES(Ont), that is, a sequence of states γ_t ($t \in$ T) in STATES(Ont). The set of all traces over ontology Ont is denoted by TRACES(Ont).

The set of *dynamic properties* DYNPROP(Ont) is the set of temporal statements that can be formulated with respect to traces based on the state ontology Ont in the following manner. Given a trace γ over state ontology Ont, a certain state at time point t is denoted by state(γ, t). These states can be related to state properties via the formally defined satisfaction relation, indicated by the infix predicate |=, and comparable to the Holds predicate in the situation calculus. Thus, state(γ, t) |= p denotes that state property p holds in trace γ at time t. Likewise, state(γ, t) |≠ p denotes that state property p does not hold in trace γ at time t. Based on these statements, dynamic properties can be formulated in a formal manner in a sorted predicate logic, using the usual logical connectives such as ¬, ∧, ∨, ⇒ and the quantifiers ∀, ∃ (*e.g.*, over traces, time and state properties). The set DYNPROP(Ont, γ) is the subset of DYNPROP(Ont) consisting of formulae with γ occurring in which it is either a constant or a variable without being bound by a quantifier.

TTL has some similarities with the situation calculus [16] and the event calculus [17], which are two well-known formalisms for representing and reasoning about temporal domains. However, a number of important syntactic and semantic distinctions exist between TTL and both calculi. In particular, the central notion of the situation calculus—a situation—has different semantics than the notion of a state in TTL. That is, by a situation is understood a history or a finite sequence of actions, whereas a state in TTL is associated with the assignment of truth values to all state properties (a "snapshot" of the world). Moreover, in contrast to the situation calculus, where transitions between situations are described by actions, in TTL, actions are in fact properties of states. Moreover, although a time line has been recently introduced to the situation calculus [15], still only a single path (a temporal line) in the tree of situations can be explicitly encoded in the formulae. In contrast, TTL provides more expressivity by allowing explicit references to different temporally ordered sequences of states (traces) in dynamic properties. Regarding event calculus, TTL does not employ the mechanism of events that initiate and terminate fluently whereas event calculus does. Events in TTL are considered to be functions of the external world that can change states of components, according to specified properties of a system. Furthermore, similar to the situation calculus, in the event calculus, only one time line is considered. TTL can also be related to temporal languages that are often used for verification (*e.g.*, propositional temporal logic [PTL] and linear-time logic [LTL] [18–20]). It is possible to translate properties expressed in these languages in TTL. However, the translation is not bidirectional, that is, it is not always possible to translate TTL expressions into LTL expressions. See Bosse et al. [14] for more details.

2.3   ORGANIZATIONAL PROPERTIES

In order to model the change process as observed in honey bee colonies, a first start is made by considering desired behavior of the organization as a whole by means of organizational properties, including the relationships between these organizational properties. The formal specifications in TTL of the dynamic properties discussed in the following can be found in Box 2.2. The highest level requirement for the organization as a whole as inspired by the biological domain knowledge is survival of the population given a fluctuating environment; in other words, the population size always needs to reach a level above a certain threshold M.

OP1(M) Surviving Population

For any time t, a time point t′ ≥ t exists such that at t′ the population size is at least M.

BOX 2.2   FORMAL SPECIFICATIONS OF DYNAMIC ORGANIZATION PROPERTIES IN TTL

OP1(M) Surviving Population

$\forall t \; \exists t' \geq t,\; v: state(\gamma, t') \models total\_living\_population\_count(v) \; \& \; v \geq M$

OP2(X, P1, P2) Organization Aspect Maintenance

$\forall t,\; v: state(\gamma, t) \models has\_value(X, v) \Rightarrow P1 \leq v \leq P2$

OP3(X, W1, W2) Sufficient Aspect Effort

$\forall t,\; v: state(\gamma, t) \models provided\_effort(X, v) \Rightarrow W1 \leq v \leq W2$

OP4(X, B, d) Adaptation Flexibility

$\forall t,\; v1 \; [[state(\gamma, t) \models provided\_effort(X, v1) \; \& \; v1 < B] \Rightarrow$
$\exists t' \geq t,\; v2: [t' \leq t{+}d \; \& \; state(\gamma, t') \models provided\_effort(X, v2) \; \& \; v2 \geq B]]$

RP(R, d, W) Worker Contribution

$\forall t \; \exists t' \geq t,\; v: [t' \leq t{+}d \; \& \; state(\gamma, t') \models work\_contribution(R, v) \; \& \; v \geq W]]$

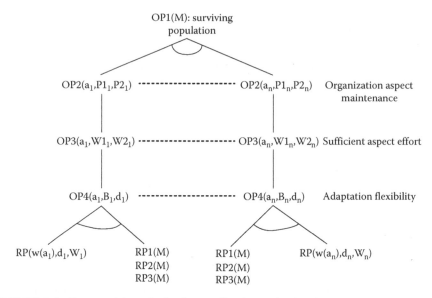

FIGURE 2.3   Property hierarchy for decentralized organizational change.

This high-level requirement is refined by means of a property hierarchy, depicted as a tree in Figure 2.3. At the highest level, OP1 is depicted, which can be refined into a number of properties (in Figure 2.3, n properties) each expressing that for a certain aspect the society is in good condition, characterized by a certain *value* for a variable (the *aspect variable*) that is to be maintained. The property template for an aspect X is as follows:

OP2(X, P1, P2) Organization Aspect Maintenance
For all time points t
If v is the value of aspect variable X at t, then v is between P1 and P2

Sometimes, one of the two bounds is omitted, and it is only required that value v is at least P1 (resp., at most P2). For the honey bee society, the aspects considered are *well-fed brood*, *safety*, *food storage*, and *cleanness* (addressed, respectively, by brood care, patroller, forager, and undertaker roles as explained in Section 2.1). For each of these aspects, a variable was defined to indicate the state of the society for that aspect. For example, for well-fed brood, this variable concerns relative larvae hunger, indicated by the larvae pheromone rate.

In order to maintain the value of an aspect variable X, a certain effort is needed all the time. To specify this, a property that expresses the *effort* made by the organization on the aspect is introduced. Notice that the notion of provided effort at a time point t can be taken in an *absolute* sense (*e.g.*, effort as the amount of feeding work per time unit), but it can also be useful to take it in a *relative* sense with respect to a certain overall amount, which itself can vary over time (*e.g.*, effort as the fraction of the amount of feeding work per time unit divided by the overall number of larvae). Below the latter, relative form will be taken. The general template property for aspect effort is as follows:

OP3(X, W1, W2) Sufficient Aspect Effort

For all time points t, the effort for aspect X provided by the organization is at least W1 and at most W2.

For the bee colony, the brood care workers take care that the larvae are well fed. The effort to maintain the hunger of larvae at a certain low level is feeding the larvae. Here, provided effort for brood care is defined as the brood care work per time unit divided by the larvae population size. Brood care work is taken as the amount of the (average) brood care work for one individual brood carer times the number of brood carers.

Whether the refined properties given earlier will always hold depends on the flexibility of the organization. For example, in the honey bee colony, if the number of larvae or enemies increases, the number of brood care workers or patrollers, respectively, should also increase. If the adaptation to the new situation takes too much time, the property Brood Care Effort will not hold for a certain time. In principle, such circumstances will damage the success of the organization. Therefore, an adaptation mechanism is needed that is sufficiently flexible to guarantee the properties such as Brood Care Effort. For this reason, the adaptation flexibility property is introduced, which expresses that when the effort for a certain organization aspect that is to be maintained is below a certain value, then within a certain time duration d it will increase to become at least this value. The smaller this parameter d is, the more flexible is the adaptation; for example, if d is very large, the organization is practically not adapting. The generic property is expressed as follows:

OP4(X, B, d) Adaptation Flexibility

At any point in time t, if the effort for aspect X provided by the organization is lower than B, then within time duration d the effort will become at least B.

An assumption underlying this property is that not all aspects in the initial situation are critical, otherwise the adaptation mechanism will not work. OP3 expressing that sufficient effort is being provided directly depends on this adaptation mechanism is shown in Figure 2.1. OP4 depends on role properties at the lowest level of the hierarchy, which are addressed in the next section.

2.4 ROLE PROPERTIES

Roles are the engines for an organization model: they are the elements in an organization model where the work that is done is specified. The properties described in Section 2.3 in a hierarchical manner have to be grounded in role behavior properties as the lowest level properties of the hierarchy. In other words, specifications of role properties are needed that entail the properties at the organizational level described in Section 2.3. In the behavioral model, two types of roles are distinguished: worker roles that provide the effort needed to maintain the different aspects throughout the organization and member roles that have the function to change worker roles. Each member role has exactly one shared allocation with a worker role. The role behavior for the worker roles within the organization is shown in Section 2.4.1, whereas Section 2.4.2 specifies the behavior for the member roles. Note that the level of groups has not been used throughout this model, but one can easily see that certain roles can be grouped according to the tasks that are being performed.

2.4.1   WORKER ROLE BEHAVIOR

Once a certain worker role exists as an active role, it performs the corresponding work. What this work exactly is depends on the application: it is not part of the organization model. The property directly relates to OP4, which specifies the overall effort provided, as shown in Figure 2.1. Note that Figure 2.1 only shows the generic form of the role property (depicted as $RP(w(a_i),d_i,W_i)$, where a_i is the specific aspect and $w(a_i)$ the worker role belonging to that aspect), whereas in an instantiated model, a role property is present for each instance of the worker role providing the effort for the specific aspect. In a generic form, this is specified by

RP(R, d, W) Worker Contribution

For all t there is a t' with $t \leq t' \leq t+d$ such that at t' the worker role R delivers a work contribution of at least W.

Here work_contribution is used as part of the state ontology for the output of the role (see Box 2.2). For each of the specific roles, it can be specified what the work contribution is in terms of the domain-specific state ontology (*e.g.*, the number of larvae to be fed for the brood carer role).

2.4.2   MEMBER ROLE BEHAVIOR

By a member role M, decisions about taking up or switching between worker roles are made. As input of this decision process, information is used about the well-being of the organization, in particular about the different aspects distinguished as to be maintained; these are input state properties indicating the value of an aspect variable X: has_value(X, v). Based on this input, the member role M generates an intermediate state property representing an indication of the aspect that is most urgent in the current situation. In the model, the decision mechanism is indicated by a priority relation priority_relation $(X_1, v_1, w_1,..., X_n, v_n, w_n, X)$, indicating that aspect X has priority in the context of values v_i and norms w_i for aspects $X_1,..., X_n$. This priority relation can be specialized to a particular form, as shown by an example specialization in the last paragraph of this section. For formal specifications in TTL of these role properties, see Box 2.3.

RP1(M) Aspect Urgency

At any t, if member role M has norms w_1 to w_n for aspects X_1 to X_n and receives values v_1 to v_n for X_1 to X_n at its input, and has a priority relation that indicates X as the most urgent aspect for the combination of these norms and values, then at some $t' \geq t$ it will generate that X is the most urgent aspect.

**BOX 2.3   FORMAL SPECIFICATIONS OF
DYNAMIC ROLE PROPERTIES IN TTL**

RP1(M) Aspect Urgency

$\forall t, v1,.., vn, w1,..., wn, X$

$state(\gamma, t) \models has\_value(X_1, v_1) \& \ldots \& has\_value(X_n, v_n) \&$

$has\_norm(X_1, w_1) \& \ldots \& has\_norm(X_n, w_n) \&$

$priority\_relation(X_1, v_1, w_1,..., X_n, v_n, w_n, X)$

$\Rightarrow \exists t' \geq t \; state(\gamma, t') \models most\_urgent\_aspect(X)$

RP2(M) Role Change Determination

$\forall t, X, R1, R2 \; state(\gamma, t) \models most\_urgent\_aspect(X) \&$

$role\_responsible\_for(R2, X) \& role\_reserved\_for(R2, M) \&$

$state(\gamma, t) \models has\_shared\_allocation(M, R1) \& R1 \neq R2$

$\Rightarrow \exists t' \geq t \; state(\gamma, t') \models shared\_allocation\_change(M, R1, R2)$

RP3(M) Role Reallocation

$\forall t, R1, R2$

$state(\gamma, t) \models shared\_allocation\_change(M, R1, R2)$

$\Rightarrow \exists t' \geq t \; state(\gamma, t') \models not \; has\_shared\_allocation(M, R1) \&$

$has\_shared\_allocation(M, R2)$

Based on this, the appropriate role for the aspect indicated as most urgent is determined. If it is not the current role sharing an allocation with M, then another intermediate state property is generated expressing that the current worker role sharing an
allocation with M should be changed to the role supporting the most urgent aspect.
In other words, the shared allocation of member role M in the change group should
change from one (the current) worker role R1 in worker group WG1 to another one,
worker role R2 in worker group WG2:

RP2(M) Role Change Determination

At any time t, if member role M generated that X is the most urgent aspect, worker
role R2 is responsible for this aspect, R1 is the current worker role sharing an allocation with M, and R1 ≠ R2, then at some t′ ≥ t it will generate that role R2 has to
become the worker role sharing an allocation with M, instead of R1.

Based on this intermediate state property, the member role M generates output indicating which role should become a shared allocation and which should not anymore:

RP3(M) Role Reallocation

At any t, if member role M generated that worker role R2 has to share an allocation
with M, instead of worker role R1, then at some t′ ≥ t it will generate the output that
role R1 will not share an allocation with M and R2 will share an allocation with M.

All three role properties for the member roles are depicted in Figure 2.3. The adaptation step property OP4 for all organizational aspects is dependent upon it, so each

of the OP4 branches depends upon RP1, RP2, and RP3, which have therefore been depicted two times in the figure.

The generic description for the member role behavior can be specialized one step further by incorporating a specific decision mechanism. This gives a specific definition of the priority relation priority_relation(X_1, v_1, w_1,..., X_n, v_n, w_n, X) as has been done for the following decision mechanism based on norms used as thresholds (see, *e.g.*, [10] for a more elaborate description of this mechanism within honey bees):

1. For each aspect X to be maintained, a norm w(X) is present. For the worker role R1 for X sharing an allocation with member role M, each time unit the norm has a decay described by fraction r.
2. For each X, it is determined how far the current value is unsatisfactory, expressed in a degree of urgency u(X) for that aspect.
3. For each aspect with urgency above the norm, that is, with u(X)>w(X), the relative urgency is determined: u(X)/w(X).
4. The most urgent aspect X is the one with highest relative urgency.

2.5   SIMULATION RESULTS

This section discusses the results of simulations that have been performed based on the organizational model, in particular the role properties presented in Section 2.4 have been put in an executable format and have been instantiated with domain-specific information for bee colonies (as the properties expressed in Sections 2.3 and 2.4 were highly generic and, hence, reusable for other domains as well).

To validate the instantiated simulation model, the high-level dynamic properties from Section 2.3 were used (in accordance with biological experts). Proper functioning of such an organization in nature is not self-evident; therefore, two simulation runs are compared: one using the adaptation mechanism and one without. The choice has been made to compare the result of using adaptation with no adaptation due to the fact that comparing with centralized change will obviously result in better performance of the centralized change model. Having a complete picture of the different aspects and their urgencies gives a major advantage. Whether decentralized adaptation is more successful than no adaptation is however not a trivial matter. In case there is merely coordination by means of observables in the world (which is the case in honey bee colonies), adaptation might even be counterproductive. Note that the results presented here are the results of a simulation of the instantiated organizational model, abstracting from allocated agents. Performing such high-level simulations of an executable organizational model enables the verification of properties against these simulation runs. Hence, it can be checked whether or not the model satisfies the properties or goals considered important. When such properties are indeed satisfied, by allocating agents to the roles that comply to the role properties, the multiagent system delivers the desired results as well. In the two simulations, several parameters have been set to certain values, where the circumstances are kept identical for both simulations.

External world: Initially, 15 larvae and 10 workers are present for which the initial type of the latter is randomly assigned. The natural mortality age is set to 500 time steps, whereas a larva is grown up after 250 time steps. Every 20 time steps, a new larva is added to the population. The initial food stock is set to 40 units of food. Once every 100 time points, an attack of 40 enemies occurs, who stay there until a patroller defeats them. In case over 200 enemies are present in the hive, each individual in the organization is removed with a probability of 0.05 per time step. In case more than 20 dead bodies are present in the hive, individuals are removed with the same probability. Food used by larvae is 0.5 per feed, and for workers 1 unit of food per time step.

Larvae: Larvae have an initial pheromone level of 0.5, increasing 0.006 per time step. In case pheromone emissions exceed 0.95, the larva dies. After being fed, the emission level is set to 0.1.

Foragers: Foragers each collect three food units per time step.

Brood carers: Feed 1 larva per 8 time steps, and only feed the larvae with a pheromone level above 0.55.

Undertaker: Carry 1 body per 12 time steps.

Patroller: Defeat 1 enemy per time step.

   In the adaptation simulation, the member thresholds are randomly generated, being somewhat above or below the average observed value of the various triggers. In Figures 2.4 and 2.5, the results of 20 runs of the model using the adaptive approach presented in Sections 2.2 and 2.3 versus no adaptation are shown. Figure 2.4 shows the population size over time, whereas Figure 2.5 shows one specific aspect of the

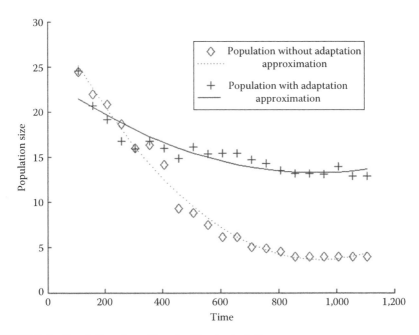

FIGURE 2.4 Population size with and without adaptation.

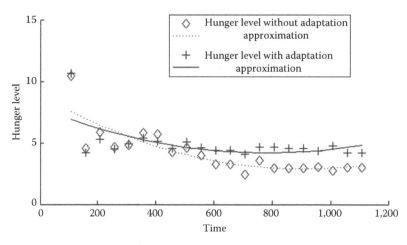

FIGURE 2.5   Hunger level with and without adaptation.

population, namely, the hunger level. Approximations of the trends are also shown, which are based upon a second-order polynomial function.

Figure 2.4 clearly shows that the approach whereby adaptation is used results in a bigger population due to the fact that the various aspects can be sustained for these higher population levels. In Figure 2.5, it can be seen that the overall hunger level in the case with adaption is slightly higher, and the level is however sufficiently maintained as can be seen by the fact that the population itself has a more healthy size.

In order to understand the change mechanism in a bit more detail, a representative example run is also explained in depth. Figure 2.6 shows results of the performance on this typical run, also compared with a typical run without adaptation. Figure 2.6A shows the overall population size over time. The population size of the simulation with adaptation remains relatively stable, whereas without adaptation it drops to a colony of size 3, which is equal to the number of larvae living without being fed. Figure 2.6B and C shows information regarding brood care: First, the average pheromone level, which is the trigger to activate the allocation to brood carers. Furthermore, the number of active brood carers in the colony is shown. In the case with adaptation, their number increases significantly in the beginning of the simulation, as the amount of pheromones observed is relatively high. Therefore, a lot of the brood carer roles are allocated. For example, at time point 300, 15 out of a population of 28 are brood carers.

Despite the fact that the overall pheromone level is not decreasing rapidly, the number of brood carer roles drops significantly after time point 300. This is due to the fact that member roles can only share an allocation with one worker role at a time. When another role receives a higher urgency (*e.g.*, there is a huge attack, demanding many patrollers), a switch of worker role takes place. Figure 2.6D shows the number of worker roles of the different types (except the resting workers) within the bee colony for the setting with adaptation. The number of brood carers decreases after time point 300 due to an increase in the number of shared allocations to the undertaker and forager roles. This results in an increase in pheromone level again,

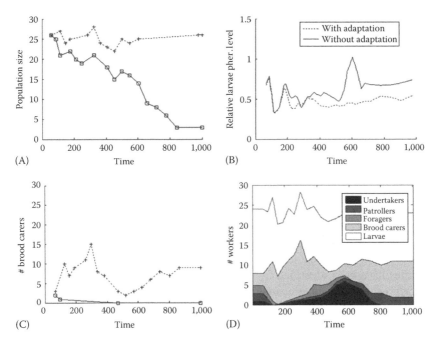

FIGURE 2.6 Results of simulating the bee colony with and without adaptation. (A) Shows the population time over time, (B) shows the relative larvae pheromone level, (C) the number of brood carers whereas (D) shows the number of workers of different types. Note that (D) only shows the worker types for the adaptive case.

causing a higher delta for brood care again, resulting in more brood carers, etc. The pheromone level finally stabilizes around 0.5 in the organizational model with adaptation. For the setting without adaptation, the brood carers simply cease to exist due to the fact that none of the larvae are growing up. The pheromone level stabilizes at a higher level. The properties from Section 2.3 have been checked by the automated TTL checker using these representative example runs. With the following parameter settings, the properties were validated and confirmed for the organizational model with adaptation and falsified for the one without adaptation: OP1(20), OP2(broodcare, 0, 0.9), OP3(broodcare, 0.15, 10,000), OP4(brood care, 0.3, 200).

2.6 CONCLUDING REMARKS

The organizational model for decentralized organizational change as seen in honey bee colonies has been formally specified by means of a methodology that describes the behavior of an organization on multiple aggregation levels [21]. The model is inspired by mechanisms observed in nature but is more general: these adaption algorithms can be very effective in many more types of applications; in Bonebeau and Theraulaz [10], examples of the effectiveness in other domains are provided. The model can therefore support organizational modelers and analysts working with multiagent organizations in highly dynamic environments, without a central

authority directing change, in general in designing and analyzing such an organization. In Hoogendoorn and Treur [22], a more generic approach toward modeling
these decentralized adaptations is presented, in which the honey bee colony is used
as an example to perform these decentralized cases. The formal specification of
the behavior in the organizational model presented in this chapter is described by
dynamic properties at different aggregation levels. Once the lowest level properties
within the organization are specified in an executable form, the organizational model
can be used for simulation abstracting from agents (to be) allocated. Such low-level
properties can be indicative for the behavior of the agent allocated to that particular
role. The possibility also exists to specify the role properties at the lowest aggregation level in a more abstract manner, in a nonexecutable format. Hierarchical relations between the properties can be identified to show that fulfillment of properties at
a lower level entails the fulfillment of the higher level properties. Simulations using
agents can be performed and checked for fulfillment of these properties. Properties
for the behavior of roles regarding decentralized organizational change have been
specified on an executable level to be able to perform simulation, and higher-level
properties have been identified as well. The model has been evaluated by looking
at an aggregate behavior over multiple runs (using certain parameter settings), but
also by investigating in more detail a single run to understand the functioning of
the mechanism itself. The results showed that, given the external circumstances, it
was effective, given overall properties put forward by biological experts. Next to the
domain of honey bee colonies, the formal approaches put forward in this chapter
have also been used for modeling other biological processes, for instance, in Jonker
et al. [23], a model of the living cell is presented.

REFERENCES

1. G. Theraulaz, E. Bonabeau, and J.L. Deneubourg, *Response thresholds reinforcement
   and division of labor in insect societies*, Proc. Royal Soc. Lond. Ser. B: Biol. Sci. 265
   (1998), pp. 327–332.
2. M.L. Winston and E.N. Punnet, *Factors determining temporal division of labor in
   honeybees*, Can. J. Zool. 60 (1982), pp. 2947–2952.
3. S. Camazine, J.L. Deneubourg, N.R. Franks, J. Sneyd, G. Theraulaz, and E. Bonabeau,
   Self-Organization in Biological Systems, Princeton University Press, Princeton, NJ, 2001.
4. H. Mintzberg, *The Structuring of Organisations*, Prentice Hall, Englewood Cliffs, NJ,
   1979.
5. V. Furtado, A. Melo, V. Dignum, F. Dignum, and L. Sonenberg, *Exploring congruence
   between organizational structure and task performance: A simulation approach*, in
   Proc. of the 1st OOOP Workshop, O. Boissier, V. Dignum, E. Matson, and J. Sichman,
   eds., Utrecht, the Netherlands, 2005.
6. P. Giorgini, J. Müller, and J. Odell, *Agent-Oriented Software Engineering IV*, Lecture
   Notes in Computer Science, vol. 2935, Springer-Verlag, Berlin, Germany, 2004.
7. C.M. Jonker and J. Treur, *Relating structure and dynamics in an organisation model*,
   in *Multi-Agent-Based Simulation II: Proc. of the Third Int. Workshop on Multi-Agent
   Based Simulation, MABS'02*, J.S. Sichman, F. Bousquet, and P. Davidson, eds., Lecture
   Notes in Artificial Intelligence, vol. 2581, Springer-Verlag, Berlin, Germany, 2003,
   pp. 50–69.

8. M. Hoogendoorn, C.M. Jonker, M. Schut, and J. Treur, *Modeling centralized organization of organizational change*, Comput. Math. Org. Theory 13 (2007), pp. 147–184.

9. M. Hoogendoorn, M.C. Schut, and J. Treur, *Modeling decentralized organizational change in honeybee societies*, in *Advances in Artificial Life: Proc. of the 9th European Conference on Artificial Life, ECAL'07*, F.A. Costa, L.M. Rocha, E. Costa, I. Harvey, and A. Coutinho, eds., Lecture Notes in Computer Science, vol. 4648, Springer-Verlag, Berlin, Germany, 2007, pp. 615–624.

10. E. Bonebeau and G. Theraulaz, *Swarm smarts*, Sci. Am. 282 (2000), pp. 72–79.

11. J. Ferber and O. Gutknecht, *A meta-model for the analysis and design of organizations in multi-agent systems*, in *Proceedings of the Third International Conference on Multi-Agent Systems (ICMAS '98)*, IEEE Computer Society, Paris, France, 1998, pp. 128–135.

12. J. Ferber, F. Michel, and J.A. Baez-Barranco, *AGRE: Integrating environments with organizations*, in *Proceedings of the First International Workshop on Environments for Multi-Agent Systems, E4MAS'04*, D. Weyns, H. Van Dyke Parunak, and F. Michel, eds., Lecture Notes in Computer Science, vol. 3374, Springer-Verlag, Berlin, Germany, 2005, pp. 48–56.

13. A. Lomi and E.R. Larsen, *Dynamics of Organizations: Computational Modeling and Organization Theories*, AAAI Press, Menlo Park, CA, 2001.

14. T. Bosse, C.M. Jonker, L. van der Meij, A. Sharpanskykh, and J. Treur, *Specification and verification of dynamics in agent models*, Intern. J. Coop. Inf. Syst. 18 (2009), pp. 167–193.

15. A. Sharpanskykh and J. Treur, *A temporal trace language for formal modelling and analysis of agent systems*, in *Specification and Verification of Multi-Agent Systems*, M. Dastani, K.V. Hindriks, and J.J.Ch. Meyer, eds., Springer-Verlag, New York, 2010, pp. 317–352.

16. R. Reiter, *Knowledge in Action: Logical Foundations for Specifying and Implementing Dynamical Systems*, MIT Press, Cambridge, MA, 2001.

17. R. Kowalski and M. Sergot, *A logic-based calculus of events*, New Gen. Comput. 4 (1986), pp. 67–95.

18. J.F.A.L. van Benthem, *The Logic of Time: A Model-Theoretic Investigation into the Varieties of Temporal Ontology and Temporal Discourse*, Reidel, Dordrecht, the Netherlands, 1983.

19. E.M. Clarke, O. Grumberg, and D.A. Peled, *Model Checking*, MIT Press, Cambridge, MA, 2000.

20. R. Goldblatt, *Logics of Time and Computation*, 2nd edn., CSLI Lecture Notes, Stanford, CA, 7, 1992.

21. C.M. Jonker and J. Treur, *Compositional verification of multi-agent systems: A formal analysis of pro-activeness and reactiveness*, Int. J. Coop. Inf. Syst. 11 (2002), pp. 51–92.

22. M. Hoogendoorn and J. Treur, *A generic architecture for redesign of organizations triggered by changing environmental circumstances*, Comput. Math. Org. Theory 17 (2011), pp. 119–151.

23. C.M. Jonker, J.L. Snoep, J. Treur, H.V. Westerhoff, and W.C.A. Wijngaards, *The living cell as a multi-agent organisation: A compositional organisation model of intracellular dynamics*, Trans. Comput. Collect. Intell. 1 (2010), pp. 160–206.

3 Illustrating the Contrasting Roles of Self-Organization in Biological Systems with Two Case Histories of Collective Decision Making in the Honey Bee

Brian R. Johnson

CONTENTS

ABSTRACT

Early self-organizing models in biology tended to focus on the role played by self-organization alone and downplayed alternative pattern formation mechanisms. In this chapter, I use two examples of how self-organization is used by honey bees to illustrate when self-organization should be favored, or disfavored, relative to other pattern formation mechanisms in biological systems. First, I argue that composite patterns, those that are not one pattern, but actually many simple patterns superimposed over one another in time or space, can have much simpler mechanistic bases than their overall complexity suggests. This means that simple templates and recipes can be important for generating such patterns. This is particularly true when patterns grow slowly in time and only become functional when they achieve a certain level of complexity. Second, I argue that processes in which information must be integrated across a large complicated system require coordination of action that can only be achieved via centralized or decentralized control structures (self-organization). Following Seeley, I argue that decentralized control evolves in contexts in which centralized control is prohibited by constraints or is not favored due to its lack of robustness in the face of perturbations to key components. When these are not true, centralized control should be favored as it leads to higher fitness. Accordingly, and contrary to some discussions of pattern formation in biology, centralized control (via nervous systems) is widespread in biology and both centralized and decentralized control are major principles in biological pattern formation.

KEYWORDS

Self-organization, Pattern formation, Templates, Recipes, Centralized control, Decentralized control, Agent-based modeling

3.1 INTRODUCTION

Pattern formation in biological systems is a rich subject that spans the development of organelles within the cell to the behavioral rules governing collective decision making in groups of animals [1–8]. While the mechanisms can be starkly different between contexts, there are common conceptual principles that span the diversity of biological pattern formation. Chief among these are the notions of self-organization, templates, and recipes (reviewed in [2]). Each plays a major role in biological pattern formation and each can operate at any biological level (from molecules to whole organisms). Self-organization refers to processes in which local interactions generate group-level behaviors that would be difficult to predict based on observations of the individual-level behavior alone [9–16]. In other words, they are processes in which a holistic approach (one at the group level) must follow a reductionistic description of the system. Examples include Benard convection cells in physics, the construction of some organelles in cell biology, and the behavioral basis of most collective decision making in ants and bees [6,14,15,17–19]. Templates refer to patterns in the environment that contain information useful for an organism attempting to generate a pattern [2,20–22]. Examples include the construction of some bird nests [2] and part of the basis of nest construction in ants [20]. Simple gradients lain down by a source beyond the system are also sometimes considered templates (reviewed in [23]). Finally, recipes are cookbook procedures in which individual-level behavior gives rise to very predictable group-level patterns [2]. Examples include web building in spiders and nest construction in many wasps.

The purpose of this work is to explore when the various forms of pattern formation mechanism are likely to be used within biological systems and why. In general, I will argue that it is useful to distinguish between structure and process and between patterns that are composites of many simpler patterns and patterns that must be constructed and function all at once. I will also discuss the varying information processing demands in each of these cases [24]. Since real pattern formation mechanisms tend to be highly complex, such that all the mechanisms are used at some point in the process, understanding the roles played by each mechanism involves explaining the general principles underlying when one mechanism is likely to be favored over another for a particular pattern formation task. I will illustrate such principles by reviewing two case histories of collective decision making in the honey bee. The first is one in which self-organization plays a vital, but relatively small, role [22], and the second is one in which self-organization is the primary pattern formation mechanism [25]. Hence, the first illustrates those cases in which self-organization is likely to be of little (or sometimes no) importance and the second illustrates a context in which self-organization is the only viable mechanism of pattern formation. I will conclude with a discussion that fleshes out how generally applicable those principles illustrated with collective decision making are to biology as a whole.

3.2 CASE HISTORY 1: PATTERN FORMATION ON HONEY BEE COMBS

Honey bee nests, like those of many insect societies, are large complex structures [20,21,26,27]. The insects that make these structures, in contrast, are small and have access to only limited information. How such creatures make such complex patterns has long fascinated biologists [17,28–30]. Because many solitary wasps and bees use rigid recipes to construct their nests, and because some insect nests contain gradients that can be used as templates [20,21], this is a case where all three pattern formation mechanisms can operate [20–22,29,30]. In this section, we will first describe the nature of the pattern, before discussing a model for how it can be generated. I will argue that early self-organizing models, though seminal in their demonstration of the importance of self-organization, were too simple to account for this case (and many other cases) of pattern formation. In contrast, I will argue that all three mechanisms (self-organization, templates, and recipes) work together in ways that can be demonstrated with simple models and likely in many other ways still beyond our understanding.

3.2.1 NATURAL HISTORY OF THE PATTERN

In many cases of pattern formation at the molecular level, we have only a very limited understanding of the full extent of the pattern. Pattern formation in developing embryos, for example, is of such complexity that layers of the overall complexity are peeled away decade after decade. It is therefore an unfortunate reality that we do not know how well our models of such processes explain the totality of the pattern (reviewed in [23]). At the macroscopic level, however, it can be relatively straight-forward to describe the pattern in great detail. If this is done, then the explanatory power of a given model can be compared to the totality of the problem it attempts to solve. Hence, I will first go into some depth about the complexity of the pattern that will be modeled. As we will see, the pattern emerges over a long period of time, in response to many social needs and physical constraints. Further, the nature of the pattern at the end is radically different from that at the beginning, meaning that the nature of information present in the pattern changes over time.

Honey bee nests are large three-dimensional structures built in the cavities of trees or sometimes under rock overhangs [26]. Figure 3.1 shows a typical honey bee nest. The nest has three major sections. There is a honey zone at the top where the bees keep their honey stores, a brood zone at the bottom where bees rear their young, and there is sometimes a band of pollen in between the brood and honey. Even when the band exists, however, there is considerable pollen stored throughout the brood zone. This is the basic pattern, but there are many critically important details. First, the cells at the top of the nest, at and near the attachment to the tree, are much thicker and more robust than the honeycomb elsewhere [26]. This is because the comb must support over 20 kg of weight with very thin wax, and hence the connection to the tree must be solid. Hence, the cells are deep at the top and taper off toward the base. Second, the queen leaves a pheromone trail wherever she goes [31], and because

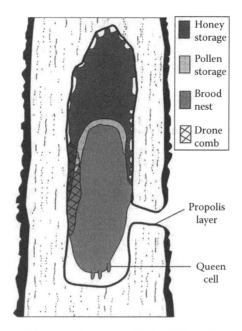

FIGURE 3.1   Wild nest of the honey bee. (From Seeley, T.D. and Morse, R.A., *Insectes Soc.*, 23, 495, 1976.)

she rarely leaves the brood zone, this means this section of the nest has a chemical signature different from that of the honey zone. Further, when brood are present they also secrete a pheromone that contributes to the olfactory properties of the brood zone [32,33]. Third, when bees emerge from cells they leave behind their cocoons, which darken eventually to a black color. Successive rounds of brood rearing result in smaller cells, as old cocoons build up within the cells [26,32]. As such, the cells in the brood zone are often a different color and size and have different olfactory properties from those in other parts of the nest [34]. Finally, nectar foragers rarely leave the bottom portion of the nest called the dance floor. This is where nectar foragers unload their nectar loads to nectar receivers, who process the nectar into honey and store it elsewhere in the nest [35].

The importance of the preceding details regarding the nest is as follows. In a newly founded nest, there is only a small comb that has not been used for brood rearing, which contains very little food. Hence, there is no spatial information in the initial comb pattern. The mature nest, however, is full of information. The two major zones differ strongly in olfactory, physical, and social cues, all of which are easily discernible to the sensory system of the bee. In short, the honey bee nest pattern is highly complex with many adaptive functions that evolve over time. The totality of this pattern is still beyond current models to explain. The goal here therefore is not to explain the entire pattern but to show that all three pattern formation mechanisms (templates, recipes, and self-organization) are important for its construction.

3.2.2 AGENT-BASED MODEL OF PATTERN FORMATION ON HONEY BEE COMBS

The model presented here builds on a previous model by Camazine [36]. Camazine described the honey bee pattern as concentric with honey on the outside, pollen in an intermediate band, and brood in the center. As reviewed in the preceding discussion, this is not the case, but it does capture part of the pattern. Camazine's self-organizing model had three simple rules: (1) the queen lays eggs in the center of the comb, (2) workers deposit nectar and pollen at random, and (3) bees preferentially remove pollen and nectar from near the brood. I made the following changes based on experimental evidence: (1) bees preferentially unload pollen next to open brood [37] and (2) bees have an upward bias in their movement pattern as they look for a place to unload nectar [35]. The first rule is a template effect as the brood provides positional information for unloading the pollen and the second is a recipe effect, in that a bias in the upward direction will naturally cause more honey to be unloaded at the top of the nest.

The model was implemented in the Netlogo agent-based modeling environment [38]. The nest surface was modeled as 14,025 cells arranged in a rectangular shape (similar to that of a natural bee nest). Four types of bees were modeled: a queen, pollen foragers, nectar receivers, and nurses. All parameter levels were consistent with studies of honey bee foraging and feeding rates [22]. Queen behavior was modeled after studies by Camazine [36]. Queens lay up to 1 egg/min in cells within four cells of another brood cell. They begin laying in the center of the nest (filling it to 80% full) before searching the rest of the nest for places to unload. Queens also avoid the nectar zone, with the simple rule that if more than 20 consecutive steps are taken on honey cells, then the queen increases her speed in a random direction until she finds the brood zone again. This is to mimic the fact that queens spend most of their time in the brood zone. Pollen foragers begin at the entrance of the nest and conduct a random search for an open brood cell. Once they find one, they unload into the first empty cell they find near it [37]. If they are unable to locate an open brood cell after a set period of time, they unload at random. Because honey bees regulate the amount of pollen they collect, a simple feedback loop was used to regulate pollen foraging. If less than 10% of the cells in the nest contain pollen, then pollen foragers forage at their maximum rate (set to 80% of the nectar foraging rate). If more than 10% of the cells contain pollen, then only 10% of the pollen foragers forage. This feedback mechanism is somewhat arbitrary; however, it leads to the accumulation of pollen stores that are sufficient to last for about 3 days, which is the typical amount found in honey bee nests [39,40]. Nectar receivers receive nectar at random locations along the bottom of the nest and then walk up a variable distance before looking for a place to unload. If receivers are unsuccessful at finding a place to unload after their first upward walk, they repeat the process until they are successful or until they reach the top of the nest at which point they begin a random search until successful. This behavior was modeled after Seeley's observations of nectar receivers [35]. Rate of nectar collection was set to 30 bees/min, the rate observed in a study of a colony of the same size modeled [45]. Overall, 1,800 nectar receivers unloaded honey into cells per hour. Nurse bees remove (eat) pollen and nectar from near the brood (in order to produce brood food) according to a simple rule. Each nurse conducts a random search for a pollen cell beginning at a random location within the brood zone. Hence, nurses search first within the brood zone for food, before searching around it. The search is simply a random walk in which the nurses inspect cells as they walk about until they find pollen. This creates differential rates of removal of food from near the brood. A more complete description of the model is available in Johnson [22].

3.2.3 NEST PATTERN RESULTS

Figure 3.2 shows the emergence of the pattern. Initially, the queen lays eggs in the center of the nest, which ultimately results in a compact central brood zone. Pollen is initially unloaded at random because there is no open brood to use as a template (the queen lays eggs that take 3 days to hatch). Nectar shows a bias toward the top of the nest, the strength of which varies with the distance receivers walk upward before looking for a place to unload. In the early phases of pattern formation, there is considerable honey stored at the bottom of the nest and all around the brood. By day 4 (the timing of each of these phases depends on the parameterization of the model), the top of the nest is solidly full of nectar and pollen. The queen, who has no bias toward the top or the bottom of the nest, is only able to find places to lay in the bottom of the nest, which begins to fill with brood. Pollen begins to accumulate throughout the brood zone at this point (because eggs are hatching into open brood) with a strong bias toward the bottom of the nest because pollen foragers unload next to the first open brood cell they find. By day 7, the bottom of the nest is mostly full of brood and pollen and the queen is only able to occasionally lay eggs (as cells are emptied by nurses). By day 14, the pattern is fully present. Although the pollen band

Day 1   Day 4   Day 7   Day 10   Day 12   Day 14

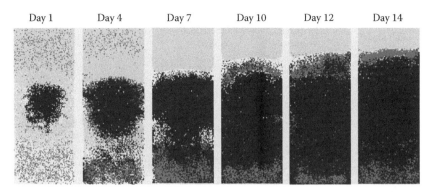

FIGURE 3.2   (See color insert.) Agent-based model of pattern formation during a period of high nectar intake. Honey cells are yellow, pollen cells are red, and brood cells are black. Initially, pollen is scattered throughout the nest, but once open brood is present, pollen accumulates at the bottom of the nest. Nectar shows an immediate bias toward being unloaded at the top of the nest due to the upward movement of nectar receivers. By day 7, the complete pattern is almost present with the exception of the pollen band, which forms by day 14. (From Johnson, B.R., *Proc. Roy. Soc. B Biol. Sci.*, 276, 255, 2009.)

between the honey and brood is not as strong as initially described by Camazine [36], this band is only sometimes seen in real colonies, and large sections of pollen within the brood zone are common in real nests. Although I will not discuss it here, it is also the case that fluctuations in the availability of food (due to bouts of rain during which bees continue to eat but do not forage) can create a strong pollen band between the honey and brood [22].

Figure 3.3 sheds light on the mechanisms underlying the pattern shown in Figure 3.2. It shows the final pattern for runs of the model with varying combinations of behavioral rules. Figure 3.3a shows the full model (with rainy days, which improves the clarity of the pattern). Figure 3.3b shows the effect of removing the template used by pollen foragers (unloading near open brood). In this case, pollen stays scattered throughout the honey zone. This simple template therefore greatly improves the clustering of pollen near the brood (where it will ultimately be used). Figure 3.3c shows the effect of differential removal of food (honey and pollen) from near the brood. As Camazine [36] showed, the effect of this is to create an open band between the brood and honey, which is then filled with pollen (which has a faster turnover rate than honey and for which there are fewer places to unload outside of the open band). It also helps divide the honey and brood zones into distinct sections, as the differential removal ensures that all the cells in the brood zone are eventually filled with brood (and not honey). A solid brood zone is important because it facilitates thermoregulation of the brood, which must be kept at 33°C–36°C [41,42]. Figure 3.3d shows the effect of removing the recipe of upward movement during unloading of nectar. In this case, the vertical pattern does not form. In nature, the vertical pattern is important, because approximately 20 kg of honey must be stored in the nest, and storing this material at the top of the nest, in the thickest, most robust cells, is necessary. It is also necessary to have the

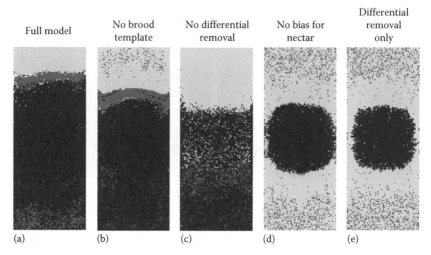

FIGURE 3.3 **(See color insert.)** Contrasting roles played by each mechanism underlying pattern formation. The pattern is shown at 14 days in each image. (a) The full model. (b) Without the queen-based template for unloading pollen near open brood, pollen is scattered throughout the honey zone. (c) Without differential removal of food from near the brood (the SO mechanism), the pattern is far from complete in that pollen does not localize and the sections of the nest are not distinct. (d) Without the recipe's effect of upward bias in unloading of nectar, the pattern is not vertical (honey on top brood on the bottom). (e) The original self-organization model of Camazine (differential removal of food from near brood alone) does not lead to complete pattern formation. (From Johnson, B.R., *Proc. Roy. Soc. B Biol. Sci*, 276, 255, 2009.)

honey zone as one compact region, since during winter when bees consume it, they must be able to move up a contiguous region of honey. Finally, Figure 3.3e shows Camazine's [36] original model without the additional template or recipe. Here, there is not a vertical pattern but only a very faint pollen band. It is possible to create a larger pollen band by tuning the parameters, but it is not possible to localize most of the pollen near the brood if bees store stockpiles of pollen sufficient to last them a few days without forage.

Figures 3.2 and 3.3 are about the mechanisms of pattern formation, the how questions. Now we switch gears to try to illuminate the why question. In other words, why might it be selectively advantageous to use a mechanism like the one the bees use? Figure 3.4 shows the growth rates of colonies that vary in the degree of upward bias shown by the receivers as they walk up looking for a place to unload nectar. Essentially, the larger the bias in unloading toward the top, the higher the growth rate of the colony. The reason is quite intuitive. If there is little or no bias in the deposition of nectar, then the whole nest quickly gets blanketed in a light layer of honey and there is no place for the queen to lay eggs. If the nectar foraging rate is high enough, then this light layer can become a thick layer and the queen gets boxed into a small central brood zone (which grows slowly as food is eaten away at its edges). If, however, the bees bias their deposition upward, then cells are left open for egg laying. At the extreme, if the entire bottom is left often, then the queen is unhindered as she lays and the growth rate is maximized. However, since a compact brood zone is useful,

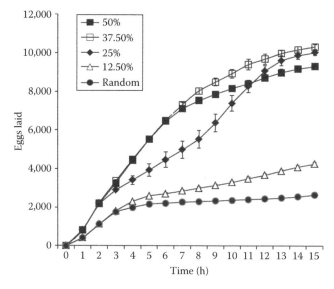

FIGURE 3.4 Exploration of the adaptive benefits of using a recipe for generating the pattern on the honey bee comb. The different lines represent different degrees of upward bias (in percent of the nest the receiver walks up before trying to unload nectar). Colonies have a higher growth rate when bees have a strong upward bias in deposition of nectar. This is because if nectar is unloaded at random, then it coats the nest and blocks the queen from laying eggs. (From Johnson, B.R., *Proc. Roy. Soc. B-Biol. Sci.*, 276, 255, 2009.)

it is probably an intermediate level of open space for egg laying that is best, and hence a moderate amount of bias upward during unloading may be favored.

3.3   CASE HISTORY 2: TASK ALLOCATION IN MIDDLE AGE HONEY BEES

Task allocation is the study of how colonies allocate the correct number of individuals to tasks that change in location and demand [43–45]. It is an intriguing problem because workers tend to only have access to local information, yet the demand and locations of tasks change at a much larger global level. Honey bee foragers, for example, visit flowers within 12 km of their nest and optimally allocate themselves to flower patches within this area [46]. They accomplish this feat by sharing information with the waggle dance, which can be used to indicate the location of a food site and can be modulated such that a good site is more heavily danced for than a lesser site [35]. Ants likewise allocate themselves to food sources located over an immense area (relative to the size of an ant) using communication systems dependent on signals, pheromone trails [13,47]. Task allocation in the middle age caste of honey bees (MABs) presents a particularly interesting question of optimal allocation of workers to tasks, because these bees, unlike the foragers, do not make use of signals [45]. Hence, they do not encode their sense of the need or value of a particular task into a signal that is then used to generate a collective decision.

To explore how MABs solve their task allocation problem, I employed a classic experimental and theoretical approach. I first described the pattern (at the individual and group levels), and then I used these data to construct an individual-oriented model to explore how these bees allocate themselves to tasks.

3.3.1 EXPERIMENTAL PATTERNS

MABs do all the work within the nest except for brood care [48–50]. This encompasses building the nest, maintaining and repairing it, processing nectar into honey, and guarding the nest, to just name a part of their task repertoire. In all, they do about 18 different tasks spread across the nest. Given that a nest is vast compared to the size of a single bee, it is clear that task demand is changing over a spatial scale far too large for a bee to have immediate direct experience of changes in task demand at the global level. Two data sets provided the key to how bees might collectively solve this problem. These data focus on the nature of task switching and individual-level movement patterns.

Figure 3.5 shows the hourly behavioral state (task) of a group of 47 bees, all of whom were observed working wax at time period one [45]. It was discovered that after a single hour, the rate of wax working in this group had fallen down to nearly the background level of wax working for the whole caste [45]. In addition, 1 h after being observed working wax, many bees were either searching for new work or had switched tasks. This data set matched somewhat the early reports of Lindauer [51] who watched a single bee for its whole lifetime and found that the bee was a generalist who would perform nearly any task within a short period of time. Johnson [45] confirmed that bees are generalists for tasks within their task repertoire, but he did not confirm whether they would perform any task (e.g., they do not perform nursing tasks).

Figure 3.6 shows the path of a bee conducting a patrol [49]. Patrols are long, seemingly random, walks through the nest [49–51]. Over the course of observing

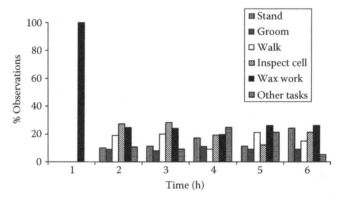

FIGURE 3.5 Hourly scans of 47 MABs observed working wax at time period one. After a single hour, wax working fell to nearly the background rate, suggesting most bees in the MAB caste are generalists who switch tasks often. (From Johnson, B.R., *Behav. Ecol. Sociobiol.*, 51, 188, 2002.)

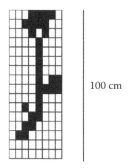

100 cm

FIGURE 3.6   Path of a MAB on patrol. The grid represents a grid drawn on the surface of a four-frame observation hive in which marked bees are being observed. Each black grid cell is a cell the bee visited during a 30 min period of observation. This bee was observed to quit working at the top of the nest and then walk down a meandering path to the bottom. Such patrols were found to be common in this caste of bees. (From Johnson, B.R., *Naturwissenschaften*, 95, 523, 2008.)

50 bees for 30 min using focal animal observations, it was found that nearly all MABs conduct patrols and that patrols cover 60 cm (total displacement) on average [49]. To illustrate the importance of this movement behavior, consider Figure 3.1, which shows the honey bee nest. The nest has a strong spatial structure, with different tasks being conducted in different regions. If one considers this in light of Figure 3.6, then it is clear that after a long patrol a bee has changed to a location with a new set of tasks. The honey zone, the brood zone, and the dance floor (where nectar is unloaded), in particular, are areas in which different tasks with strongly variable demands are performed.

To review the experimental work, it is likely that after a patrol a bee that begins to search for work will find a new task, contributing to the task switching patterns found in Figure 3.5. This suggests a simple self-organizing model for how the MABs can allocate themselves to variable tasks over a large area.

3.3.2 MODEL

An agent-based model was constructed to determine if the simple behaviors recorded in Figures 3.5 and 3.6 are sufficient to explain MAB task allocation [25,38]. Like most agent-based models, the model has a landscape and a variable number of agents who interact with one another and with the patches. The landscape of the present model has three work zones (the honey zone, the brood zone, and the dance floor). Each zone is composed of patches on which one type of work is conducted. Each patch has a stimulus level that corresponds to the amount of work available there. If the stimulus is five, for example, then when a worker finds work on that patch, the stimulus falls to four until the worker quits working. The nest surface overall had 528 patches. The number of patches of each type was varied depending on the simulation (details in the results section).

The simulated bees are autonomous agents that interact with the patches but not with each other. That is to say, they only effect one another's behavior through their

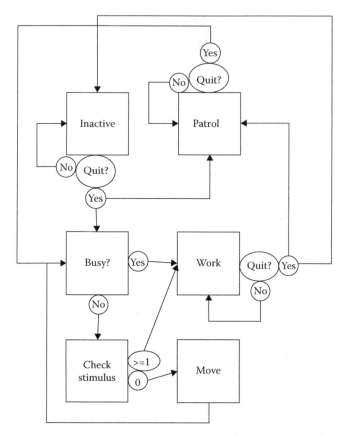

FIGURE 3.7 Flow diagram of the behavioral rules governing worker behavior in the model for task allocation in MABs. (From Johnson, B.R., *Am. Nat.*, 174, 537, 2009.) Workers can be in one of three behavioral states: working, patrolling, or inactive. Working bees conduct a simple search for work, patrollers conduct a correlated random walk, and inactive bees are simply inactive.

interactions with the patches. With respect to movement, they move over the land-scape according to a step length that is independent of patches. That is to say, workers do not jump from patch to patch but take a step that may or may not cause them to leave a patch in one time step. Figure 3.7 shows the simple behavioral algorithm used by workers based on experimental data. Bees can be in one of three behavioral states at a given period of time: working, patrolling, or inactive [49]. Each time step (10 s), a worker either stays in its current state or switches to another (choosing at random) based on a quitting probability parameterized with the data collected in the experimental studies [49]. While in the working behavioral state, bees conduct a random search for work. This means they perform a random walk during which they periodically stop to check if work is available on the patch they are currently within. If so, they take up work there and stay. If not, they move on. If a worker starts to work on a patch, the stimulus level on that patch falls by one and stays at that level until the worker quits the working state, at which point it increases by one. Patrolling bees conduct a continuous correlated random walk until they switch to another behavioral state. Patrolling bees do not inspect cells or interact with the nest in any way. Inactive bees are simply stationary and do nothing until they quit their state.

The model was parameterized as follows. Patrollers move at a rate of 3.15 cm/min (net displacement). This means a step length of 1.5 patches per time step with successive turn angles within a 180° arch. Bees searching for work had a step length of 0.95 patches per time step. Working bees had a quitting probability of 1/1,800 s, inactive bees one of 1/1,000 s, and patrollers one of 1/800 s. This distribution gave a ratio of workers in these states that matches those found in experimental studies [45,49].

3.3.3 RESULTS

Model validation, comparing the models' output to real data, was performed successfully, and the results are available in Johnson [25]. Here, I focus on the nuts and bolts of how the model works. Figure 3.8 shows the basic behavior of the model. For the first day of this simulation, the stimulus levels for each task were fixed, and the model quickly produced an allocation of workers to tasks that matched the variation in stimulus levels (task 1 had the highest stimulus level and task 2 the lowest). When I simulated a change in the environment, such that the demand for one task rapidly fell off and that for the other two increased, there was a rapid response by the bees and a new pattern of labor allocation appeared. Hence, the model is able to solve the sort of task allocation problems faced by MABs; that is, it can allocate bees to tasks that unpredictably change in demand over a large area. We now explore how the model accomplishes this feat.

The model functions by coupling two opposing forces (Figure 3.9). Figure 3.9a illustrates the first, which is a localizing mechanism. This is the result of the random search for work that workers perform when in the working state. This serves to localize

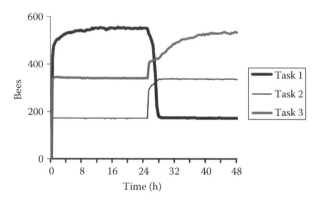

FIGURE 3.8 Output from the model of task allocation in MABs. Shown are the numbers of bees working on three tasks with different levels of demand (stimulus levels). Stimulus levels are held constant for the first day and a stable allocation of workers to tasks that matches variation in demand is achieved. After a change in the environment at 24 h, the model quickly reallocates bees to tasks in accordance with the new ratios of stimulus levels. (From Johnson, B.R., *Am. Nat.*, 174, 537, 2009.)

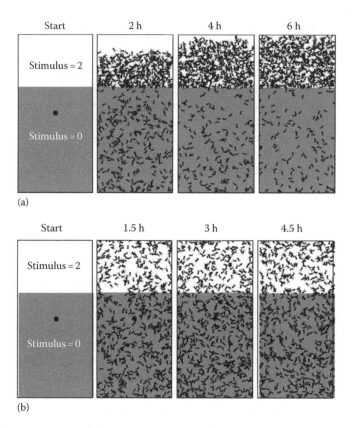

FIGURE 3.9 Two mechanisms underlie task allocation in MABs. In each simulation, work is only available at the top of the nest (zero stimulus levels elsewhere). In the beginning of each simulation, bees are at the center of the nest. (a) Bees search for work but do not conduct patrols. (b) Bees patrol but do not search for work. Searching for work localizes bees to tasks, while patrolling randomizes worker location over the surface of the nest. (From Johnson, B.R., *Am. Nat.*, 174, 537, 2009.)

workers in accordance with variation in task demand. Figure 3.9b illustrates the opposing force, which is the result of patrolling behavior. Patrolling serves to randomize worker location without reference to variation in task demand. Hence, it causes a constant rate of diffusion away from regions of high workload to regions of low workload in a constant environment. Figure 3.10 shows how the two forces work together to balance productivity and flexibility. Productivity is a measure of how many workers have been assigned tasks (how much work is being produced), while flexibility refers to how quickly workers can reallocate themselves after a change in the environment that necessitates a shift in labor allocation. Figure 3.10a shows a situation with low productivity, but high flexibility, while Figure 3.10c shows the opposite. What differs between these figures is the ratio of time spent patrolling to working (or localizing to diffusing). When workers spend a large amount of time in the working state, and little in the patrolling state, they have high productivity but low flexibility and vice versa

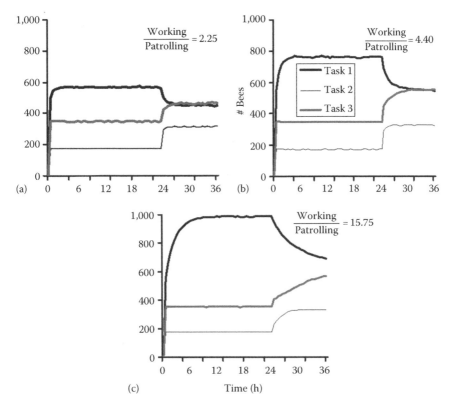

FIGURE 3.10   (a) Optimizing productivity and flexibility in task allocation. Simulations are shown in which variable ratios of time are spent in the working and patrolling states. (b) High rates of working lead to high productivity but low flexibility, while high rates of patrolling lead to high flexibility but low productivity. (c) An optimal balance between the two forces allows a colony to maximize productivity without compromising flexibility in the face of a variable environment. (From Johnson, B.R., *Am. Nat.*, 174, 537, 2009.)

when the ratio is the opposite. Hence, there is an optimal balance between the time spent localizing to tasks and diffusing over the surface of the nest that maximizes productivity without compromising flexibility (for a given set of parameter values).

3.4   DISCUSSION AND CONCLUSIONS

3.4.1   CASE HISTORY 1: COMPOSITE PATTERNS

The pattern on the surface of a honeycomb evolves slowly over time according to many rules. Unlike in the simplified simulations, the real pattern starts small and grows in three dimensions. As the comb ages, different portions of the nest take on new forms (old versus new comb), and the workers modulate their behavior based on this information [32]. Therefore, a pattern emerges slowly over time according to many rules, some of which interact, some of which do not, and some of which only appear at certain stages of development. In such cases, the pattern is not one pattern at all, but many patterns superimposed over one another in time and/or space. Such composite patterns are highly complicated structures that may or may not need self-organizing principles for their construction.

A composite pattern that is the result of many simple patterns superimposed over one another can have a simple explanation that is amenable to a reductionist approach. It can be the case that each pattern is the result of simple template effects and that a recipe-type program controls when each template is used. "Recipe" is a term for an inflexible cookbook set of instructions that are carried out in a sequence with no feedback between the different steps. A treasure map, for example, is a recipe. The searcher does not reference any positional information as it moves; it simply walks according to instructions. Any errors in the implementation of instructions cannot be corrected with external information, and small errors early in the process can completely negate the possibility of ultimate success. This sort of process with zero feedback or interaction is not self-organization. It can be useful for solving many problems in biology, however, if the initial state can be precisely defined (analogous to starting in the right spot in the context of the treasure map) and/or a fixed sequence of events can get the system from one state to another.

An example will be useful to illustrate how composite patterns could be formed with or without the use of self-organization. This example is loosely based on embryogenesis in *Drosophila* [23,52–55] but is not meant to be a literal discussion of the topic. If we imagine an egg with a set of gradients imposed on the system by the mother, then the initial conditions are anything but random. They are in fact precisely initialized by an external agent. Hence, at the start of the process there are templates (externally imposed gradients of chemicals that can provide information about position within the system) that have a precise relationship to one another. If all that happens after this is that the templates are used to turn on different genes according to a combinatorial relationship between concentration gradients, then the ultimate pattern that forms is based on many simple template effects superimposed in time and space. If the initial phases of gene expression trigger downstream cascades that are entirely fixed processes with no feedback, then a recipe provides the control for the process. This should be true no matter how many steps there are in the recipe, as long as there is no interaction between template mechanisms. (Many rules can be used at once or different rules can be used at different times, but no coordination between them occurs.)

If such a recipe making use of templates was all that occurred in development, then self-organization would likely not play much of a role. However, development is mainly about the differential growth of separate components of the system such that a complex three-dimensional structure is formed [56–59]. To the extent that differential growth within the embryo requires coordination (or responsive control) in order for pieces that are initially apart to come together or to take particular spatial relationships to one another, simple templates and recipes are not sufficient to explain these cases of pattern formation. Further, it is easy to imagine how templates can coordinate pattern formation within fixed space if the system is not growing, but a more dynamic changing pattern is difficult to reconcile with a recipe and template basis. This is because small errors early in the process would not be buffered to produce a robust mechanism. Hence, for dynamic patterns, particularly those that require coordination between elements, some mechanism that is interactive is necessary. As Turing pointed out in his now classic paper, the reaction diffusion mechanism [9] is one such mechanism for generating differential growth. It does not necessarily apply to every pattern formation situation, but some such Turing-like process is often at play when coordination between growing structures is necessary at the molecular level [3,23,56–60]. Self-organization can even be necessary when pattern formation occurs in a fixed space, if shifts of pattern are necessary relative to the initial concentration gradients, or if robustness is necessary in the context of imprecise gradients (reviewed in [23]).

In general, it is likely that just as the honey bee pattern we explored is based on a combination of pattern formation mechanisms, so are other composite patterns in biology. Given the large number of changes in conformation requiring sophisticated coordination that occur in developmental biology, it may well be the case that self-organizing processes are the most common pattern formation mechanisms in that context. However, this should be taken with two strong caveats. First, templates are easy to lay down within a subsystem by elements of the larger whole system (e.g., the mother). Templates should therefore play a large role whenever pattern formation problems are solvable with such mechanisms (i.e., when they occur in a fixed volume and can be based on simple concentrate gradients). Second, the steps of a recipe can be any type of mechanism. Hence, if the pattern formation process depends on first laying down a grid on the surface and then growing differentially into a new three-dimensional structure, then one recipe can control the entire process. All that is required is that at the end of each step there is an indicator that is used to trigger cessation of the current step and initiation of the next. Each step can either be based on templates or some form of self-organizing process. In general, genetically based recipes in which the steps are mostly self-organizing, but with significant use of templates, is a good null hypothesis for how pattern formation of composite structures occurs.

3.4.2   CASE HISTORY 2: PROCESSES THAT INTEGRATE INFORMATION IN TIME

Unlike the case for a pattern that emerges slowly over time and/or space and only eventually takes on some function (like many structures used for support or protection), some patterns reflect optimal solutions to current problems. In the case of the MABs trying to optimally allocate themselves to tasks, the goal is to find the current best allocation of workers to tasks given the state of the environment. Information from all relevant sources must be used and a fast and accurate response must ensue [7,43]. Hence, the pattern is complex; it emerges quickly and is functional immediately. There are two mechanisms used in biological systems for solving this type of problem: centralized organization and decentralized control or self-organization [24].

It is worthwhile to illustrate the difference between the two mechanisms by describing a hypothetical situation facing a factory at which people are assigned to many jobs that can rise and fall in demand. The way a centralized control structure solves this problem is for managers to collect information from each sector of the factory and integrate this information with the overall needs of the factory before assigning individual to tasks. As such, when 20 more workers are necessary for a job, they are taken from other sectors that are less important or overstaffed. If we imagine a hypothetical decentralized control structure inspired by social insects instead, it might look something like the following. Each hour a worker draws a random number (1–100) from a hat. If the number is less than 10, then that worker quits his or her current job and chooses a new job within the factory at random. If that job does not need more workers, then they choose again at random until they find one that does. In this case, when the demand for a job suddenly rises, it will attract more workers because all those workers who choose that job will find work there and stay, while for many other jobs the workers will not find work there and need to choose again. Ultimately, labor would be allocated according to work demand, but some labor would be wasted by individuals inefficiently looking for work at random according to the lottery process. In order for this to work, it is also necessary that the ratio of workers to total work is appropriate, because if there is infinite work available for each job, then workers will not be assigned to tasks with respect to variation in demand. This hypothetical example is quite close to what the MABs are faced with and how they use decentralized control to solve their task allocation problem.

The hypothetical example illustrates the basic difference between centralized and decentralized control. In centralized control, efficient decisions are made by managers with the ability to integrate large amounts of information. For decentralized control, a simple rule of thumb accomplishes the same result, but with a cost in inefficiency. Bees are unlikely to be able to process all the information that would be necessary to direct one another to tasks in an efficient directed manner, and, hence, they make do with a simple, robust, though inefficient, algorithm. Of course, such mechanisms can be made increasingly efficient by adding to the rules (such that tasks are not chosen at random, e.g., but with some bias toward important tasks or those that tend to change in demand most), but it is unlikely they can ever reach the efficiency of a well-orchestrated system of centralized control.

In biological systems, centralized control is implemented quite commonly within the bodies of metazoan animals. Here the brain is a central processor, which takes in information from throughout the system, integrates it, compares it to global needs and goals, and makes decisions that are sent back out into the system. As one looks across the diversity of animals, there is a clear trend toward the most sophisticated organisms having central processors of increasing complexity. This pattern is so striking and obvious that a strong explanation for why animals should ever use decentralized control is necessary.

A simple hypothesis for when animals should use centralized versus decentralized control was proposed by Seeley [24]. Simply put, animals use decentralized control when either the evolution of centralized control is prevented by strong constraints or decentralized control confers higher fitness. Constraints can take many forms. Chiefly, centralized control may be nearly impossible to implement given the physical character of the system. Contrast the nature of a nervous system, for example, with the internal dynamics of cell biology, which depend on collisions facilitated by molecular crowding [61,62]. It is difficult to imagine that a well-ordered system, like the nervous system, could be implemented within a cell no matter what the selective pressure. It is no surprise then that cell biology seems to be the branch of biology in which self-organization plays the largest role. With respect to when decentralized control confers higher fitness than centralized control, centralized control is powerfully efficient and accurate but costly in terms of metabolic power and risky in terms of not being robust in the case of damage to the central processor. Decentralized control is less efficient and accurate, but it is robust to damage to any part of the system and is cheap to implement in metabolic terms. In addition, for centralized control to work well, much information must be collected and transmitted quickly to a central organizer. If the scope of the problem is vast, compared to the information processing capability of the individuals, then this may be impossible. Both of these contexts are important in social insect biology, for example, where any insect (even the queen) is likely to be killed and in which workers have limited information processing capabilities relative to the scope of the problems they must solve. Beyond the social insects, it is also the case that decentralized control may be favored because the loss of efficiency with decentralized control is less than the cost of managers with centralized control. Given that managers do no work other than to direct others and can be hugely costly to produce and maintain, it is likely that this selective context exists for some human social organizations (even if action is rarely taken as a result of it).

ACKNOWLEDGMENTS

I thank Cameron Jasper, James Devillers, and William Smith for careful reading of the manuscript and James Devillers for the invitation to contribute to this edition. Work discussed here was supported by a postdoctoral fellowship to the author from the National Science Foundation of the United States.

REFERENCES

1. T. Misteli, *The concept of self-organization in cellular architecture*, J. Cell Biol. 155 (2001), pp. 181–185.
2. S. Camazine, J.L. Deneubourg, N.R. Franks, J. Sneyd, G. Theraulaz, and E. Bonabeau, *Self-Organization in Biological Systems*, Princeton University Press, Princeton, NJ, 2001.
3. I. Salazar-Ciudad and J. Jernvall, *A gene network model accounting for development and evolution of mammalian teeth*, Proc. Natl. Acad. Sci. USA 99 (2002), pp. 8116–8120.
4. I.D. Couzin, J. Krause, R. James, G.D. Ruxton, and N.R. Franks, *Collective memory and spatial sorting in animal groups*, J. Theor. Biol. 218 (2002), pp. 1–11.
5. I.D. Couzin, J. Krause, N.R. Franks, and S.A Levin, *Effective leadership and decision-making in animal groups on the move*, Nature 433 (2005), pp. 513–516.
6. R.E. Carazo-Salas and P. Nurse, *Self-organization of interphase microtubule arrays in fission yeast*, Nat. Cell Biol. 8 (2006), pp. 1102–U1194.
7. D.J.T. Sumpter, *The principles of collective animal behaviour*, Philos. T. Roy. Soc. B 361 (2006), pp. 5–22.
8. T.D. Seeley, P.K. Visscher, T. Schlegel, P.M. Hogan, N.R. Franks, and J.A.R. Marshall, *Stop signals provide cross inhibition in collective decision-making by honeybee swarms*, Science 335 (2012), pp. 108–111.
9. A.M. Turing, *The chemical basis of morphogenesis*, Philos. T. Roy. Soc. B 237 (1952), pp. 37–72.

10. A. Gierer and H. Meinhard, *Theory of biological pattern formation*, Kybernetik 12 (1972), pp. 30–39.

11. G. Nicolis and I. Prigogine, *Self-Organization in Nonequilibrium Systems: From Dissipative Structures to Order through Fluctuations*, Wiley, New York, 1977.

12. S.A. Kauffman, *The Origin of Order: Self-Organization and Selection in Evolution*, Oxford University Press, Oxford, UK, 1993.

13. E. Bonabeau, G. Theraulaz, J.L. Deneubourg, S. Aron, and S. Camazine, *Self-organization in social insects*, Trends Ecol. Evol. 12 (1997), pp. 188–193.

14. A. Kurakin, *Self-organization vs watchmaker: Stochastic gene expression and cell differentiation*, Dev. Genes Evol. 215 (2005), pp. 46–52.

15. E. Karsenti, *Self-organization in cell biology: A brief history*, Nat. Rev. Mol. Cell Bio. 9 (2008), pp. 255–262.

16. B.R. Johnson and S.K. Lam, *Self-organization, natural selection, and evolution: cellular hardware and genetic software*, Bioscience 60 (2010), pp. 879–885.

17. J.L. Deneubourg and S. Goss, *Collective patterns and decision making*, Ethol. Ecol. Evol. 1 (1989), pp. 295–311.

18. T. Surrey, F. Nedelec, S. Leibler, and E. Karsenti, *Physical properties determining self-organization of motors and microtubules*, Science 292 (2001), pp. 1167–1171.

19. J.A.R. Marshall and N.R. Franks, *Colony-level cognition*, Curr. Biol. 19 (2009), pp. R395–R396.

20. A.S. Mikheyev and W.R. Tschinkel, *Nest architecture of the ant* Formica pallidefulva: *Structure, costs and rules of excavation*, Insectes Soc. 51 (2004), pp. 30–36.

21. W.R. Tschinkel, *The nest architecture of the Florida harvester ant,* Pogonomyrmex badius, J. Insect Sci. 4, (2004), p. 19.

22. B.R. Johnson, *Pattern formation on the combs of honeybees: Increasing fitness by coupling self-organization with templates*, Proc. Roy. Soc. B-Biol. Sci. 276 (2009), pp. 255–261.

23. J. Jaeger, *Modelling the* Drosophila *embryo*, Mol. Biosyst. 5 (2009), pp. 1549–1568.

24. T.D. Seeley, *When is self-organization used in biological systems?* Biol. Bull. 202 (2002), pp. 314–318.

25. B.R. Johnson, *A self-organizing model for task allocation via frequent task quitting and random walks in the honeybee*, Am. Nat. 174 (2009), pp. 537–547.

26. T.D. Seeley and R.A. Morse, *The nest of the honey bee* (Apis mellifera L.), Insectes Soc. 23 (1976), pp. 495–512.

27. E. Bonabeau, G. Theraulaz, J.L. Deneubourg, N.R. Franks, O. Rafelsberger, J.L. Joly, and S. Blanco, *A model for the emergence of pillars, walls and royal chambers in termite nests*, Philos. T. Roy. Soc. B 353 (1998), pp. 1561–1576.

28. E.O. Wilson, *The Insect Societies*, Harvard University Press, Cambridge, MA, 1971.

29. C.D. Michener, *The Social Behavior of the Bees*, Harvard University Press, Cambridge, MA, 1974.

30. C. Jost, J. Verret, E. Casellas, J. Gautrais, M. Challet, J. Lluc, S. Blanco, M.J. Clifton, and G. Theraulaz, *The interplay between a self-organized process and an environmental template: Corpse clustering under the influence of air currents in ants*, J. R. Soc. Interface 4 (2007), pp. 107–116.

31. Y. Lensky and Y. Slabezki, *The inhibiting effect of the queen bee* (Apis mellifera L) *footprint pheromone on the construction of swarming queen cups*, J. Insect Physiol. 27 (1981), pp. 313–323.

32. J.B. Free and I.H. Williams, *Factors determining food storage and brood rearing in honeybee* (Apis mellifera *L.*) *comb*, J. Entomol. Ser. A. 49 (1974), pp. 47–63.

33. T. Pankiw, *Brood pheromone regulates foraging activity of honey bees (Hymenoptera: Apidae)*, J. Econ. Entomol. 97 (2004), pp. 748–751.

34. M.L. Winston, *The Biology of the Honeybee*, Harvard University Press, Cambridge, MA, 1987.
35. T.D. Seeley, *Social foraging in honey bees: How nectar foragers assess their colony nutritional status*, Behav. Ecol. Sociobiol. 24 (1989), pp. 181–199.
36. S. Camazine, *Self-organizing pattern formation on the combs of honeybee colonies*, Behav. Ecol. Sociobiol. 28 (1991), pp. 61–76.
37. C. Dreller and D.R. Tarpy, *Perception of the pollen need by foragers in a honeybee colony*, Anim. Behav. 59 (2000), pp. 91–96.
38. U. Wilensky, NetLogo. http://ccl.northwestern.edu/netlogo/. Center for Connected Learning and Computer-Based Modeling, Northwestern University, Evanston, IL, 1999
39. K. Crailsheim, *Interadult feeding of jelly in honeybee* (Apis mellifera L) *colonies*, J. Comp. Physiol. B 161 (1991), pp. 55–60.
40. K. Crailsheim, *The flow of jelly within a honeybee colony*, J. Comp. Physiol. B 162 (1992), pp. 681–689.
41. E.E. Southwick and G. Heldmaier, *The temperature control in honey bee colonies*, Bioscience 37 (1987), pp. 395–399.
42. B. Heinrich, *The Hot-Blooded Insects: Strategies and Mechanisms of Thermoregulation*, Harvard University Press, Cambridge, MA, 1993.
43. D.M. Gordon, *The organization of work in social insect colonies*, Nature 380 (1996), pp. 121–124.
44. S.N. Beshers and J.H. Fewell, *Models of division of labor in social insects*, Annu. Rev. Entomol. 46 (2001), pp. 413–440.
45. B.R. Johnson, *Reallocation of labor in honeybee colonies during heat stress: The relative roles of task switching and the activation of reserve labor*, Behav. Ecol. Sociobiol. 51 (2002), pp. 188–196.
46. T.D. Seeley, S. Camazine, and J. Sneyd, *Collective decision making in honey bee: How colonies choose among nectar sources*, Behav. Ecol. Sociobiol. 28 (1991), pp. 277–290.
47. R. Beckers, J.L. Deneubourg, and S. Goss, *Trail laying behavior during food recruitment in the ant* Lasius niger (L.), Insectes Soc. 39 (1992), pp. 59–72.
48. T.D. Seeley, *Adaptive significance of the age polyethism schedule in honeybee colonies*, Behav. Ecol. Sociobiol. 11 (1982), pp. 287–293.
49. B.R. Johnson, *Global information sampling in the honey bee*, Naturwissenschaften 95 (2008), pp. 523–530.
50. B.R. Johnson, *Division of labor in honeybees: Form, function, and proximate mechanisms*, Behav. Ecol. Sociobiol. 64 (2010), pp. 305–316.
51. M. Lindauer, *Ein Beitrag zur Frage der Arbeitsteilung im Bienenstaat*. Z. Vgl Physiol. 34 (1952), pp. 299.
52. D. Morisato and K.V. Anderson, *Signaling pathways that establish the dorsal-ventral pattern of the* Drosophila *embryo*, Annu. Rev. Genet. 29 (1995), pp. 371–399.
53. C.S. Thummel, *From embryogenesis to metamorphosis: The regulation and function of* Drosophila *nuclear receptor superfamily members*, Cell 83 (1995), pp. 871–877.
54. R. Rivera-Pomar and H. Jackle, *From gradients to stripes in* Drosophila *embryogenesis: Filling in the gaps*, Trends Genet. 12 (1996), pp. 478–483.
55. J.W. Bodnar and M.K. Bradley, *Programming the* Drosophila *embryo 2—From genotype to phenotype*, Cell Biochem. Biophys. 34 (2001), pp. 153–190.
56. H. Meinhardt, *Model of pattern formation in insect embryogenesis*, J. Cell Sci. 23 (1977), pp. 117–139.
57. H. Meinhardt, *Models for the generation and interpretation of gradients*, Cold Spring Harb. Perspect. Biol. 1, 2009.

58. J. Jaeger, M. Blagov, D. Kosman, K.N. Kozlov, Manu, E. Myasnikova, S. Surkova, C.E. Vanario-Alonso, M. Samsonova, D.H. Sharp, and J. Reinitz, *Dynamical analysis of regulatory interactions in the gap gene system of* Drosophila melanogaster, Genetics 167 (2004), pp. 1721–1737.

59. A.D. Lander, *Pattern, growth, and control*, Cell 144 (2011), pp. 955–969.

60. N.M. Stroeymeyt, M. Giurfa, and N.R. Franks, *Improving decision speed, accuracy and group cohesion through early information gathering in house-hunting ants*, Plos One 5 (2010), p. e13059.

61. R.J. Ellis, *Macromolecular crowding: An important but neglected aspect of the intracellular environment*, Curr. Opin. Struct. Biol. 11 (2001), pp. 114–119.

62. R.J. Ellis, *Macromolecular crowding: Obvious but underappreciated*, Trends Biochem. Sci. 26 (2001), pp. 597–604.

4 Models for the Recruitment and Allocation of Honey Bee Foragers

Mary R. Myerscough, James R. Edwards, and Timothy M. Schaerf

CONTENTS

ABSTRACT

Honey bee foragers recruit other bees to visit productive patches of flowers by advertising, on their return to the hive, their source of nectar or pollen by a waggle dance to indicate the location and quality of the source. The distribution of foragers among sources, generated by this waggle dance recruitment, has been modeled with differential equations. Differential equation models either represent each stage of foraging—dancing, visiting forage sites, waiting as an unemployed forager, following a dance—or more simply divide the foraging force into pools of workers where each pool exploits a different site. Other models include receiver bees who work in the hive and modulate the foragers' dance response to the quality of forage sites. Simulation and individual-oriented models for honey bee foraging are briefly discussed as well as foraging in other social insects. A brief description of the application of ideas from honey bee foraging to create bee-based computer algorithms concludes the chapter.

KEYWORDS

Self-organization, Collective decision making, Nature-inspired algorithms, Differential equation model

4.1 INTRODUCTION

The effectiveness of food collection by workers in social insect colonies is one of the key determinants of whether a colony will thrive or die. Foraging will be most efficient if workers forage at easily accessible food sources that are rich in required nutrients. Honey bees forage in landscapes that provide flower patches of variable nutritional quality, of variable accessibility, both in terms of distance from the nest and ease of exploitation and of variable availability; many flowers only provide nectar and pollen at certain times of the day or they may be made unavailable to foragers by the presence of predators. The task of the hive is to find resources and optimize resource use by scouting, by dance communication, and by intercaste communication inside the hive.

In this chapter, we examine models for how the hive allocates its available foragers among known patches of flowers. We will principally focus on differential equation models for forager allocation, but we will also mention a variety of other types of model. We will conclude by briefly reviewing how honey bees' strategies for optimizing food collection have inspired computer algorithms.

4.2 BIOLOGY OF HONEY BEE FORAGING

Honey bee colonies forage for nectar, pollen, water, and substrates for building internal structures in the hive. There is evidence that different bees may specialize in foraging for one or more of these substances. This may be genetically programmed or due to the past experience of the individual. Here, we shall concentrate on nectar foraging; this has been well studied as it is easy to observe, and nectar foraging can be manipulated or imitated by providing feeders for bees that are stocked with sugar syrup.

Individual foragers leave the hive to scout for a flower patch that provides suitable nectar. Once a bee finds a suitable patch of flowers, she returns to the hive, and if the forage source provides sufficiently high-quality nectar and there is a need for nectar in the hive, then she will perform a waggle dance on the comb near the entrance of the hive to communicate the location and quality of the source that she has found (see Seeley [1] for a more complete description). A waggle dance consists of a number of waggle runs. In each waggle run, the dancer runs forward, rapidly waggling her abdomen from side to side. At the end of the run, she turns either to the left or to the right and returns to the start of the run and repeats the process. Usually the dancer turns left and right alternately at the end of each run so that the whole dance traces out a figure-of-eight pattern. The length of the waggle run indicates the distance of the forage source from the hive, and the angle between the waggle run and the vertical indicates the direction of the source relative to the sun [2].

Potential foragers may follow the dancer as she advertises the source that she has found. They follow close behind her and may follow for several cycles. Sometimes several potential foragers may be following a dancer at once, but it is not clear that they can all gather the dance information at the same time.

A bee that has become an active forager through following a dance leaves the hive to seek the advertised forage source. As well as the information from the dance, she may also use the scent from the flowers that the dancer has been visiting to find the advertised forage source. On her return to the hive, she may also advertise the source by waggle dancing or she may continue to forage without dancing or cease to forage.

Whether or not a returning forager performs a waggle dance depends not only on her assessment of the profitability of the nectar source that she is visiting but also on whether she is able to rapidly unload the nectar that she gathers. Receiver bees, who work in the hive, collect nectar from incoming foragers and take it to the comb where it is stored and concentrated to become honey. If a forager has to wait for an extended time to unload her nectar, then she is less likely to dance before returning to the forage sources [3].

Foragers who encounter a predator, overcrowded forage sources, or the presence of an alarm signal may produce a stop signal at the hive. They apply this signal to other foragers who are foraging at nectar sources that carry the same scent [4,5]. This stop signal is produced by the forager butting her head against the bee that she is communicating with and making a short buzzing noise. The recipient of the signal is unlikely to continue dancing and so the dangerous source is not advertised as much and the number of foragers that visit it may decline.

4.3   SCALE AND TYPE OF MODELS

Social insects typically exist at intermediate scales. Depending on the behavior that is being modeled, a colony of bees can be regarded as a varying density of bees that can be modeled with differential equations, as a probability distribution, as undergoing a stochastic process, or as a number of discrete individuals that interact with one another according to predetermined rules. Most models for foraging fall into the large-scale category, which uses ordinary differential equations [6], or the small scale, which uses simulations of individual bees [7]. Models at different scales have different strengths and will give different types of insight into foraging and so different types of models have the capacity to inform and illuminate each other, although this has been more evident in models for nest site selection [8] than models for foraging.

In this chapter, we focus on differential equation models for honey bee foraging— these are principally concerned with communication by waggle dancing—but we will also refer to simulation models as appropriate.

4.4   FUNDAMENTAL MODEL: LINKING INDIVIDUAL
AND COLLECTIVE BEHAVIOR

How does a colony of several thousand bees ensure that it is exploiting the best available forage sources and not spending undue time on poor sources? In fact, before this question can be answered, it is important to know how an individual bee assesses the quality of a source and whether that assessment is done by foragers or by receiver bees or by some other mechanism. Experiments by Seeley et al. [9] showed that foragers assess the quality of a source based on both the sugar concentration in the nectar and the distance of the source from the nest. Other factors such as crowding at the source and ease-of-access to nectar in the flowers may also be relevant, but their experiments did not consider this.

Further, Seeley et al. [9] showed that the number of waggle runs that a forager danced and the rate that she made her foraging trips (foraging tempo) increased with the quality of the forage source, and the rate that foragers ceased to forage from a source decreased as quality increased. This suggested that individual behavior leads to a distribution of foragers where there are high forager numbers at high-quality sources and low forager numbers at low-quality sources.

Using the observations of Seeley et al. [9], Camazine and Sneyd [6] constructed a differential equation model to represent the flow of foragers between different behavioral classes.

Like all subsequent differential equation models for foraging, this is a compartment model. The model traces the transitions that foragers make between resting, following dances, foraging, unloading, and dancing. The flow of foragers between different states is illustrated in Figure 4.1. The model assumes that when a bee begins foraging, she follows dances on the dance floor on a comb inside the entrance of the

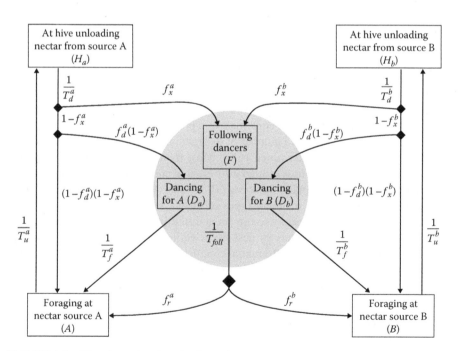

FIGURE 4.1 Flowchart showing the flow of bees between compartments in the model of Camazine and Sneyd (1991) where there are two sources and using the notation of Equations 4.1 through 4.4. (Adapted from Camazine, S and Sneyd, J., *J. Theor. Biol.*, 149, 547, 1991.)

hive. If there are dances for several sources in the hive, she will choose one to follow and learn the location of that source. She then flies to that source and gathers nectar before returning to the hive and unloading. Once she has unloaded, she has the choice as to whether to continue to forage at that source or to follow another dance. If she does continue to forage, then she has the choice of whether to dance or not. The rate of recruitment by dancing is dependent on both the number of foragers dancing for a source and the number of waggle runs each dancer performs, which is, in turn, a function of the quality of the source. The rate of abandonment and failure to dance are also functions of source quality.

Let A be the number of bees occupied in actively foraging at source A and B the number of bees foraging at source B. Let H_a be the number of bees that are at the hive unloading nectar from source A and D_a the number of bees that are dancing for source A; H_b and D_b are similar classes for source B. Finally, let F be the number of bees that are on the dance floor and seeking to follow dances. Camazine and Sneyd's model tracks the number of bees in each class (compartment) through the following differential equations:

Rate of change of followers:

$$\frac{dF}{dt} = \underbrace{\frac{f_x^a}{T_d^a}A + \frac{f_x^b}{T_d^b}B}_{\substack{\text{rate that} \\ \text{foragers abandon foraging} \\ \text{at sources A and B and rejoin} \\ \text{follower class}}} - \underbrace{\frac{1}{T_{foll}}F}_{\substack{\text{rate that} \\ \text{followers are} \\ \text{recruited to} \\ \text{sources A or B}}} \tag{4.1}$$

where

T_d^a and T_d^b are the times it takes for a bee to decide to abandon source A or B, respectively, or continue once it has unloaded;

f_x^a and f_x^b are the probability that a bee will abandon source A or B, respectively; and

T_{foll} is the time that it takes a bee from the start of following dances until it is recruited to a source.

As this is a deterministic differential equation model, these times could be regarded as means of a distribution of times. Clearly, not all bees will take exactly the same time to transition from one compartment to another.

Rate of change of the number of foragers who are actively foraging at source A:

$$\frac{dA}{dt} = \underbrace{\frac{(1-f_x^a)(1-f_d^a)}{T_d^a}H_a}_{\substack{\text{rate that bees return} \\ \text{directly to foraging after} \\ \text{unloading without dancing}}} + \underbrace{\frac{1}{T_f^a}D_a}_{\substack{\text{rate of} \\ \text{returning to} \\ \text{foraging after} \\ \text{dancing}}} + \underbrace{\frac{f_r^a}{T_{foll}}F}_{\substack{\text{rate of} \\ \text{recruitment} \\ \text{of followers} \\ \text{by dancers}}} - \underbrace{\frac{1}{T_u^a}A}_{\substack{\text{rate that} \\ \text{foragers return} \\ \text{and start} \\ \text{unloading}}}$$

$$\tag{4.2}$$

where

$(1 - f_x^a)$ is the probability that a bee does not abandon source A immediately after unloading;

$(1 - f_d^a)$ is the probability that a bee returns to source A without becoming a dancer;

T_f^a is the time that a dancer takes to dance for source A before returning to forage;

T_u^a is the time that a bee takes to forage at source A before unloading in the hive; and

f_r^a is the probability that a follower will be recruited to source A and this probability will be a function of the number of dancers at each source, D_a and D_b.

Rate of change of the number of bees, who are committed to source A and who are unloading in the hive:

$$\frac{dH_a}{dt} = \underbrace{\frac{1}{T_u^a} A}_{\substack{\text{rate that foragers} \\ \text{return and start} \\ \text{unloading}}} - \underbrace{\frac{1}{T_d^a} H_a.}_{\substack{\text{rate that bees finish} \\ \text{unloading}}} \qquad (4.3)$$

Rate of change of numbers of dancers for source A:

$$\frac{dD_a}{dt} = \underbrace{\frac{f_d^a(1 - f_x^a)}{T_d^a} H_a}_{\substack{\text{rate that bees start} \\ \text{dancing for source A} \\ \text{after unloading}}} - \underbrace{\frac{1}{T_f^a} D_a.}_{\substack{\text{rate that bees cease} \\ \text{dancing for source A and} \\ \text{return to source A to forage}}} \qquad (4.4)$$

Here, $f_d^a(1 - f_x^a)$ is the probability that a bee who has finished unloading will not abandon source A but will dance for it.

A set of equations similar to (4.2) through (4.4) apply to B, H_b, and D_b. This model assumes that the total number of bees that are available to become foragers is constant and that bees cycle through the tasks of following, foraging, unloading, and dancing and that all bees who engage in foraging must be committed to either source A or B or be actively following dances.

If source A is better quality than source B, then this can be expressed mathematically in this model by setting $f_r^a > f_r^b$, $f_d^a > f_d^b$ and $f_x^a/T_d < f_x^b/T_d^b$. Using these ideas, Camazine and Sneyd were able to show that the model could reproduce the experimental results.

Seeley et al. [9] had shown that if bees had the choice of two experimental feeders, one (the south feeder) with high-quality nectar and the other (the north feeder) with acceptable but lower-quality nectar, then the high-quality feeder would recruit more foragers than the lower-quality feeder. If the feeders were swapped after 4 h, so that the north feeder had high-quality nectar and the south feeder had lower-quality nectar, then the new high-quality source would recruit more foragers and the new low-quality source would lose foragers (see Figure 4.2a). Using parameter values

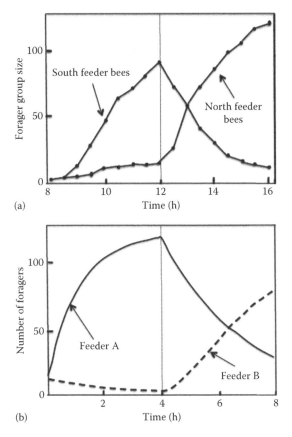

(a)

(b)

FIGURE 4.2 Comparison of (a) the experimental results obtained by Seeley et al. [9] with (b) the results of the model of Camazine and Sneyd. (Plot a: Redrawn from Seeley et al., *Behav. Ecol. Sociobiol.*, 28, 277, 1991; plot b: Redrawn from Camazine, S and Sneyd, J., *J. Theor. Biol.*, 149, 547, 1991.)

obtained experimentally, Camazine and Sneyd [6] were able to reproduce this behavior with their model (Figure 4.2b).

This model is the earliest mathematical modeling for foraging. It linked individual behavior—expressed mathematically by the likelihood that a bee would choose a state as given in the probabilities $f_x^a, f_x^b, f_d^a, f_d^b, f_r^a$ and f_r^b, and in the average times that bees remained in different states, $T_{foll}, T_d^a, T_d^b, T_f^a, T_f^b, T_u^a$ and T_u^b—with collective hive behavior expressed in the way that the hive distributed its foragers over forage sources.

4.5 FROM DIVERSE TASKS TO DIVERSE BEES

Camazine and Sneyd's model [6] with different behavioral compartments was important as it allowed observations to be explicitly incorporated into the model. However, the model can be streamlined by collecting the bees who are dancing, foraging, or unloading into a single group that is committed to a source [10]. This gives scope for relatively uncomplicated models for large number of sources or subclasses within the hive population.

A forager completes a foraging cycle—flying to the source, gathering nectar, returning to the hive, unloading, and, perhaps, dancing—over a timescale of several minutes, but recruiting significantly more dancers or abandoning foraging at a source occurs on a longer timescale that is closer to an hour. Therefore, it makes sense, mathematically, to group all committed foragers into a single class, regardless of their behavior and monitor only recruitment or abandonment. With a simpler model, it is also easy to include independent discovery of a source by scouting.

The model of Cox and Myerscough [10] can be written purely in terms of the number of foragers for each source j, given by F^j (this is equivalent to $A + H_a + D_a$ or $B + H_b + D_b$ in Camazine and Sneyd's model [6]) and the number of unemployed foragers U. In the simplest terms, the equation for the rate of change of foragers committed to source j is

$$\frac{dF^j}{dt} = B_1^j U - B_2^j F^j + B_3^j U \tag{4.5}$$

where
the first term on the right-hand side represents recruitment by dancers of new foragers for source j;
the second term represents abandonment of source j by foragers; and
the third term represents unemployed bees who find source j by independent foraging and become committed to that source.

A flowchart illustrating the dynamics of the model is given in Figure 4.3. The coefficients B_1^j, B_2^j, and B_3^j are not constant but depend on bee behavior.

To model the effect of dancing for source j, we need to know the proportion of foragers committed to source j that are dancing at any one time. If we assume that foragers are evenly spread throughout the foraging cycle, then the proportion of foragers that are dancing for source j is the same as the proportion of time in the foraging cycle that each forager spends dancing, multiplied by the proportion of foragers p_d^j that are committed to source j who perform a dance rather than merely return to foraging after unloading. The proportion p_d^j is the same as the probability that any given scout who is committed to source j will dance at a given return to the hive. If T_d^j is the average time that a single dancer spends dancing for source j on a given return to the hive and r^j is the average visit rate for source j, so that the average total time spent in the foraging cycle for source j is $1/r^j$, then the total number of dancers for source j at any given time is $D^j = F^j p_d^j T_d^j r^j$. If we assume that the probability that an unemployed bee seeks to follow a dance is p_f and that the rate of following a dance for source j is given by the proportion of dances that are for source j, then

$$B_1^j = p_f \frac{D^j}{\sum_{k=1}^m D^k} \tag{4.6}$$

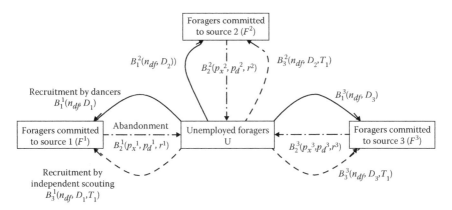

FIGURE 4.3 A flowchart showing the flow of bees between compartments in the model of Cox and Myerscough. (From Cox, M.D. and Myerscough, M.R. *J. Theor. Biol.*, 223, 179, 2003.) Here all foragers committed to a source are grouped into a single compartment and the transition rates are functions of the number of dancers advertising a site D_j; the probability of abandoning source j, p_x^j; the probability of dancing for source j, p_d^j; the average visit rate for source j, r^j; and the number of bees on the dance floor, n_{df}. More detailed explanations of the symbols are given in the text.

where m is the total number of sources in the model. The probability that an unemployed forager finds a dance to follow will depend on how many dances are taking place on the dance floor and how many nondancing bees are present. The model assumes that these nondancing bees inhibit potential followers from finding dancers. If we assume that there are n_{df} bees on the dance floor (most of whom will not be dancers but resting foragers or receiver bees or other hive bees), then the probability of contacting a dancer to follow is

$$p_f = \frac{1}{n_{df}} \sum_{k=1}^{m} D_k \tag{4.7}$$

so that

$$B_1^j = \frac{1}{n_{df}} D_j = \frac{1}{n_{df}} F^j p_d^j T_d^j r^j \tag{4.8}$$

If we assume that foragers that abandon sources do not dance, then the rate that foragers abandon source j in any given foraging cycle is proportional to the probability that a forager does not dance and that a nondancing forager abandons the source in any given foraging trip. Hence,

$$B_2^j = p_x^j (1 - p_d^j) r^j \tag{4.9}$$

where p_x^j is the probability of a nondancing forager abandoning the source.

We assume that an unemployed bee undertakes independent scouting when she cannot find a dance to follow within a period of time T_l. Hence,

$$B_3^j = \frac{1}{m}\frac{1}{T_\ell}\left(1 - \frac{1}{n_{df}}\sum_{k=1}^{m}D^k\right) \tag{4.10}$$

where the term in brackets is the probability of only finding nondancing bees. This assumes that independent scouts have an equal probability of finding each of the m available sources.

If the total number of bee available to become foragers is fixed, that is,

$$U + \sum_{k=1}^{m}F^k = N_T \tag{4.11}$$

where N_T is a constant, then the rate of change of the number of unemployed bees is given by

$$\frac{dU}{dt} = \sum_{k=1}^{m}\left(B_2^k F^k - (B_1^k + B_3^k)U\right) \tag{4.12}$$

Many of the parameters of this model such as the visit rate r^j for source j and the probabilities of dancing p_d^j (and so of recruiting by dancing) and of abandoning a source j are strongly related to the profitability of that source. Cox and Myerscough [10] explicitly incorporated source profitability into their model and used the model to explore how foragers distributed themselves over multiple sources of varying profitability and then to explore how a forager workforce that had varying degrees of phenotypic variability [11,12] affected a hive's ability to gather food. While Cox and Myerscough [10] considered genetic factors as the source of variability, other factors such as age are just as likely to introduce variation among foragers [13].

Because this model simplifies bee behavior and collects all the behaviors in the foraging cycle into one class, it makes it conceptually and computationally easier for other types of variations to be included.

Of course, as foragers are divided into more and more compartments in a model, the number of foragers in each compartment becomes smaller and smaller, and so the assumption that underlies the use of ordinary differential equation models breaks down. That is, there are not enough individuals in each class to justify the assumption that a smooth average adequately represents their behavior. In this case, it is necessary to go to individual-based models where each bee is represented individually and rules are created for how individuals interact with one another and with their environment. These models will be briefly described in Section 4.8.

4.6 ROLE OF HIVE BEES: MODELING INTERACTIONS BEYOND FORAGERS AND DANCERS

Foragers are, of course, not isolated from the rest of the hive and interact with the other bees in the colony in other ways besides dancing. The main class of hive bees that interact with active foragers are receiver bees that collect nectar from incoming nectar foragers and store it in the combs of the hive. There is some evidence that a forager's tendency to rest, rather than return to the forager source and to dance or not, depends on the time that it takes her to find receiver bees and unload [1,3]. If it takes a long time for a forager to offload the nectar that she has collected, then she is less likely to dance and less likely to return to forage. Receiving nectar may take longer when there is an unusually large amount of nectar being brought into the hive or when the combs are nearly full so that receivers have to search for cells to store the nectar that they are carrying, which means that they are away from the hive entrance for a long period of time. In either case, the need to forage is less urgent when there is plenty of nectar being brought into the hive or when the supplies of stored nectar are high. Edwards and Myerscough [14] have constructed a model to explore how a forager's search time for nectar receivers may be a vehicle for communicating the hive's need for nectar to the foraging work force and inhibiting foraging if it is not immediately necessary. Figure 4.4 illustrates how search time information allows foragers and receivers to transfer information from one class to the other.

Edwards and Myerscough devised an ordinary differential equation model, which includes the effects of feedback from search time on foragers' behavior and nectar

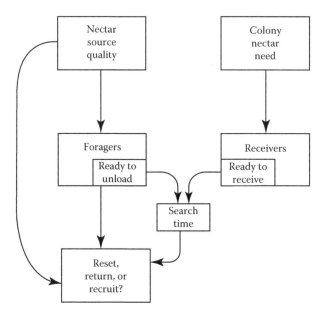

FIGURE 4.4 Diagram illustrating how foragers and receiver bees implicitly exchange information via the search time. (From Edwards, J.R. and Myerscough, M.R., *J. Theor. Biol.*, 271, 64, 2011.)

collection. There are two classes of bees in this model: foragers, numbering F, and receiver bees with numbers denoted by R. Each of these classes contains a subclass of bees that are active in nectar exchange: $\bar{F}$, the number of foragers actively unloading and $\bar{R}$, the number of receivers actively receiving or waiting to receive. The model depends on the key variable, search time, $S(t)$, which is given by

$$S(t) = s\frac{Q_M}{Q}\frac{\bar{F}(t) + \bar{R}(t)}{\bar{R}(t)} \tag{4.13}$$

where $S(t)$ is the time taken to find a receiver bee and unload, which is assumed to depend on the ratio of the quality of the nectar to be unloaded, Q, to some scaling factor Q_M, which may be regarded as the maximum quality that the bees expect. The search time also depends on the fraction of bees in the entrance that are receivers, $\bar{R}(t)/(\bar{F}(t) + \bar{R}(t))$. This assumes that the only bees that a returning forager will encounter are receivers or other foragers. The parameter s is a constant that reflects the time that a returning forager takes to determine if a bee that she encounters is willing to receive her nectar. We assume that the actual unloading time is small compared to the search time.

We assume that the probability that a forager will dance or abandon foraging and return to resting depends on the search time and on the quality of the source. If search time is high, that is, a forager takes a long time to find a receiver willing to receive her load of nectar, and the quality of the source that she is exploiting is low, then she is unlikely to dance and prone to rest rather than return to foraging. A forager who is exploiting a high-quality source and is able to rapidly unload her nectar is most likely to dance and to continue foraging. Under these assumptions, the equation for the rate of change of forager numbers F is

$$\frac{dF}{dt} = f_r F(t)\Gamma_1(S(t), Q) - f_s F(t)\Gamma_2(S(t), Q) \tag{4.14}$$

The first term on the right-hand side represents recruitment of new foragers to the source and the second term represents the rate that foragers cease to be active and return to rest. We assume that scouting is not important in this model. The constants f_r and f_s control the rate of recruitment and return to rest, respectively. The functions $\Gamma_1(S,Q)$ and $\Gamma_2(S,Q)$ are sigmoid functions shown in Figure 4.5 and given explicitly in [14].

The rate of change of the number of foragers in the entrance of the hive waiting to unload is given by

$$\frac{d\bar{F}}{dt} = \frac{F - \bar{F}}{f_a} - \frac{\bar{F}}{S} \tag{4.15}$$

The first term on the right-hand side gives the rate that foragers arrive at the hive and start to seek to unload. Here, $F - \bar{F}$ is the number of foragers that are not waiting to

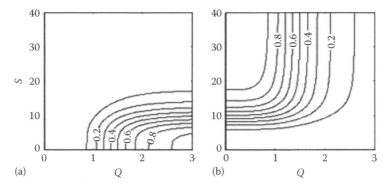

FIGURE 4.5   Contour plots of functions (a) $\Gamma_1(S,Q)$ and (b) $\Gamma_2(S,Q)$. These functions both have ranges between 0 and 1; $\Gamma_1(S,Q)$ is high for sources of high quality when waiting time is short and $\Gamma_2(S,Q)$ is high for sources of low quality when waiting times are long.

unload, and f_a is the time that a forager takes to complete one foraging cycle, so that $1/f_a$ is the rate of arrival of foragers at the hive, assuming that foragers are evenly distributed through the foraging cycle.

If we assume that the total number of receiver bees R_0 is constant, then the equation for the rate of change of receiver bees in the entrance is

$$\frac{d\bar{R}}{dt} = \frac{R_0 - \bar{R}}{r_s} - \frac{\bar{F}}{S} \tag{4.16}$$

where the first term represents the rate of return of receiver bees to the entrance of the hive and to readiness to receive and the second term represents the rate that foragers find and unload to a receiver, assuming that this can be well-approximated by a relationship of one forager to one receiver bee. The constant r_s represents the time that it takes a single receiver to offload the nectar that she carries into a cell in the honey comb.

Using this model, Edwards and Myerscough [14] showed that tuning dancing and resting rate using information from search time limited the number of foragers that were recruited during times when high-quality nectar was abundant (Figure 4.6) but allowed foragers to recruit freely when nectar was less abundant. Essentially the forager workforce was limited by the limited number of receiver bees.

In a real hive, when foragers are faced with long waiting times before they unload, they perform a tremble dance, which may recruit more receivers and discourage other foragers from visiting the same source [15]. If the tremble dance is included in the model so that the total number of receivers $R(t)$ is no longer constant, then the forager numbers continue to increase but more slowly than the initial rapid increase.

There is good reason why a hive might be parsimonious with their foragers; there is a limit to how much nectar a hive needs to store, and foragers are the most vulnerable of all the behavioral castes in the hive. If many bees go outside to forage

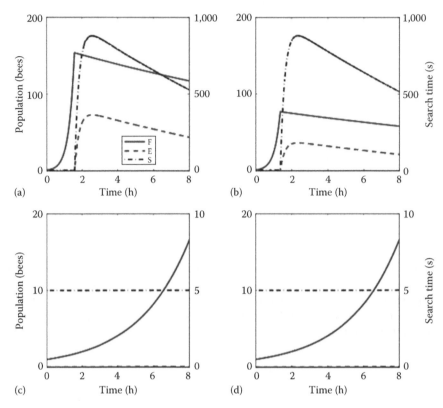

FIGURE 4.6 Results from the model of Edwards and Myerscough, showing how the limited number of receiver bees curtails the recruitment of foragers when nectar is abundant, but not when nectar is scarce. In (a) and (b), the hive is exploiting a high-quality source with $Q=3$, but in (a) the hive has 100 receivers and in (b) 50 receivers. In (c) and (d), the hive is exploiting a relatively low-quality source with $Q=1.3$ with 100 receivers in (c) and 50 receivers in (d). (From Edwards, J.R. and Myerscough, M.R., *J. Theor. Biol.*, 271, 64, 2011.)

and there is a rapid loss of foragers, then the hive itself may be endangered [13]. It is also possible that the availability or otherwise of receivers may reflect the capacity of the hive to store nectar. A receiver who has to search for space to unload her nectar in a honey comb that is almost full will take longer than if there are many cells with space for nectar storage. In this way, the interaction between receivers and foragers may be able to provide feedback to the foragers about the hive's need for nectar.

The model of Edwards and Myerscough [14] can be extended to multiple sources. In this model, when different sources have different quality nectar, the foragers that exploit those sources experience different search times. This means that recruitment to good-quality sources is even stronger relative to recruitment to poor-quality sources than in Camazine and Sneyd's model [6] (Figure 4.7). This is particularly evident in the model results for the feeder swapping experiment, when tremble dancing is included.

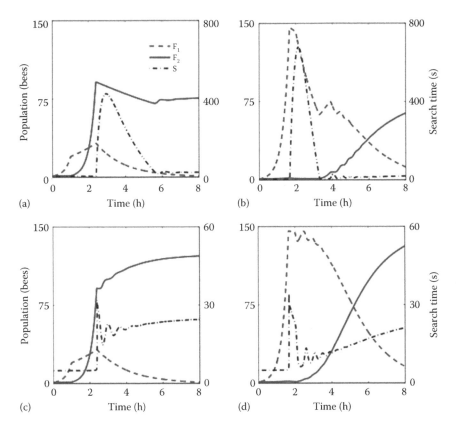

FIGURE 4.7 Diagrams showing the effect of tremble dancing and the effect of the time that the feeders are swapped in the Edwards and Myerscough model. In (a) and (b), there is no tremble dancing—the number of receivers is fixed, but in (c) and (d), more receivers are recruited by foragers who tremble dance in response to long search times (see Edwards and Myerscough, 2011 for details). In (a) and (c), the feeders are swapped after 1 h, but in (b) and (d), the feeders are swapped after 2 h. The increase in the number of receivers due to tremble dancing in (c) and (d) means that more foragers are in the field, and so more foragers are recruited to the second site after swapping. The search time that is plotted is for the more profitable site. (From Edwards, J.R. and Myerscough, M.R., *J. Theor. Biol.*, 271, 64, 2011.)

4.7 GENERALIZED ODE FORAGING MODELS: ARE BEES LIKE ANTS?

Honey bees are not the only social insects that forage collectively by passing information about forage sources from one member of the colony to another. Ants also share information, sometimes through trail laying and sometimes through knowledgeable individuals leading novices to the food source.

Sumpter and Pratt [16] devised a generalized framework for writing ordinary differential equation models for social insects. Their framework relies on breaking down the foraging cycle into various behavioral compartments such as waiting, searching, exploiting, and recruiting. In this sense, this framework is closer to the model of Camazine and Sneyd [6] than those of Cox and Myerscough [10] or Edwards and Myerscough [14]. By choosing appropriate functional coefficients in their framework, they were able to write example models for ant foraging using pheromone trails and for honey bee foraging with both dancing and independent scouting. Their honey bee model agrees better with experimental results in the initial phase of foraging expansion than Camazine and Sneyd's model, possibly because Sumpter and Pratt introduce a class of bees that have followed a dance and are now engaged in searching for the advertised source. This inherently will slow down recruitment to the class of bees that are exploiting the source. They note, however, that their model does not represent the experimental results for the swap between sources as well as Camazine and Sneyd's model.

Sumpter and Pratt [16] comment that ant foraging is like honey bee foraging where ants are leading new foragers to the forage source by group recruitment or tandem running. When ants use pheromone trails, however, their foraging recruitment is qualitatively different to honey bee dance recruitment as the recruitment takes place not by direct individual-to-individual communication but with the environment—that is the pheromone trail—as an intermediary in the process.

4.8 INDIVIDUAL-BASED MODELS AND DIFFERENTIAL EQUATION MODELS

Ordinary differential equation (ODE) models possess a conceptual clarity and force the modeler to consider what are the key processes in the system that they are representing. The choices that a modeler makes in constructing an ODE model depend, to a large extent, on the purpose of their model and the scale that they want to explore with their model system. For example, a modeler who wanted to look at the distribution of individuals in different parts of the foraging cycle needs to use a model similar to those of Camazine and Sneyd [6] or Sumpter and Pratt [16], but if they want to look at the distribution of individuals among many different sources, it may be better to use the model of Cox and Myerscough [10], which collects all individuals foraging at the same source into a single class regardless where they are in the foraging cycle.

Dividing the hive or the forager workforce into many different categories will inevitably lead to a situation where many classes might only have one or two individuals each. This introduces a "graininess" to the problem and violates the assumption that there are enough individuals in each class to think of foraging in terms of average numbers or rates. A different sort of model is needed to represent the cases where there are many different classes or where particular distinct characteristics are imputed to individuals.

Individual-based models represent the colony or the part of the colony of interest as a number of individuals who each have their own set of characteristics and at each time step behave according to a set of rules imposed by the modeler. These rules may involve interactions with other hive members or with the environment. Individual-based models can be very directly constructed from observational data to give an *in silico* colony, but it is not always easy to tease apart how different behaviors and interactions contribute to the computed colony-wide behavior.

A very detailed individual-based model was first proposed by de Vries and Biesmeijer [7]. This model included details such as the individual bees' positions on flying out to a source or scouting and a bee's experience of foraging, including a knowledge of sources where she had previously foraged. They compared their model to Seeley et al.'s [9] empirical data and obtained a reasonable match but not as close as expected, given the detailed model and care with experimental parameters. Their model did suggest, however, that the foraging history of individual foragers is important; bees remember the odor and location of previously profitable sources and may undertake reconnaissance flights to those sources before seeking other sources, either through dancing or independent scouting. This type of behavior would be very difficult to incorporate into a differential equation model. A later extension of this model included monitoring of energy harvest rates and cross-inhibition between sources [17].

When a hive forages over a large number of patches, each with a distinct location and quality, this is best represented by an individual-based model. Dornhaus et al. [18] used individual-based models to examine when recruitment to foraging was advantageous compared to independent scouting and concluded that recruitment was most advantageous in environments where there were few patches of forage of poor or variable quality.

Schmickl et al. [19] use an individual-based model to examine a similar question to that of Edwards and Myerscough [14]: How does feedback from nectar receivers affect the foraging effort? Their model (Figure 4.8) is much more complicated than

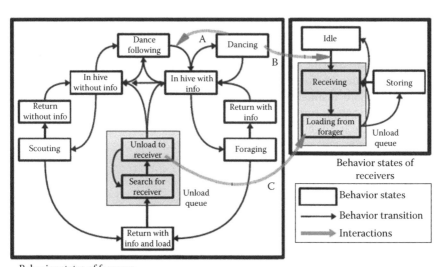

FIGURE 4.8   Flowchart for the individual-oriented model of Schmickl et al. [19]. The black arrows represent the flow of individuals from one compartment to another and the gray arrows represent interactions among different classes of bees. (With kind permission from Springer Science+Business Media: *Neural Comput. Appl.*, Swarm-intelligent foraging in honey bees: Benefits and costs of task-partitioning and environmental fluctuations, 21, 2012, 251–268, T. Schmickl, R. Thenius, and K. Crailsheim.)

that of Edwards and Myerscough but reaches much the same conclusion, that limiting the receiver numbers limits the number of foragers that are out in the field.

There are definite benefits in applying individual-based models to certain types of problems in honey bee foraging, but depending on the problem, it may be more worthwhile to avoid the very complicated and time-consuming computations that these models need and to think carefully about abstracting the model system into a differential equation model.

4.9   ALGORITHMS INSPIRED BY HONEY BEE FORAGING

The cooperative foraging behavior of honey bees has been a source of inspiration for numerical techniques for solving a wide variety of problems. Indeed much of the literature in this area references Camazine and Sneyd's model [6]. Intuitively, algorithms based on foraging seem well suited to problems of efficient task allocation and optimal use of resources, but the list of applications for these algorithms extends well beyond these problems. Some important foraging inspired techniques include *BeeHive* [20], *Bee Colony Optimization* (BCO) [21], the *Artificial Bee Colony* (ABC) algorithm [22], and the *Bees Algorithm* (BA) [23]. *BeeHive* has been used principally for studies of efficient routing of data packets through a network and has been extended to include an Artificial Immune System security framework [24] inspired by the principles of the human immune system. *Bee Colony Optimization* has been used to examine complex transportation problems [21], job shop scheduling (the problem of distributing n tasks of varying duration between m machines such that the total duration to complete all tasks is minimized) [25,26], and constrained portfolio optimization (the problem in finance of constructing a portfolio of shares, and possibly derivatives, that will likely maximize profit for the portfolio holder subject to constraints based on the portfolio holder's aversion to risk) [27]. The *Artificial Bee Colony* algorithm and the *Bees Algorithm* cover the widest range of applications, including standard mathematical problems of finding the optimal value of function of many variables, training or optimizing neural networks, and data clustering. A comprehensive list of numerical techniques inspired by honey bees up until 2008 can be found in the review of Karaboga and Akay [28].

REFERENCES

1. T.D. Seeley, *The Wisdom of the Hive*, Harvard University Press, Cambridge, MA, 1995.
2. K. von Frisch, *The Dance Language and Orientation of Bees*, Harvard University Press, Cambridge, MA, 1967.
3. T.D. Seeley and C.A. Tovey, *Why search time to find a food-storer bee accurately indicates the relative rates of nectar collecting and nectar processing in honey-bee colonies*, Anim. Behav. 47 (1994), pp. 311–316.
4. J.C. Nieh, *A negative feedback signal that is triggered by peril curbs honey bee recruitment*, Curr. Biol. 20 (2010), pp. 310–315.
5. C.W. Lau and J.C. Nieh, *Honey bee stop-signal production: Temporal distribution and effect of feeder crowding*, Apidologie 41 (2010), pp. 87–95.
6. S. Camazine and J. Sneyd, *A model of collective nectar source selection by honey-bees-Self-organisation through simple rules*, J. Theor. Biol. 149 (1991), pp. 547–571.

7. H. de Vries and J.C. Biesmeijer, *Modelling collective foraging by means of individual behavior rules in honey-bees*, Behav. Ecol. Sociobiol. 44 (1998), pp. 109–124.
8. P.K. Visscher, *Group decision making in nest-site selection among social insects*, Ann. Rev. Entomol. 52 (2007), pp. 255–275.
9. T.D. Seeley, S. Camazine, and J. Sneyd, *Collective decision-making in honey-bees—How colonies choose among nectar sources*, Behav. Ecol. Sociobiol. 28 (1991), pp. 277–290.
10. M.D. Cox and M.R. Myerscough, *A flexible model of foraging by a honey bee colony: The effects of individual behavior on foraging success*, J. Theor. Biol. 223 (2003), pp. 179–197.
11. B.P. Oldroyd, T.E Rinderer, and S.M. Buco, *Intracolonial variance in honey bee foraging behavior: The effects of sucrose concentration*, J. Apicult. Res. 30 (1991), pp. 137–145.
12. B.P. Oldroyd, T.E. Rinderer, S.M. Buco, and L.D. Beaman, *Genetic variance in honey bees for preferred foraging distance*, Anim. Behav. 45 (1993), pp. 323–332.
13. D.S. Khoury, M.R. Myerscough, and A.B. Barron, *A quantitative model of honey bee colony population dynamics*, PLoS One 6 (2011), e18491.
14. J.R. Edwards and M.R. Myerscough, *Intelligent decisions from the hive mind: Foragers and nectar receivers of* Apis mellifera *collaborate to optimise active forager numbers*, J. Theor. Biol. 271 (2011), pp. 64–77.
15. T.D. Seeley, S. Kuhnholz, and A. Weidenmuller, *The honey bee's tremble dance stimulates additional bees to function as nectar receivers*, Behav. Ecol. Sociobiol. 39 (1996), pp. 419–427.
16. D.L.S. Sumpter and S.C. Pratt, *A modelling framework for understanding social insect foraging*, Behav. Ecol. Sociobiol. 53 (2003), pp. 131–144.
17. H. de Vries and J.C. Biesmeijer, *Self-organization in collective honeybee foraging: Emergence of symmetry breaking, cross inhibition and equal harvest-rate distribution*, Behav. Ecol. Sociobiol. 51 (2002), pp. 557–569.
18. A. Dornhaus, F. Klugl, C. Oechslein, F. Puppe, and L. Chittka, *Benefits of recruitment in honey bees: Effects of ecology and colony size in an individual-based model*, Behav. Ecol. 17 (2006), pp. 336–344.
19. T. Schmickl, R. Thenius, and K. Crailsheim, *Swarm-intelligent foraging in honeybees: Benefits and costs of task-partitioning and environmental fluctuations*, Neural Comput. Appl. 21 (2012), pp. 251–268.
20. H.F. Wedde, M. Farooq, and Y. Zhang, *BeeHive: An efficient fault-tolerant routing algorithm inspired by honey bee behavior*, in *Ant Colony, Optimization and Swarm Intelligence: 4th International Workshop, ANTS 2004*, Brussels, Belgium, September 5–8, 2004 Proceedings, Lect. Notes Comput. Sci. 3172 (2004), pp. 83–94.
21. D. Teodorović and M. Dell'Orco, *Bee colony optimization—A cooperative learning approach to complex transportation problems*, in *Proceedings of 10th EWAT Meeting, Poznan, Poland,* (2005), pp. 51–60.
22. D. Karaboga, *An idea based on honey bee swarm for numerical optimization*, Technical Report TR06, Computer Engineering Department, Engineering Faculty, Erciyes University, Kayseri, Turkey, 2005.
23. D.T. Pham, A. Ghanbarzadeh, E. Koc, S. Otri, S. Rahim, and M. Zaidi, *The bees algorithm*, Technical Report, Manufacturing Engineering Centre, Cardiff University, Cardiff, UK, 2005.
24. H.F. Wedde, C. Timm, and M. Farooq, *BeeHiveAIS: A simple, efficient, scalable and secure routing framework inspired by artificial immune systems*, in *Parallel Problem Solving from Nature—PPSN IX*, Lect. Notes Comput. Sci. 4193 (2006), pp. 623–632.
25. C.S. Chong, A.I. Sivakumar, M.Y.H. Low, and K.L. Gay, *A bee colony optimization algorithm to job shop scheduling*, in *WSC'06: Proceedings of the 38th Conference on Winter Simulation*, Winter Simulation Conference, Monterey, CA, (2006), pp. 1954–1961.

26. C.S. Chong, M.Y.H. Low, A.I. Sivakumar, and K.L. Gay, *Using a bee colony algorithm for neighbourhood search in job scheduling problems*, in *21st European Conference on Modeling and Simulation (ECMS 2007)*, Prague, Czech Republic, (2007), pp. 2050–2058.

27. V. Vassiliadis and G. Dounais, *Nature inspired intelligence for the constrained portfolio optimization problem*, in *Artificial Intelligence: Theories, Models and Applications*, Lect. Notes Comput. Sci. 5138 (2008), pp. 431–436.

28. D. Karaboga and B. Akay, *A survey: Algorithms simulating bee swarm intelligence*, Artif. Intell. Rev. 31 (2009), pp. 61–85.

5 Infectious Disease Modeling for Honey Bee Colonies

Hermann J. Eberl, Peter G. Kevan,
and Vardayani Ratti

CONTENTS

ABSTRACT

Mathematical models have been successfully used to study the progression of various infectious diseases for many years. In this chapter, we demonstrate how the susceptible–infected–removed (SIR) modeling framework can be applied to study diseases of honey bee colonies. We focus on the acute bee paralysis virus, which is transmitted by the parasitic mite *Varroa destructor* as a vector. The resulting model consists of four nonautonomous, nonlinear ordinary differential equations, which we study with analytical and computational techniques. Our results indicate that, depending on model parameters, in the absence of the virus, a mite infestation can lead either to extinction of the bee colony or to an endemic infestation that allows the bee colony to survive. However, when the mites also carry the virus, this infection is very difficult to be fought off without remedial intervention and might lead to the extinction of the colony.

KEYWORDS

Acute bee paralysis virus, Honey bees, Mathematical model, Ordinary differential equations, SIR model, *Varroa* mite, Vector-borne disease

5.1   INTRODUCTION

Honey bees, like all animals, have diseases and pests that can destroy individuals or the entire colony if not treated in time. Diseases in the honey bee colony are mainly caused by pests and parasites, bacteria, fungi, and viruses [1]. This chapter focuses on viral diseases in honey bee colonies for which the parasitic mite *Varroa destructor* acts as a vector. Besides vectoring diseases, *varroa* mites also affect the life span of honey bees by feeding on the bee's hemolymph and by piercing its intersegmental membrane [2,3]. There have been at least 14 viruses found in honey bee colonies [4,5], which can differ in intensity of impact, virulence, etc., for their host. For example, the acute bee paralysis virus (ABPV) affects the larvae and pupae that fail to metamorphose to adult stage, while in contrast the deformed wing virus (DWV) affects larvae and pupae that can survive to the adult stage [6]. *Varroa* mites are the main agents of transmission of viruses between bees. Transmission occurs when mites feed on bees and when a virus-carrying mite attaches to a healthy bee during the former's phoretic phase [3,7,8]. Thus, the previously uninfected bee becomes infected. A virus-free phoretic mite can begin carrying a virus after it moves from an uninfected to an infected bee [8,9].

The relationships between honey bees, parasitic mites, and viruses comprise a complex, interdependent system with nonlinear interactions. It cannot be understood by studying the three players, host, vector, and virus, individually and superimposing the findings. Instead, their interdependencies must be taken into account. For example, not only does the virus have an effect on the bee population, but also a decreasing bee population affects the mites, which again alters disease transmission. For several decades, mathematical models have been developed as a major tool to understand the course and complexity of infectious diseases in human populations and to aid in the development of vaccination and control strategies. In this chapter, we describe how this highly successful methodology can be adapted to study the effect of *varroa* mites and their viruses in honey bee colonies. We stress that although these mathematical models are predictive, they are primarily qualitative, rather than quantitative, in nature. Their main purpose is not to predict how many bees in a colony become infected by a certain virus and when. Their real power lies in describing the fate of a disease qualitatively and to provide insight under which conditions a disease can be eradicated or when it will overcome the colony, potentially eradicating it. Such disease models can also be valuable tools for investigating the effectiveness of remedial and disease control strategies. But this aspect is beyond what we want to achieve in this chapter.

5.2   MATHEMATICAL MODELS

Mathematical models are (sometimes simple) representations of complex real-world problems. They can increase our understanding of the processes involved and of their interplay. Mathematical models have a long tradition in most disciplines of science and engineering. The aim of all scientific study is to derive theories that enable us to make predictions. A mathematical model is the formulation of such a theory. Mathematical modeling requires a good command of mathematical tools and an understanding of the underlying physical or biological phenomena.

The formulation of a mathematical model is a careful trade-off between *complexity*, *difficulty*, and *accuracy*. Increased complexity ideally also increases accuracy, but always at the expense of added difficulty. Difficulty can take several forms. Foremost, this refers to the difficulty in obtaining a solution of the mathematical model, either analytically or computationally. Equally important, however, is the difficulty in providing the additional information required, such as parameters or other input data for the model. Often, the theoretical accuracy gain of a more complex model can be completely lost if the additionally required information cannot be obtained and must be substituted by crude assumptions.

Every mathematical model is specific to its purposes and no model fits all. Moreover, mathematical models are not unique. Often several mathematical models can be formulated for seemingly the same, say, biological or physical process. Models differ in the assumptions on which they are based and in the mathematical machinery used. As a general rule, mathematical models that contain lots of details without relevance to the question at hand are not desired. In some sense, that can be understood as a consequence of Occam's razor: Simpler explanations are, other things being equal, generally better than more complex ones. In other words, a model should be as simple as possible and as complicated as necessary.

Model formulation is an iterative, multifaceted cyclical process as shown in Figure 5.1. First, the most essential features of real-world problems are identified and described in mathematical terms. Once such a candidate mathematical model is formulated, it is analyzed with respect to its internal consistency. For most mathematical models, it is not possible to write down the solution. In fact, the primary question a modeler has to ask is whether or not his or her model actually possesses

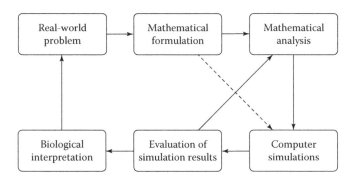

FIGURE 5.1   The modeling process as an iterative process.

a solution and, if so, whether it is unique and satisfies important properties. For example, in mathematical ecology, it is expected that population sizes predicted by a model are not negative. In some cases, these mathematical questions are relatively easy to answer using standard theory, for example, for many ordinary differential equation models. In other cases, this might involve highly nontrivial mathematics, for example, in models that consist of nonlinear systems of partial differential equations. In fact, this step in the modeling cycle can be so challenging that in many cases one postpones it until the first computational results are obtained to decide whether the model is worth pursuing.

After the mathematical model is checked for its consistency, the next step is to obtain its solution. In most cases, that needs to be done computationally, using numerical methods to approximate the unknown solutions of the model. The computer simulation results are then subjected to scrutiny. First, it is verified that the computational results indeed are good approximations of the unknown solution of the underlying model. It is also validated whether or not the computer results agree with the theoretical properties that have been established. On the other hand, computational results can also inform the mathematical analysis and lead to an improvement of theoretical results. Second, the computational results are interpreted in their biological or physical context. This might lead to further model refinement and increased complexity. In other cases, it might be determined that the model is overly complex and can be simplified. The modeling cycle begins anew. Eventually, one arrives at a mathematical model that is deemed satisfactory for the current situation but would need to be modified if used to tackle a different problem. Also, when new scientific knowledge becomes available, the mathematical model needs to undergo new scrutiny. As such, mathematical modeling is inherently Popperian: Mathematical models can be validated but never verified; falsifiability is one of the main characteristics of a mathematical model. And "He who decides one day that scientific statements do not call for any further test, and that they can be regarded as finally verified, retires from the game [of science]" [10]. On the other hand, "Once a hypothesis has been proposed and tested and has proved its mettle, it may not be allowed to drop out without 'good reason'" [10].

Once a mathematical model is accepted as sufficiently appropriate for the problem at hand, computer simulation experiments are conducted to investigate a particular scientific question. Such computer simulation experiments can resemble laboratory or field experiments, but also differ from them. The big advantage of simulations is that they are usually cheaper to carry out, often faster, and always easier to control. On the other hand, they are subject to inherent uncertainty, in particular if no experimental data are available for quantitative validation. Even if a model is quantitatively not validated, if it is grounded on accepted and well-understood basic principles, it can offer great insight into the potential behavior of the system that it models. In the words of Heinrich Hertz: "We form ourselves images of external objects and the form which we give them is such that the necessary consequents of the images in thought are always the images of the necessary consequents in nature of the things pictured" [11]. The form that we give them is determined by the mathematical expertise and experience of the modeler. The consequents of the images, the model predictions, offer potential explanations or testable hypotheses for the consequents of the things pictured. As such, they inform future experimental studies or suggest potential interventions to avoid undesirable consequences. This, in fact, is an important area for the application of mathematical models and applies to a broad class of models used in science and engineering, among which are many infectious disease models.

5.3 INFECTIOUS DISEASE MODELS

The course of infectious diseases can be predicted using mathematical models that have been developed over many years and have experienced a huge surge in activity in the last decade [12–14]. While most of these research efforts are driven by diseases of human populations, the underlying concepts can be adapted to diseases of animals as well. Because parameters change temporally and geographically, an accurate quantitative prediction of future epidemics over several years normally cannot be expected, especially for newly emerging diseases. Consequently, also a quantitative validation is only possible after the fact. However, instead of focusing on quantitative predictions, the real strength and power of the mathematical approach to understand disease dynamics are the qualitative insight that it gives into the complex interplay between host and disease. In this light, the success of infectious disease models has been phenomenal, even in the absence of accurate data that describe a particular disease quantitatively. We want to make use of this successful strategy for honey bee diseases.

The traditional modeling framework that we have adapted for virus infections in honey bee populations, is the so-called susceptible–infected–removed (SIR) models in which the population is described in terms of susceptible and infected individuals; further particularities depend on the disease under consideration. For viruses in honey bees, several of the simplifying assumptions often made in infectious disease models for humans are not possible. For example, because the bee population size fluctuates greatly with the seasons and because viral diseases can reduce the population considerably, to the point of extinction, the usual assumption that the overall population size remains relatively unaffected by the virus and can be considered a constant is not possible, leading to more algebraically involved models. Moreover, because the characteristic timescale of the disease is comparable to or even bigger than the characteristic timescale of bee biology, birth and natural death of bees cannot be neglected and models must be developed that account for the particularities of bee reproduction. Furthermore, ABPV and other *varroa*-associated viruses are to be modeled as vector-borne diseases; a possible extension of the SIR modeling framework for vector-borne diseases is Ross–Macdonald model [15].

The basic, traditional SIR model divides the host population into susceptible (S), infected (I), and removed or recovered (R) individuals. The susceptibles are individuals that are not currently, but may become, infected, and the infected group consists of individuals that are currently infected. Depending on the disease, the R group can indicate those individuals who were infected but have now recovered from the disease or those who were removed, for example, by death. The SIR model is conceptualized in Figure 5.2. Each compartment in this model is described by a relatively simple ordinary differential equation that constitutes

FIGURE 5.2 Conceptualization of the SIR modeling framework: After infection, susceptible individuals leave the S compartment and enter the I compartment. After recovery/removal, they leave the I compartment and enter the R compartment. The infection rate can depend on both, that is, the number of already infected and susceptible individuals.

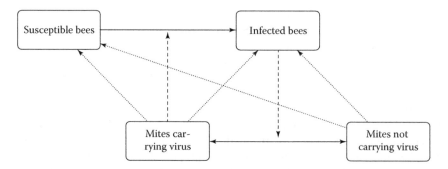

FIGURE 5.3 Role of mites in the spread of infection in a honey bee colony: Virus-carrying mites affect the transmission rate from susceptible to infected bees. All mites, virus carrying or not, affect mortality of susceptible and infected bees.

a "mass balance," that is, a balance of gain and loss in the compartment. The rate at which susceptible individuals become infected, that is, the rate of transfer from the S into the I compartment, depends on the number of already infected individuals but can also depend on the number of susceptible individuals. In the literature, the SIR model has been adapted to describe a variety of diseases. For example, it can be considered that recovered individuals become susceptible again. An extension of the model to vector-borne diseases can be achieved by making the infection rate dependent on the number of vectors that carry the virus. In many instances, this requires the inclusion of additional differential equations that describe how the vector population evolves and how vectors acquire the virus (see, e.g., Figure 5.3).

5.4 APPLICATION TO HONEY BEES, *VARROA* MITES, AND THE ACUTE BEE PARALYSIS VIRUS

We apply the traditional SIR modeling framework to describe the fate of a honey bee colony that is exposed to the ABPV. ABPV, together with the Kashmir bee virus, black queen cell virus, and the Israeli acute paralysis virus, belongs to the family Dicistroviridae. These viruses share a number of biological characteristics, such as principal transmission routes and primary host life stages [16]. ABPV is a common infective agent of honey bees that is frequently detected in apparently healthy colonies. Bees affected by this virus are unable to fly, lose hair from their bodies, and tremble uncontrollably. The virus has been suggested to be a primary cause of bee mortality. Infected pupae and adults suffer rapid death. ABPV is associated with *varroa* mites and has been implicated in colony collapse disorder. Due to the simultaneous appearance of mite and virus under field conditions, it is difficult to separate the effects of both pathogens [17]. Therefore, they should be studied together, and mathematical models of the disease dynamics should include both pathogens simultaneously.

In the SIR framework, susceptible individuals are healthy bees, infected individuals are sick bees, and the vector is the *varroa* mite. The virus is transmitted to the bees via mites that feed on their hemolymph by piercing the intersegmental membrane. The model for honey bees is with demography, that is, the birth and the death of the bees is taken into consideration. In our case, the *removed* compartment is the deceased bees. A schematic of the virus transmission is given in Figure 5.3. The mites that carry the virus affect the transmission of the disease. All mites, vector carrying or not, affect the mortality of susceptible and infected bees.

The primary dependent variables for our model are the healthy (susceptible) bees, x, and the virus infected bees, y. The transmission of the disease is by the vectoring *varroa* mites M, and the mites carrying the virus are denoted by m. The deceased bees are denoted by R. In our model, we include birth and natural death of bees, bee death due to parasitic mites, infection of healthy bees with the virus, bee death due to the disease, transfer of the virus between mites, and the evolution of the mite population. The governing equations read

$$\frac{dx}{dt} = \mu(t)g(x,t)h(m,t) - \beta_3(t)m\frac{x}{x+y} - d_1(t)x - \gamma_1(t)Mx \tag{5.1}$$

$$\frac{dy}{dt} = \beta_3(t)m\frac{x}{x+y} - d_2(t)y - \gamma_2(t)My \tag{5.2}$$

$$\frac{dm}{dt} = \beta_1(t)(M-m)\frac{y}{x+y} - \beta_2(t)m\frac{x}{x+y} \tag{5.3}$$

$$\frac{dM}{dt} = r(t)M\left(1 - \frac{M}{\alpha(x+y)}\right) \tag{5.4}$$

$$\frac{dR}{dt} = d_1(t)x + d_2(t)y - \gamma_1(t)Mx - \gamma_2(t)My \tag{5.5}$$

Here, all coefficients and parameters depend on time t to reflect seasonal changes in the bee–mite–virus population and transmission dynamics. Because the dependent variables x, y, m, and M do not depend on R, Equation 5.5 would decouple from Equations 5.1 through 5.4.

The function $g(x,t)$ in (5.1) expresses that a sufficiently large number of healthy worker bees are required to care for the brood. We think of $g(x,t)$ as a switch function. If x falls below a critical value, which may seasonally depend on time, then essential work in the maintenance of the brood cannot be carried out anymore and no new bees are born. If x is above this value, the birth of bees is not hampered. A mathematically convenient formulation of such switch-like behavior is given by the sigmoidal Hill function:

$$g\left(x,t\right) = \frac{x^n}{K^n(t) + x^n} \tag{5.6}$$

Here, the parameter $K(t)$ is the size of the bee colony at which the birth rate is half of the maximum possible rate and the integer exponent $n > 1$.

The function $h(m,t)$ in (5.1) indicates that the birth rate is affected by the presence of mites that carry the virus. This is, in particular, important for viruses like ABPV that kill infected pupae before they develop into bees. The function $h(m,t)$ is assumed to decrease as m increases and $h\left(0,t\right) = 1$ (no reduction in bee birth in the absence of the virus); [6] suggests that this is an exponential function $h(m,t) \cong e^{-mk(t)}$, where $k(t)$ is nonnegative. We use this expression in the computer simulations later on.

The parameter β_1 in (5.3) is the rate at which mites that do not carry the virus acquire it. The rate at which infected mites lose their virus to an uninfected host is β_2. The rate at which uninfected bees become infected is β_3 in bees per virus-carrying mite and time.

Finally, d_1 and d_2 are the death rates for uninfected and infected honey bees. We can assume that infected bees have shorter lives than healthy bees, thus $d_2 > d_1$. The last equation of the model is a logistic growth model for *varroa* mites. By r, we denote the maximum mite birth rate. The carrying capacity for the mites changes with the host population site, $x + y$, and is characterized by the parameter α, which indicates how many mites can be sustained per bee on average. This assumption is in agreement with [18].

Mites contribute to an increased mortality of bees. This is considered in (5.1) and (5.2) by including death terms that depend on M; the parameters $\gamma_{1,2}$ are the rate at which mites kill bees.

The mathematical model mentioned earlier is based on Sumpter and Martin [6] and Eberl et al. [19]. In Sumpter and Martin [6], the mite population in the hive was assumed to be a given model parameter and the honey bee birth rate was assumed to be a constant; moreover, the model was considered for constant coefficients only, that is, without seasonal fluctuations. Eberl et al. [19] added the brood maintenance terms $g(x, t)$ and seasonal fluctuations. In the aforementioned formulation, it was also added that the mite population is not a given model parameter but is described by a logistic growth equation with a carrying capacity that depends on the current bee population size. Therefore, this model can be considered a fourth iteration of the modeling cycle toward understanding the fate of bee colonies if infested by ABPV-carrying mites.

5.5 COMPUTER SIMULATIONS OF THE DISEASE MODEL

5.5.1 COMPUTATIONAL SETUP AND PARAMETERS

Models (5.1) through (5.4) comprise a system of four nonlinear, nonautonomous ordinary differential equations. An exact analytical solution in closed form cannot be found, but many suitable and well-tested computer algorithms are readily available for their numerical approximation. Such a simulation requires explicit numerical values for the model parameters even if the overall goal is a phenomenological qualitative understanding rather than a quantitative prediction. The parameters $\mu, k, \alpha, \beta_i, d_i, \gamma_i, g(x), h(m), r$ are assumed to be nonnegative. They can change with time. In particular, major differences may be observed between seasons. For example, the life span of a worker bee in summer is much shorter than in winter [2,20]; the birth rate for bees is higher in summer than in spring and autumn, and it drops down to 0 in winter [21]. Seasonal averages for the model parameters $\beta_{1,2,3}, \mu, d_{1,2}, k$ can be derived from the data in Sumpter and Martin [6]. These are summarized in Table 5.1 and will be used in the simulations later. Numerical values for the remaining parameters $r, \alpha, and K$ can be estimated (order of magnitude) from other data in the published literature. In our computer simulations, these are the parameters that we vary to understand the model's behavior.

For computer simulations, we use the software package MATLAB®. The ordinary differential equations are integrated by the built-in routine ode15 s, a variable order solver based on numerical differentiation formulas. In all cases, we run the computer simulations for a period of 7,000 days (approx. 20 years) or until the colony vanishes, whichever comes first.

TABLE 5.1
Seasonal Averages of Model Parameters

Parameters	Spring	Summer	Autumn	Winter
β_1	0.1593	0.1460	0.1489	0.04226
β_2	0.04959	0.03721	0.04750	0.008460
β_3	0.1984	0.1460	0.1900	0.03384
d_1	0.02272	0.04	0.02272	0.005263
d_2	0.2	0.2	0.2	0.005300
μ	500	1,500	500	0
k	0.000075	0.00003125	0.000075	N/A
r	0.0165	0.0165	0.0045	0.0045

Source: Derived from Sumpter, D.J.T. and Martin, S.J., *J. Animal Ecol.*, 73, 51, 2004.

5.5.2 ILLUSTRATIVE RESULTS

Before studying the full models (5.1) through (5.4), we start with the smaller virus-free submodel. That is, we assume that initially no sick bees and no virus is in the system, $y(0) = m(0) = 0$. It is easily verified from (5.1) through (5.4) that then $y(t) = m(t) = 0$ for all time, in accordance with biological expectation.

Illustrative simulations are shown in Figure 5.4, for two different sets of brood maintenance coefficients K: one at the lower end of the estimated range for this parameter, $K=6,000$ (spring), $K=8,000$ (summer), $K=6,000$ (fall), $K=6,000$ (winter), in panels (a) and (b), and one at the higher end $K=11,800$ (spring), $K=12,000$ (summer), $K=11,800$ (fall), $K=6,000$ (winter), in panels (c) and (d). For α, we use 0.4784 (Winter, Spring), 0.5 (summer, autumn) in both cases. We compare the development of the bee population in the absence ((a), and (c)) and in the presence of mites ((b) and (d)).

Without mites, the bee population assumes a periodic pattern after a brief transient period due to initial conditions. The bee colonies are strongest in summer and then decline through fall and winter before they increase again in spring, etc. For lower K (*i.e.*, fewer worker bees are required to maintain the brood), the population size is somewhat larger than for higher values of K. With mites, but no viruses, in the colony, the model predicts for the chosen parameters a coexistence of mites and bees. The bee colony is slightly smaller than in the mite-free case but can still be considered of healthy strength. The mite population assumes a periodic pattern as well from the second year on. These simulations suggest the possibility of coexistence of bees and mites at relatively healthy bee population levels.

That, however, mites alone, even in the absence of virus, can lead to a decline of the bee colony is suggested in a further simulation, as shown in Figure 5.5. Here, the tolerance of the bees to the mites, expressed in terms of parameter α, was reduced to $\alpha=0.1321$ in spring and summer. We repeat the simulation of bees and mites for four slightly different sets of brood maintenance coefficients K (a) $K=11,600$ (spring), $K=12,000$ (summer), $K=11,600$ (autumn), $K=6,000$ (winter), (b) $K=11,700$ (spring), $K=14,000$ (summer), $K=11,700$ (autumn), $K=6,000$ (winter), (c) $K=11,900$ (spring), $K=14,000$ (summer), $K=11,900$ (autumn), $K=6,000$ (winter), (d) $K=12,200$ (spring), $K=14,000$ (summer), $K=12,200$ (autumn), $K=6,000$ (winter).

For the first case (a), we see qualitatively the same picture as in Figure 5.4. However, for, depending cases (b) through (d), the bee colony vanishes after the fourth, third, and second winter, respectively. This suggests that the fate of a virus-free, but mite-infested, bee colony, that is, the question of whether or not it survives or vanishes, depends on the circumstances as expressed in parameters. These, in turn, could change with the years on factors such as weather and climate, etc.

In all the simulations we have shown, and many more that are not shown here, we note that according to the model presented, once mites have infested the colony, they stay there and cannot be fought off.

In the next simulations that we show, we introduce at initial time a virus in the population; that is, we assume that $m(0) > 0$. This is illustrated in Figure 5.6. The brood maintenance coefficient was set to $K=8,000$ (spring, autumn), $K=12,000$ (summer),

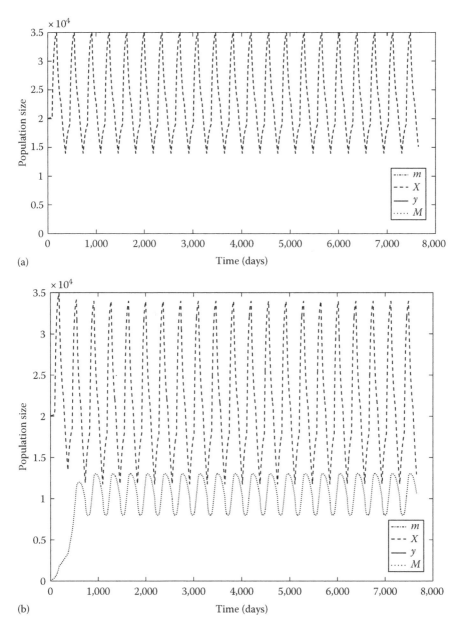

(a)

(b)

FIGURE 5.4 Simulation of bee–mite population dynamics: periodic solutions for varying brood maintenance coefficients K, low K (a), (b) versus high K (c), (d) values, without mites (a), (c) and with mites (b), (d).

(*continued*)

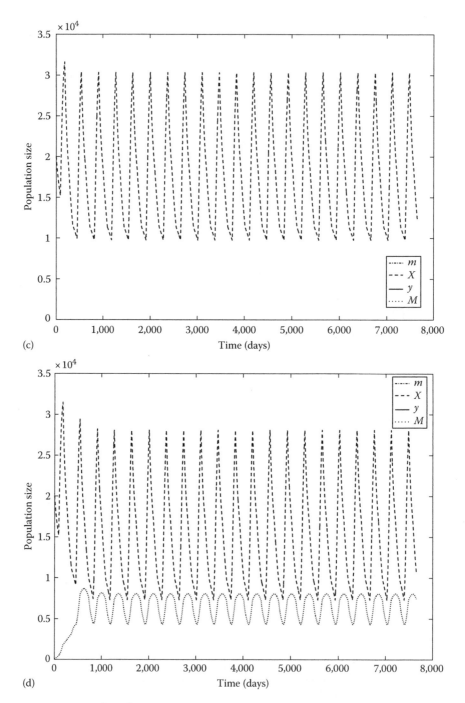

FIGURE 5.4 (continued) Simulation of bee–mite population dynamics: periodic solutions for varying brood maintenance coefficients K, low K (a), (b) versus high K (c), (d) values, without mites (a), (c) and with mites (b), (d).

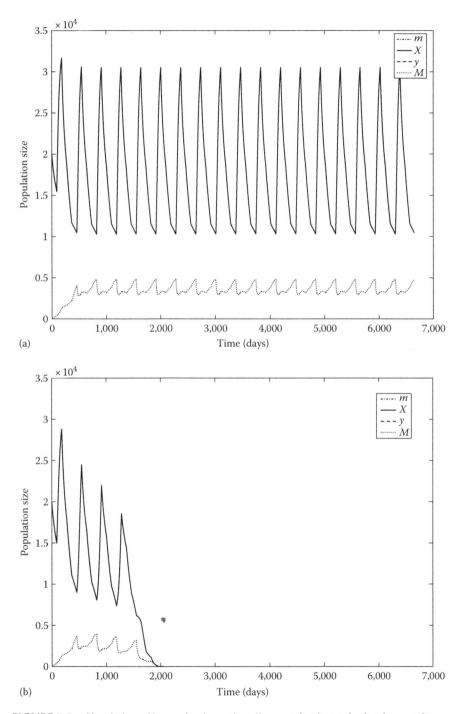

(a)

(b)

FIGURE 5.5 Simulation of bee–mite dynamics: disappearing bee colonies due to mites.

(*continued*)

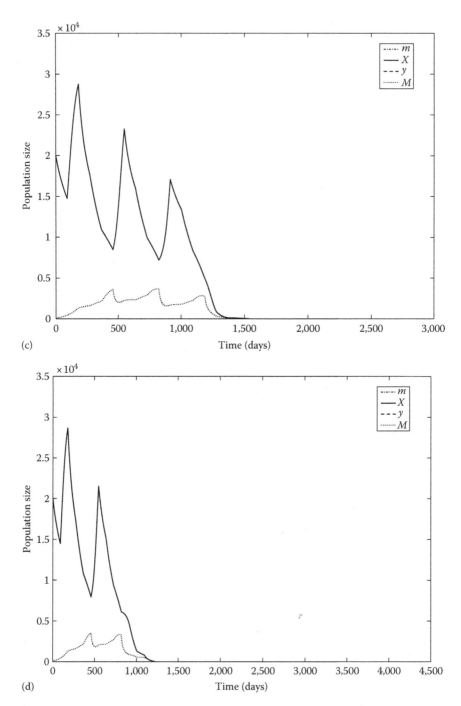

FIGURE 5.5 (continued)   Simulation of bee–mite dynamics: disappearing bee colonies due to mites.

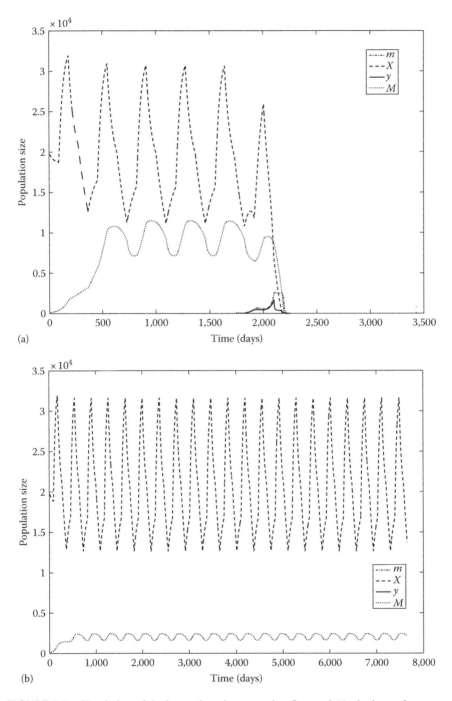

(a)

(b)

FIGURE 5.6 Simulation of the bee–mite–virus complex: In panel (a), the bee colony vanishes in the sixth year after the virus is fully established; in panel (b), with strongly reduced parameter α, the virus is fought off.

$K=6,000$ (winter), and α as in Figure 5.4. Under these conditions, in the absence of the virus, we would find bees and mites attaining a coexistence pattern (simulation data not shown) similar to that in Figure 5.4b and d. With the virus included, we observe that initially bees and mites attain a coexistence pattern. However, in the sixth year, the colony vanishes abruptly and somewhat surprisingly. This coincides with, and is a consequence of, a steep increase in the number of virus-carrying mites. Indeed, upon closer inspection, we note that the maximum strength of the bee colony in summer decreased slightly every year as did the minimum strength in winter. We repeat this simulation with the strongly reduced, more or less unrealistic, $\alpha=0.1$ for all seasons. Under these conditions, the virus can be fought off by the bee colony and bees and virus-free mites attain a periodic, stable pattern.

Our simulations seem to suggest that it depends on the environmental and ecological conditions, as expressed in the model parameters, whether the virus disease can be eradicated or whether it will lead to the disappearance of the bee colony.

5.5.3 Mathematical Analysis of the Constant Coefficient Case in an Attempt to Gain More Insight into Computer Simulations

As a four-dimensional, nonautonomous, nonlinear system of ordinary differential equations, models (5.1) through (5.4) are algebraically too involved to establish, with rigorous mathematical techniques, conditions on model parameters that allow the prediction of the fate of the colony. However, in order to explain and describe in more detail the interaction of bees, mites, and viruses, it is helpful to study the models for constant coefficients, which, for example, can be understood as investigating the seasons individually.

The mathematical techniques used for this purpose are relatively well developed and described in many textbooks in mathematical biology (see, *e.g.*, [22]). In short, we determine the steady states (equilibria points) of the underlying model and investigate their linear stability. We say an equilibrium is asymptotically stable if solutions close to equilibrium converge to equilibrium. We say it is unstable if solutions close to equilibrium are repelled by equilibrium. Unstable equilibria are difficult to observe and usually not attained (unless they are already starting conditions). This stability analysis is performed by linearizing the equation about the equilibrium and studying the behavior of solutions close to the equilibrium. The stability of an equilibrium is then obtained from the eigenvalues of the linearization matrix, the Jacobian.

We advance in three steps. First, we investigate the simple submodel with bees only (without mites and viruses). Second, we also consider mites but not viruses, and finally we consider bees, mites, and viruses together. We summarize here the result only and refer the reader to Ratti et al. [23] for the mathematical details.

5.5.3.1 One-Dimensional Healthy Bee Submodel

In the absence of parasites and viruses, models (5.1) through (5.4) become

$$\frac{dx}{dt} = \mu g(x) - d_1 x \qquad (5.7)$$

with $g(x)$ as in (5.6) for the brood maintenance function. Now μ, K, d_1 are assumed to be constant.

The dynamics of one-dimensional autonomous systems is structurally simple: either the solutions converge to an asymptotically stable equilibrium or they diverge to (positive or negative) infinity.

The trivial equilibrium $x_0^* = 0$ (no bees) exists always and is asymptotically stable. The unconditional stability of the trivial equilibrium reflects that a certain number of honey bees are required to care for the brood.

Two other nontrivial equilibria $x_1^* > 0$ and $x_2^* > 0$ exist if the maximum growth rate μ is large enough, compared to the death rate d_1. More specifically, after some routine algebra, the condition for the existence of these equilibria is obtained as

$$\frac{\mu}{d_1} > \frac{n}{n-1} K \sqrt[n]{n-1} \tag{5.8}$$

that is, it depends also on the brood maintenance coefficients, that is, the number of bees required to care for the brood.

We denote the larger of the two equilibria by x_1^* and the smaller one by x_2^*. The latter is unstable, while the former is asymptotically stable.

Because the solution $x(t)$ of (5.7) must be strictly monotonic between two equilibrium points, we obtain the following picture: If initially the bee population is small, $x(0) < x_2^*$, the solution converges to the trivial equilibrium x_0^*; that is, the colony dies out because too few worker bees are present to care for the brood. On the other hand, if $x(0) > x_2^*$, the solution converges to the asymptotically stable equilibrium x_1^*; that is, a healthy colony can establish itself attaining equilibrium x_1^*. For the data in Table 5.1, the nontrivial equilibria $x_{1,2}^*$ exist in spring, summer, and fall, but not in winter, when no new bees are born. Thus, in winter, all solutions converge to the trivial equilibrium. However, as it is not reached in finite time, the population can recover in spring if at the beginning of the season the bee population is stronger than the equilibrium x_2^* obtained for that season. In the time-dependent computer simulations that were shown earlier, this corresponds to the situation found in Figure 5.4a and c.

5.5.3.2 Two-Dimensional Bee–Mite Submodel

We investigate now how the stability of the equilibria $x_{0,1,2}^*$ of (5.8) changes when parasitic mites (but not viruses) are considered in the colony. To this end, we study the bee–mite subsystem of (5.3) and (5.4):

$$\frac{dx}{dt} = \mu g(x) - d_1(x) - \gamma_1 M x \tag{5.9}$$

$$\frac{dM}{dt} = rM\left(1 - \frac{M}{\alpha x}\right) \tag{5.10}$$

Here, we again assume (5.6) for the brood maintenance function $g(x)$.

The two-dimensional model (5.9) and (5.10) can have up to five equilibrium points, that is, points (x^*, M^*) for which $dx/dt = dM/dt = 0$. See Figure 5.7.

The first one is again a trivial equilibrium point A with $(x^*, M^*) = (0,0)$ (no bees, no mites), which can be shown to be always asymptotically stable, for the same reasons that x_0^* was stable in the one-dimensional bee-only model and because in the absence of bees mites do not have a host. Solutions that come close enough to this equilibrium will lead to the dying out of both bees and mites. This corresponds to our dynamic computer simulations in Figure 5.5b through d.

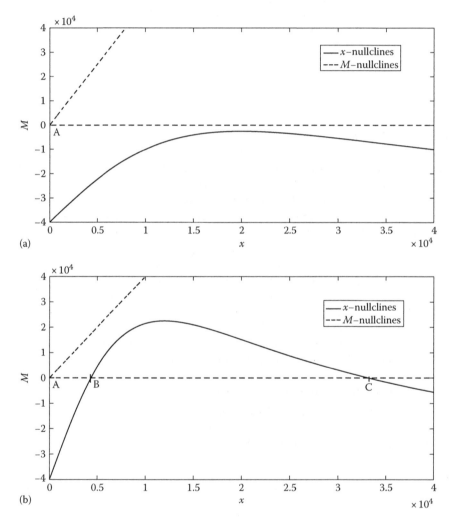

FIGURE 5.7 Nullclines and location of equilibria for the bee–mite model (5.9) and (5.10): Shown are the dashed curves along which $dM/dt=0$ and the solid curves along which $dx/dt=0$. The equilibria are the intersections of a dashed and a solid curve. (a) The trivial equilibrium A is the only equilibrium; (b) only the mite-free equilibria A, B, C exist; (c), (d) two additional equilibria D, E with $M^*=\alpha x^*>0$ exist. In (c) $x_E^* < x$ and in (d) $x_E^* > x = \sqrt[n]{n-1}\, K$.

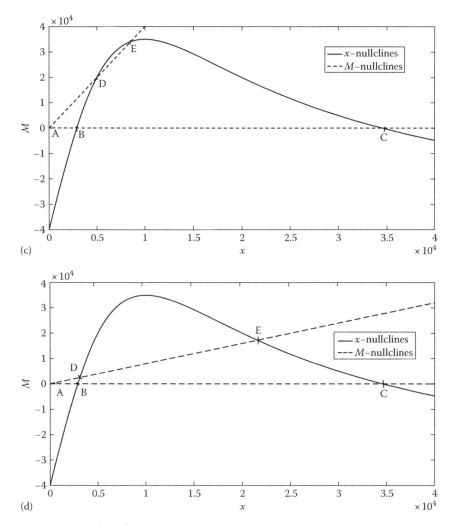

FIGURE 5.7 (continued)

Furthermore, it is easy to verify that if (5.8) is satisfied, then there exist two mite-free equilibrium points, associated with the two equilibria $x^*_{1,2}$ of the aforementioned bee-only model: $B : (x^*_2, 0)$ and $C : (x^*_1, 0)$. The former inherits the instability of x^*_2, its phase portrait is an unstable node. The latter is an unstable saddle: only solutions on the line $M = 0$ (no mites in the system) approach the equilibrium; all other equilibria are repelled. The instability of these mite-free equilibria is explained as follows: The growth of the mites is logistic. Therefore, as soon as mites enter the system, they start establishing themselves. That explains why in our computer simulations of the model we cannot find a case where the mites are fought off by the bees.

Depending on parameters, there might also be two equilibria points $D : (x^*_D, M^*_D)$ and $E : (x^*_E, M^*_E)$ for which both $x^*_{D,E}$ and $M^*_{D,E}$ are positive. Here, we have

$x_D^* < x_E^*$ and $M_D^* < M_E^*$. Using phase plane analysis techniques, we find that the equilibrium D is always unstable, while E is stable if $x_E^* > \sqrt[n]{n-1}\, K$ and unstable otherwise. See [23] for the mathematical details. The simulations in Figure 5.4b and d correspond to stable E.

5.5.3.3 Complete Bee–Mite–Virus Model

We investigate now the question whether a stable, mite-infested honey bee colony can fight off the virus. To this end, we consider the complete four-dimensional models (5.1) through (5.4) with (5.6). It is easily verified that the equilibrium E of the bee–mite model has a corresponding disease-free equilibrium $E_4 : (x_E^*, 0, M_E^*, 0)$. Of particular interest is the case where the equilibrium E is stable under (5.9) and (5.10). Linearization of (5.1) through (5.3) around this equilibrium shows that it is asymptotically stable if the following condition on the parameters is satisfied:

$$\beta_3 \beta_1 \alpha < \beta_2 \left(d_2 + \gamma_2 \alpha x_E^* \right) \tag{5.11}$$

Therefore, if the virus is introduced into a properly working mite-infested colony, the colony can fight off the virus if the sick bees die fast. This corresponds to the scenario we observed in our simulations in Figure 5.6b. However, as we have seen there, this required a relatively low choice for the parameter α, and, therefore, our results suggest that it might not be a frequent event that a colony can rid itself of the ABPV disease. As this equilibrium is the only stable equilibrium with a positive bee population size, the alternative seems the complete decline of the bee colony.

5.6   CONCLUDING REMARKS

In recent years, mathematical models have been successfully used to study the progression of infectious diseases. While typically set in the context of human populations, many of the principles are generally valid and can be applied to animal populations as well. In this chapter, we try to show how this is done for honey bee diseases in a simple example, focusing on honey bees, *varroa* mites, and the ABPV. This model, although conceptually simple, is complicated enough to not allow for a complete, rigorous mathematical analysis. Therefore, it must be studied in computer simulations, aided by analytically obtained mathematical insight. The model we describe here is but a first step from which many potential directions for further work originate. This includes more detailed descriptions of the disease, incorporation of effects such as queen failure or swarming, which we neglected, or the effect of mite control strategies on the viral disease.

ACKNOWLEDGMENT

The authors would like to thank the Ontario Ministry for Agriculture, Food and Rural Affairs (OMAFRA) and the Canadian Pollination Initiative (CANPOLIN) for financial support.

REFERENCES

1. R.A. Morse, *Honey Bee Pests, Predators, and Diseases*, Cornell University Press, New York, 1978.
2. G.V. Amdam and S.W. Omholt, *The regulatory anatomy of honeybee lifespan*, J. Theor. Biol. 216 (2002), pp. 209–228.
3. P.L. Bowen-Walker, S.J. Martin, and A. Gunn, *Preferential distribution of the parasitic mite,* Varroa jacobsoni *Oud. on overwintering honeybee* Apis mellifera L. *workers and changes in the levels of parasitism*, Parasitology 114 (1997), pp. 151–157.
4. L. Bailey and B.V. Ball, *Honey Bee Pathology*, 2nd edn. Academic Press, London, UK, 1991.
5. P.G. Kevan, E. Guzman, A. Skinner, and D. van Engelsdorp, *Colony collapse disorder in Canada: Do we have a problem?* HiveLights 20 (2007), pp. 14–16.
6. D.J.T. Sumpter and S.J. Martin, *The dynamics of virus epidemics in* Varroa-*infested honey bee colonies*, J. Anim. Ecol. 73 (2004), pp. 51–63.
7. S.J. Martin, *The role of* Varroa *and viral pathogens in the collapse of honeybee colonies: A modeling approach*, J. Appl. Ecol. 38 (2001), pp. 1082–1093.
8. S. Nordstrom, *Virus infections and* Varroa *mite infestations in honey bee colonies*, PhD thesis, Swedish University of Agricultural Sciences, Uppsala, Sweden, 2000.
9. S.J. Martin, B.V. Ball, and N.L. Carreck, *Prevalence and persistence of deformed wing virus (DWV) in untreated or acaricide treated* Varroa *destructor infested honey bee* (Apis mellifera) *colonies*, J. Apicult. Res. 49 (2010), pp.72–79.
10. K. Popper, *The Logic of Scientific Discovery*, Routledge Classics, London, UK, 2002. (first English edition published by Hutchinson & Co, 1959).
11. H. Hertz, *The Principles of Mechanics Presented in a New Form*, Dover Publications, New York, 1956 (translation of the German edition first published in 1894).
12. F. Brauer and C. Castillo-Chavez, *Mathematical Models in Population Biology and Epidemiology*, Springer, Heidelberg, Germany, 2001.
13. O. Diekmann and J.A.P. Heesterbeek, *Mathematical Epidemiology of Infectious Diseases: Model Building, Analysis, and Interpretation*, Wiley & Sons, New York, 2000.
14. Z. Ma, Y. Zhou, and J. Wu, *Modeling and Dynamics of Infectious Diseases*, World Scientific Publishers, Singapore, 2009.
15. H. Yang, H. Wei, and X. Li, *Global stability of an epidemic model for vector-borne disease*, J. Syst. Sci. Complex. 23 (2010), pp. 279–292.
16. J. de Miranda, G. Cordoni, and G. Budge, *The acute bee paralysis virus—Kashmir bee virus—Israeli acute paralysis virus complex*, J. Invert. Pathol. 103, S1 (2010), pp. 30–47.
17. R. Siede, M. Konig, R. Buchler, K. Failing, and H.J. Thiel, *A real time PCR based survey on acute bee paralysis virus in German bee colonies*, Apidologie 39 (2008), pp. 650–661.
18. A.A Ghamdi and R. Hoopingarner, *Modeling of honey bee and* Varroa *mite population dynamics*, Saudi J. Biol. Sci. 11 (2004), pp. 21–36.
19. H.J. Eberl, M.R. Frederick, and P.G. Kevan, *The importance of brood maintenance terms in simple models of the honeybee—*Varroa destructor—*Acute bee paralysis virus complex*, Electr. J. Diff. Equat. Conf. Ser. 19 (2010), pp. 85–98.
20. S. Omholt, *Relationships between worker longevity and intracolonial population dynamics of the honeybee*, J. Theor. Biol. 130 (1988), pp. 275–284.
21. J. Tautz, *The Buzz about Bees. Biology of a Superorganism*, Springer, Berlin, Germany, 2008.

22. J.D. Murray, *Mathematical Biology: I. An Introduction*, Springer, New York, 2002.
23. V. Ratti, P.G. Kevan, and H.J. Eberl, A mathematical model for population dynamics in honeybee colonies infested with Varroa destructor and the acute bee paralysis virus, preprint (2012), Accepted for publication. Available at http://uoguelph.ca/canpolin/ Publications/Ratti_et_al_PopulationdynamicsHBvarroa.pdf.

6 Honey Bee Ecology from an Urban Landscape Perspective
The Spatial Ecology of Feral Honey Bees

*Kristen A. Baum, Maria D. Tchakerian,
Andrew G. Birt, and Robert N. Coulson*

CONTENTS

ABSTRACT

Urban environments provide suitable habitats for feral honey bees, especially in
the desert Southwest where urban areas may provide a more continuous supply
of floral resources than surrounding natural areas. Studying honey bees in urban
environments can be challenging due to logistical constraints, but health and safety
concerns about Africanized honey bees have made understanding the urban ecology
of honey bees increasingly important. We describe two data sources for Africanized
honey bees in the greater Tucson metropolitan area. The first dataset consists of
honey bee colony and swarm removals from 1994 to 2001 compiled from the records

of a company specializing in Africanized honey bee control. The second dataset contains removal data from water meter boxes from 1996 to 2008 based on water company records. The first dataset represents point data, whereas the second dataset represents lattice/grid data. We discuss approaches for analyzing these different types of spatial data using these datasets as examples. We use the first dataset to evaluate spatiotemporal patterns in the distribution of Africanized honey bees and the contribution of precipitation to those patterns. We use the second dataset to identify characteristics associated with Africanized honey bee occupancy of water meter boxes. The two datasets combined provided over 14,000 records of colony and/or swarm removals for the study area. We found that after Africanized honey bees became established, colony removals were clustered in space and time. We also identified winter precipitation as a good predictor of the number of honey bee removals, with colony and swarm removals increasing following a wet winter. Water meter boxes in residential locations were more likely to be occupied than those in commercial locations. Occupancy was also associated with smaller lots, older structures, and closer proximities to vacant land. Water meter boxes were more likely to be occupied if neighboring water meter boxes had been occupied. The spatial analysis of these two nontraditional sources of data provided important insights into the biology and ecology of feral honey bees in the greater Tucson metropolitan area, with implications for developing control strategies for Africanized honey bees.

KEYWORDS

Apis mellifera, Cross-correlation, Feral colonies, Logistic model, Mantel test, Poisson model, Sonoran Desert, Swarms.

6.1   INTRODUCTION

This chapter focuses on the spatial ecology of feral honey bees in urban landscapes. We have extensive experience conducting research on various aspects of Africanized honey bee biology [1–12], with our more recent research addressing the distribution and abundance of feral colonies in urban landscapes [1,6]. This research has concentrated on two unique datasets encompassing the greater Tucson metropolitan area. The first dataset consists of data collected on honey bee colony and swarm removals from the invoices of a pest control company specializing in the removal of honey bee colonies and swarms [6]. The second dataset consists of data on the removal of Africanized honey bee colonies from water meter boxes by the Tucson Water Department, which covers most of the greater Tucson metropolitan area [1]. We provide a brief overview of the spatial aspects of honey bee colonies, followed by a review of what is known about the spatial distribution of feral colonies. We then discuss approaches for studying the spatial ecology of feral honey bees in urban landscapes, using the two datasets as examples. We specifically focus on spatiotemporal patterns in the distribution of feral honey bee colonies and swarms and on identifying landscape characteristics associated with their occurrence. We discuss the results in the context of understanding the ecology of feral honey bees in urban landscapes and the application of the results for developing control strategies for Africanized honey bees and reducing human–honey bee interactions. We also discuss additional considerations and future directions for research using similar, nontraditional data sources.

6.2 SPATIAL ASPECTS OF HONEY BEE COLONIES

Honey bee (*Apis mellifera* L.) colonies are spatially organized in numerous ways. The hive itself contains distinct spatial patterns in the structure of the comb and the distribution of brood, pollen, and honey [13–15]. Colonies consist of thousands of bees that are divided into castes and distributed throughout the hive according to task, with spatial structure within the hive potentially contributing to task specialization [16]. The implementation of tasks typically incorporates spatial information as well. For example, scout bees recruit worker bees to pollen and nectar sources by performing dances that provide information about the distance and direction of floral patches [17–19]. Spatial information is also used in the selection of nest sites. Honey bees reproduce at the colony level by swarming, where a portion of the colony leaves and forms a new colony. Honey bee swarms send out scouts to investigate potential nest sites based on a variety of factors, such as cavity size, entrance size, orientation, exposure, evidence of previous use, etc. [20–23]. The scout bees return to the swarm and perform waggle dances, which provide information about the distance and direction of potential nest sites. Nest site choice is narrowed down as additional scout bees recruit to advertised nest sites, until a single nest site option remains [22]. Nest site selection also likely reflects the suitability of broader-scale components of the surrounding environment for feral honey bee colonies, such as nectar and/or pollen availability or water resources. This chapter focuses on approaches for quantifying the spatial distribution of feral honey bee colonies and identifying landscape level features associated with the distribution and abundance of colonies and/or swarms.

6.3 FERAL HONEY BEES

Feral honey bees occur in many different habitats, such as coastal prairie [5], semi-deserts/deserts [24–27], forests [7,28,29], and urban areas [1,6,30,31]. Many studies have recorded the spatial locations of feral colonies, with some estimating colony density and others identifying environmental characteristics associated with colony locations [1,5,6,13,25,26,28–30,32–40]. However, relatively few studies have evaluated spatial and/or temporal patterns in the distribution of feral honey bee colonies [1,5,6,25,33,34,40] or identified landscape characteristics associated with nest sites [1]. The loss of many managed honey bee colonies to colony collapse disorder [41,42] and the arrival of Africanized honey bees, hybrids between European and African (*A. m. scutellata*) honey bees [11,43], have increased the need for more information about feral honey bee populations [44,45], especially in terms of understanding factors influencing the spatial and temporal distribution of feral colonies. With fewer managed colonies available to provide pollination services to agricultural crops and native plants, more data are needed on the distribution and abundance of feral colonies and their potential contribution to pollination. Furthermore, data on spatial and temporal distribution patterns can provide useful information for developing strategies for controlling and managing Africanized honey bees, as well as for controlling and managing honey bee parasites and diseases that show spatial and/or temporal patterns of infection [46,47].

Several studies have identified aggregations (*i.e.*, clumped distributions) of honey bee colonies in different locations, including Australia [33,34]; Botswana, Africa [25]; southern Texas, United States [5]; and Tucson, Arizona, United States [1,6]. Aggregations have also been documented for other honey bee species, including giant honey bees (*A. dorsata* Fabricius and *A. laboriosa* Smith) [48–50] and dwarf honey bees (*A. florea* Fabricius and *A. andreniformis* Smith) [40,51]. Several biological and/or ecological reasons have been proposed for why colonies may form aggregations [33,40,52]. For example, short swarm dispersal distances may lead to aggregations, whereas long dispersal distances may lead to other distribution patterns. However, reports of swarm dispersal distances are variable, with both relatively short and long dispersal distances reported [33,53–58]. These conflicting results may reflect the influence of site-specific characteristics on swarm dispersal distances, such as differences in resource availability or genetic differences among populations [54,59]. For example, swarms may select nearby cavities when cavities are abundant and disperse longer distances when cavities are scarce [20,55–57]. Also, Africanized honey bee swarms have been recorded dispersing farther than swarms of European origin [54,59]. Other potential benefits of aggregations include increased predator detection and/or defenses and increased mating efficiency [40,50,52,60]. Alternatively, swarms may select nest sites near existing colonies if the presence of colonies is used by the honey bees as an indicator of environmental suitability, such as high floral resource availability [33,40,50], or if swarms employ similar behavioral rules and make similar decisions during the process of evaluating potential nest sites [40,61]. Potential disadvantages of aggregations include the attraction of predators, increased transmission of pests and/or pathogens, and competition for floral resources [40,46,47,50]. Therefore, identifying spatial and temporal patterns in the distribution of honey bee colonies and identifying landscape characteristics associated with these patterns can provide a framework for evaluating potential biological and ecological mechanisms behind aggregations, as well as developing additional hypotheses to explain these patterns.

We have studied the spatial distribution of feral honey bee colonies in natural and urban settings [1,5,6] using different types of spatial data, including point data and lattice data. Point data consist of spatial coordinates for colony locations, which can be recorded using global positioning system (GPS) technology [5] or by assigning coordinates to street addresses where colonies have been recorded [6]. In the latter case, street addresses can be linked to line segments in a street network coverage using a commercial geographic information system (GIS) product to generate address-specific spatial coordinates. Lattice data consist of data associated with a lattice or grid, where a data point is recorded for each grid cell. Lattice data could be obtained in many ways, such as placing a trap within each grid cell for sampling. Depending on the question of interest, separate analyses of spatial locations may be needed for colonies and swarms, since colonies represent the selection of somewhat permanent nest site locations and swarms only occupy temporary locations while in the process of selecting a more permanent nest site.

6.4 APPROACHES FOR STUDYING THE SPATIAL ECOLOGY OF FERAL HONEY BEE COLONIES IN URBAN LANDSCAPES

A variety of approaches can be used for studying the spatial ecology of honey bees. We focus our discussion on approaches relevant for feral honey bee colonies and/or swarms, with an emphasis on urban landscapes. We start by describing two nontraditional datasets on feral honey bees in the greater Tucson metropolitan area. We then explain different approaches for evaluating the spatial ecology of feral colonies and/or swarms, providing an example of how each approach can be applied using these datasets.

6.4.1 DESCRIPTION OF DATASETS

Our research on feral honey bees in urban settings has focused on two unique datasets encompassing the greater Tucson metropolitan area (Figure 6.1). The first dataset consists of data collected on honey bee colony and swarm removals from 1994 to 2001 from invoices obtained from a pest control company specializing in the removal of honey bee colonies (BeeMaster, Inc.; [6]). The second dataset consists of data on the removal of Africanized honey bee colonies from water meter boxes from 1996 to 2008 from records provided by the Tucson Water Department, which covers most of the greater Tucson metropolitan area [1]. Both datasets indicate that honey bee colonies were rarely found in water meter boxes or other urban cavities until after the arrival of Africanized honey bees. For example, only 14 colonies and swarms were removed by BeeMaster, Inc. in 1994, the year after Africanized honey bees were first recorded in Arizona, compared to a minimum of 458 colonies and swarms recorded in subsequent years [6]. Africanized honey bee colonies are typically smaller than colonies composed of bees of European origin, and they will use smaller cavities and a wider range of nest sites, including water meter boxes, garbage cans, flower pots, etc. [23,62–64]. This reduced selectivity in nest site selection (compared to European honey bees) could place Africanized honey bees in closer proximity to humans [1,6]. This may be especially true in southeastern Arizona (and the greater Tucson metropolitan area) where cavities are relatively uncommon in natural areas where colonies nest in rock crevices [29,55]. Combined with their stronger defensive behaviors, these characteristics raise concerns about the interactions between Africanized honey bees and humans. Furthermore, feral honey bee colonies in areas of the United States dominated by Africanized honey bees are primarily Africanized [10,63], including the greater Tucson metropolitan area in Arizona [65,66].

For the pest removal dataset, we collected data from the invoices of a pest control company (BeeMaster, Inc.), including the dates and street addresses for colony and/or swarm removals [6]. We converted each street address into spatial coordinates

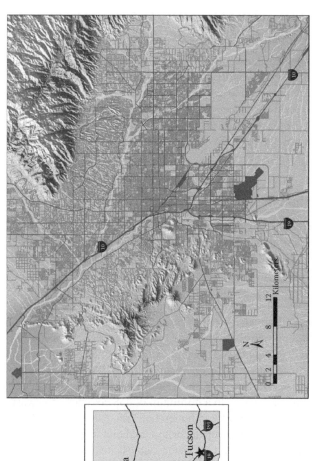

FIGURE 6.1   Location of Tucson within Arizona, United States, and elevation in a shaded relief view of the greater Tucson metropolitan area.

by linking each address to line segments in the Pima County Department of Transportation street network coverage using ArcGIS®. Therefore, the pest removal dataset represents point data because it is based on specific spatial locations where honey bee colonies and/or swarms were removed. The dataset is limited by socio-economic constraints, as the data depend on the ability and/or willingness to pay for honey bee removal services. The dataset is also limited by colony and swarm detectability, which could vary based on numerous social, economic, and environmental considerations [6]. However, we expect these biases are minimal in the study area because of public concern about Africanized honey bees (aka "killer bees").

The water meter dataset also represents point data because it is based on records of water meter boxes from which colonies were removed. We converted these data into lattice data by adding locations of water meter boxes where honey bee colonies were not recorded from records available from the Pima County Tax Assessor's Office. Because water meter boxes are checked regularly for billing purposes, this additional information allowed us to incorporate both presence and absence data. These data are necessary for identifying landscape characteristics associated with the occurrence of honey bee colonies and to predict where honey bee colonies will likely be found. We made several necessary assumptions about the dataset. First, we assumed the dataset was complete and that all colonies were recorded and removed by the water company (*i.e.*, no one else found and removed colonies between water meter checks). We also assumed that water meter boxes were checked once every 30 days and that no new colonies occurred within that 30-day time period. Finally, we assumed that water meter boxes were checked regularly after the construction date for a building, although water meters may have been turned off for various amounts of time or not turned on until after the construction date. An additional limitation of this dataset relates to the uneven distribution of water meter boxes throughout the greater Tucson metropolitan area. Furthermore, the availability of other types of cavities and their relative suitability for honey bee colonies may have varied throughout the study area. However, we expect cavity availability to be high throughout the study area and that cavities are not limiting for honey bees, especially given the diversity of nest sites used by Africanized honey bees [23,62,64,66]. Regardless of these limitations, the water meter dataset provides a unique perspective on the spatial ecology of feral honey bees because it simulates an array of traps distributed across the study area, which were checked monthly for honey bee colonies [1].

The pest removal dataset indicated that colony and swarm removals in the greater Tucson metropolitan area typically exhibited a bimodal intra-annual distribution with peaks in March through May and again in October (Figure 6.2). Some years (*e.g.*, 1995, 1996, 1997, and 2001) exhibited both peaks in activity, whereas some years showed only an early peak (*e.g.*, 1999, 2000) and others showed a pronounced late peak (*e.g.*, 1998). Colony and swarm removals also showed pronounced fluctuations among years. For example, the number of swarms removed increased from 1995 to 1996 (+41%), 1997 to 1998 (+230%), and 2000 to 2001 (+336%), and decreased from 1996 to 1997 (−10%), 1998 to 1999 (−10%), and 1999 to 2000 (−49%). Similar patterns were observed for the water meter dataset over the same time period. Most water meter boxes were located in residential areas (82%) and most water meter boxes occupied by honey bee colonies occurred in residential areas (87%).

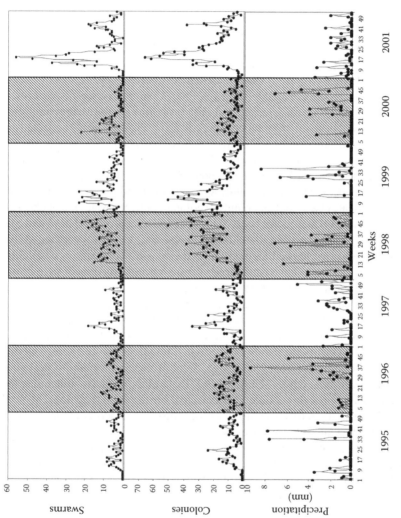

FIGURE 6.2 The number of swarms removed, number of colonies removed, and weekly precipitation (mm) from 1995 through 2001 for the pest removal dataset. (Reprinted from *Landscape Urban Plann.*, 85, Baum, K.A., Tchakerian, M.D., Thoenes, S.C., and Coulson, R.N., Africanized honey bees in urban environments: A spatio-temporal analysis, 123–132, Copyright [2008], with permission from Elsevier.)

Overall, less than 3% of the water meter boxes in any given land use type (*e.g.*, residential, commercial/public, recreation, agriculture, etc.) ever contained a honey bee colony, and only 1.28% of the water meter boxes in the overall study area. Over the 12-year study period, 8,211 colonies were removed from 5,640 water meter boxes, with an average of 632 colonies removed per year. Of the occupied water meter boxes, 1,350 were occupied more than once over the 12-year study period.

6.4.2 Overview of Analytical Approach

Although working with nontraditional datasets (*i.e.*, data not collected from a designed study) generates many challenges, they provide a unique opportunity to increase our understanding of feral honey bees in urban landscapes. We used the pest removal dataset to evaluate spatiotemporal patterns in the distribution of feral honey bee colonies and swarms using a Mantel test. We used a cross-correlation analysis to evaluate the contribution of precipitation to the observed patterns. We developed logistic and Poisson models to identify landscape characteristics that contribute to the nest site locations of feral honey bee colonies using the water meter dataset. The results of these analyses provide insight into the ecology of feral honey bees in urban landscapes and also suggest potential control strategies for Africanized honey bees in urban areas.

6.4.3 Spatiotemporal Patterns

A Mantel test can be used to evaluate the similarity or dissimilarity between two distance matrices [67], including distance in space and distance in time. If distance units are not comparable (as is the case with space and time units), the data can be ranked prior to analysis [13,68]. A randomization test (*i.e.*, random reshuffling of the data values) is typically used to evaluate if values are significant, with significance occurring when the Mantel test statistic is more extreme than the randomly generated reference distribution as determined by Spearman's rank correlation coefficient. We used a Mantel test to evaluate if honey bee colony and swarm removals located close together in space were also located close together in time in the pest removal dataset [6]. We found that during the initial years following the arrival of Africanized honey bees (*i.e.*, 1995–1998), colony locations and a combination of colony and swarm locations were not spatiotemporally aggregated (Table 6.1). However, significant spatiotemporal clustering occurred during the last 3 years for which data were available (*i.e.*, 1999–2001). During 1999 and 2000, colony and swarm removals tended to be locally concentrated, with removals located close together in space also being located close together in time. These years were also dry years, and there was a pronounced decline in the number of colonies and swarms removed from 1999 to 2000. In 2001, which was a wet year, approximately 3× as many colonies and swarms were removed compared to the previous year. Removals were also broadly distributed throughout the greater Tucson metropolitan area (*i.e.*, spatiotemporal clustering was negative). The extreme variation in the number of colonies and swarms removed among years suggests that climatic conditions may contribute to inter- and intra-annual fluctuations in the size of the feral honey bee population.

TABLE 6.1

Mantel Test Results to Evaluate if Colony Removals or Colony and Swarm Removals Located Close together in Space Were also Located Close together in Time for the Pest Removal Dataset

Year	# Colonies	R	p	# Colonies and Swarms	R	p
1994	14	−0.071	0.376	14	−0.071	0.376
1995	323	0.009	0.286	458	0.004	0.385
1996	474	0.011	0.187	665	0.009	0.203
1997	445	0.022	0.096	617	0.021	0.075
1998	1,012	0.019	0.061	1,407	0.017	0.051
1999	871	0.027	0.036	1,225	0.025	0.024
2000	353	0.039	0.021	525	0.046	0.005
2001	1,035	−0.029	0.009	1,613	−0.033	0.000

Source: Modified from *Landscape Urban Plann.*, 85, Baum, K.A., Tchakerian, M.D., Thoenes, S.C., and Coulson, R.N., Africanized honey bees in urban environments: A spatio-temporal analysis, 123–132, Copyright [2008], with permission from Elsevier.

Temperature and precipitation may influence the distribution of Africanized honey bees. However, Tucson is in southeastern Arizona in the northern Sonoran Desert. Therefore, it is unlikely that winter temperatures would limit the distribution of Africanized honey bees in the study area. However, honey bees depend on floral resource availability, which may be limited by precipitation. Average annual rainfall in this area is 29.7 cm, with rainfall events concentrated during the summer monsoon period (July to mid-September) and to a lesser extent during the winter (December to early March). Therefore, nectar and pollen availability are relatively high in natural areas in the spring and later summer/fall [69,70] and lower throughout the remainder of the year. African honey bees are adapted to rapidly exploit the seasonal flushes of floral resources that occur in tropical habitats and are expected to respond more strongly to precipitation patterns than European honey bees, which are adapted to exploit the more consistently available floral resources of temperate habitats [66,71,72]. However, floral resource availability may be higher in urban areas where landscaping (including the use of exotic ornamentals) and associated irrigation may extend flowering periods. Therefore, urban environments may provide a more spatially and temporally consistent supply of floral resources compared to surrounding natural areas (as well as more abundant cavities; see earlier text). Also, floral resource availability is not expected to be directly correlated with precipitation, as there is typically a time lag between precipitation events and flower production.

A cross-correlation analysis provides a useful approach for evaluating the relationship between colony and swarm removals and climatic conditions across a range of time lags. Cross-correlation values range from −1 to 1. When strong positive correlation occurs, cross-correlation values approach 1, and when strong negative correlation occurs, values approach −1, with 0 indicating the absence of correlation. Significance is evaluated using 95% upper and lower confidence limits. We used a cross-correlation analysis to evaluate the contribution of precipitation to the abundance of colonies and swarms in the pest removal dataset using daily precipitation data. Colony removals were positively cross-correlated with precipitation in 1997, 1998, and 2001 and negatively correlated in 1996 and 1998 (Figures 6.2 and 6.3). Swarm removals were positively cross-correlated with precipitation in 1997, 1998, 2000, and 2001 and negatively cross-correlated with precipitation in 1998 (Figures 6.2 and 6.3). Relatively short lag times likely reflect frequent rainfall and high or early pollen availability, whereas longer lag times likely reflect low or delayed pollen availability. For example, 1998 was an El Niño year and the wettest year based on overall precipitation, and colony and swarm removals showed relatively short, positive lags (15 and 12 weeks, respectively), suggesting pollen availability was high. Furthermore, 1998 also had the greatest number of days with rain, with colony and swarm removals also exhibiting very short, negative lags (4 and 0 weeks, respectively), as would be expected with a temporary decrease in honey bee activity associated with rainfall. Lag times in 1997 were intermediate in length (*e.g.*, 31, 32, and 34 weeks for colony and swarm removals) and long in 2001 (28–30 weeks for colony removals and 24, 26, and 28 weeks for swarm removals), suggesting intermediate and low levels of pollen availability, respectively. Lag times for colonies and swarms would be expected to be similar, given that colony activity levels are expected to be high during periods of high pollen availability, which would also correspond to swarm production. These high levels of activity would increase the likelihood of detection (and thus inclusion in the pest removal dataset), depending on the cavity selected, and the timing would likely be associated with the developmental periods of honey bees.

In the greater Tucson metropolitan area, spring flowering (and swarming activity) is associated with rainfall arriving late in the previous year [55]. To identify expected periods of swarming activity, it is necessary to consider the developmental periods of honey bees. For European honey bees, worker bees develop in approximately 21 days and queens in about 18 days. Africanized honey bees develop 1–2 days faster than European honey bees. Therefore, one would expect early spring swarming to occur approximately 24–26 weeks after precipitation in October that generates early spring flowering in mid-February (which would correspond to 6–8 weeks after early spring flowering begins). Precipitation during the growing season would be expected to generate shorter lags, as floral resources (and subsequently honey bees) would exhibit a more immediate response. Negative correlations could be associated with detection probabilities, with the expectation that fewer swarms and colonies would be detected and/or removed during rainy periods when honey bees are less active (and less likely to be noticed) and people may spend less time outside. Positive correlations also would be associated with detection probabilities and related to activity periods of honey bees and humans. Seasonal absconding could obscure these patterns if colonies leave their nests in areas of low resource availability (common in the mountains surrounding Tucson during the fall and winter; [73]) in search of nest sites with more suitable conditions, such as areas where agricultural and landscaping practices increase floral resource availability [72–74].

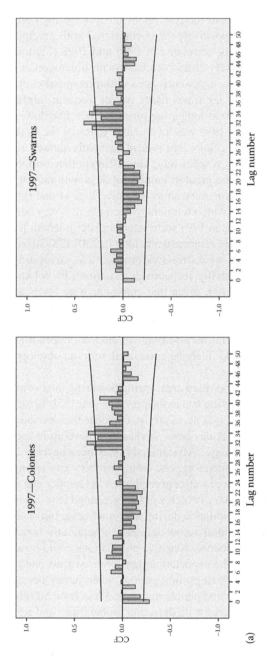

FIGURE 6.3 Cross-correlation coefficients for precipitation and colony and swarm removals for (a) 1997, (b) 1998, and (c) 2001 in the pest removal dataset. Lags were calculated from 0 to 50 weeks and 95% upper and lower confidence limits were used to evaluate significance, with coefficients at or above the confidence limits indicating significant lags.

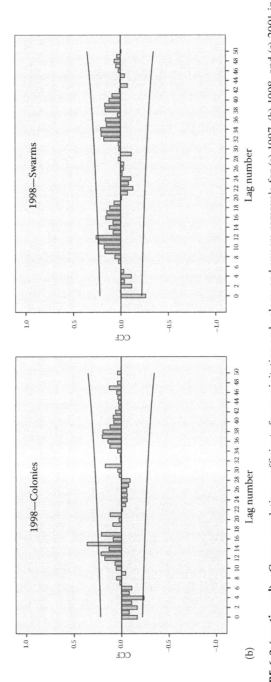

FIGURE 6.3 (continued) Cross-correlation coefficients for precipitation and colony and swarm removals for (a) 1997, (b) 1998, and (c) 2001 in the pest removal dataset. Lags were calculated from 0 to 50 weeks and 95% upper and lower confidence limits were used to evaluate significance, with coefficients at or above the confidence limits indicating significant lags.

(continued)

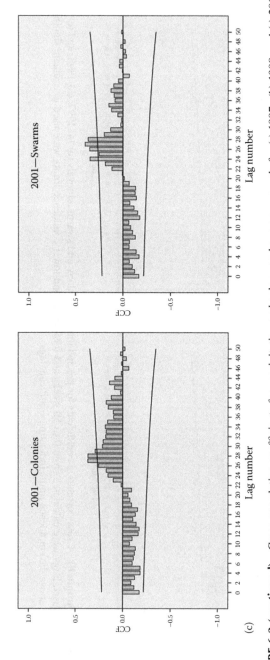

FIGURE 6.3 (continued) Cross-correlation coefficients for precipitation and colony and swarm removals for (a) 1997, (b) 1998, and (c) 2001 in the pest removal dataset. Lags were calculated from 0 to 50 weeks and 95% upper and lower confidence limits were used to evaluate significance, with coefficients at or above the confidence limits indicating significant lags. (Modified and reprinted from *Landscape Urban Plann.*, 85, Baum, K.A., Tchakerian, M.D., Thoenes, S.C., and Coulson, R.N., Africanized honey bees in urban environments: A spatio-temporal analysis, 123–132, Copyright [2008], with permission from Elsevier.)

6.4.4 LANDSCAPE ASSOCIATIONS

Landscape characteristics also can influence the distribution and abundance of feral honey bee colonies. We used the water meter dataset to evaluate the influence of land use, lot characteristics, and structure characteristics on the spatial distribution of feral honey bees in the greater Tucson metropolitan area [1]. We developed logistic regression and Poisson models to predict the occupancy of water meter boxes based on landscape characteristics (*i.e.*, distance to different land use types based on distance to nearest edge) and lot/structure characteristics (*i.e.*, construction year, lot size, and indicator variables for the presence of a swimming pool within 0.8 km and the presence of a residential location) (Table 6.2). Indicator variables are binary variables, receiving a value of 1 when the indicator is present and a value of 0 when it is not present (*e.g.*, a value of 1 when the location was a residential location and a value of 0 when the location was not a residential location). We also included the percent occupancy of nearby water meter boxes, which was calculated as the percent of water meter boxes within 0.8 km of a given water meter box that were occupied at least once during the 12-year study period (Table 6.2). This distance was selected to represent typical foraging distances and swarm dispersal distances based on data reported in the literature [53,55,58,75]. In both models, the location (*i.e.*, landscape characteristics) of every water meter box in the study area was used to identify the variables that could influence water meter box occupancy by feral honey bee colonies. Therefore, an unbiased analysis of factors most relevant to colony occupancy required that the colony removal dataset (positive counts only) be combined with a dataset including the locations (and landscape characteristics) of all the water meter boxes in the study area, many of which were never occupied. For both the logistic and Poisson models, this generated a dataset of 275,877 data points. In the case of the logistic model, we used a binary variable (1 or 0) to indicate whether a water meter box was ever occupied by a honey bee colony. Thus, the quantity estimated from the model can be interpreted as the probability of the water meter box being occupied over the 12-year study period. For the Poisson model, we used a similar approach to regress the landscape characteristics of each water meter box location against the number of times, over the 12 year × 1 month sampling period, that the water meter box was occupied. Therefore, the output of the model is a Poisson rate that can be used to assess a probability that a given number of occupancies will occur in a given water meter box (dependent on its landscape characteristics) over the 12 year × 1 month sampling period.

In practice, both models were developed by calculating a large number of potential independent variables (*i.e.*, landscape characteristics) and then assessing the significance of each variable when included in either the logistic regression or Poisson models. This stepwise approach to model development contributes to the ecological interpretation of results. Although it is not reasonable to expect that landscape characteristics alone can predict feral honey bee colony occurrence (*e.g.*, many other nonlandscape or smaller scale factors may also influence colony nest site selection), the stepwise procedure allowed us to evaluate factors that we thought might influence feral honey bee colony behavior. It elucidated the significance of each potential variable, allowing us to identify those that do affect colony nest site location as well as eliminate those that do not.

TABLE 6.2

Description of the Variables Included in the Logistic Regression and Poisson Models for the Water Meter Box Dataset, Including How Each Variable Was Measured, and Corresponding Means and Standard Deviations (SD)

Variable	Mean	SD	Description
% Nearby water meter boxes with colonies	2.46	2.47	The percent of nearby water meter boxes within 0.8 km that were occupied at least once over the 12-year study period for every water meter box.
Construction year	1976	18.73	The year the structure associated with a water meter box was built.
Lot size (ha)	0.466	2.27	The land area of a parcel of land associated with a water meter box.
Distance from "water"[a] (km)	1.42	1.27	Distances calculated as the distance from the center of the lot associated with each water meter box to the nearest edge of the specified land use type.
Distance from recreational land (km)	3.73	3.74	
Distance from natural land (km)	2.42	2.51	
Distance from vacant land (km)	1.44	1.67	
Distance from agricultural land (km)	10.92	6.90	
Distance from transportation land (km)	1.36	1.70	
Indicator of pool within 0.8 km	0.737	0.440	Binary variable for the presence or absence of a pool within 0.8 km (*i.e.*, 1 indicates a pool is present and 0 indicates no pool is present). The mean is the percentage of water meter boxes with a pool within 0.8 km.
Indicator of residential location	0.821	0.383	Binary variable for residential location (*i.e.*, 1 indicates a residential location and 0 indicates a nonresidential location). The mean is the percentage of water meter boxes that were in residential locations.

Source: Modified from *Landscape Urban Plann.*, 100, Baum, K.A., Kolodziej, E.Y., Tchakerian, M.D., Birt, A.G., Sherman, M., and Coulson, R.N., Spatial distribution of Africanized honey bees in an urban landscape, 153–162, Copyright (2011), with permission from Elsevier.

[a] Most lots classified as water only temporarily held water and were dry for part of the year; some were washes that only contained water during the monsoon season [78].

Variables were either significant in both models or not significant in either model. Variables that were significant in both models included percent of occupied neighboring water meter boxes, construction year, lot size, distance from vacant land, and the indicator of residential location (Tables 6.3 and 6.4). Colonies were more likely to occupy water meter boxes associated with higher percentages of neighboring water meter boxes with colonies, older structures, smaller lots, closer distances to vacant land, and residential (versus commercial) locations. The importance of occupied neighboring water meter boxes supports the view that colonies tend to form aggregations (see earlier discussion on the potential biological advantages and disadvantages of aggregations). Older structures likely provide more potential cavities, whereas residential locations may provide higher floral resource availability compared to other land use types based on urban landscaping practices [76], and residential locations tend to be located on smaller lots.

Variables that were not significant in either model included distances from water, recreational land, natural land, agricultural land, and transportation land, as well as the indicator variable for a nearby swimming pool (Tables 6.3 and 6.4). These results suggest that water was not limiting in the study area. The distribution of the

---

TABLE 6.3

Parameter Estimates, Standard Errors (SE), and Two-Tailed P-Values from the Logistic Regression Model to Predict the Occupation of Water Meter Boxes

Parameter	Parameter Estimates	SE	P-Values
(Intercept)	31.62	3.98	<0.0001*
Percent neighbors with colonies	23.12	2.39	<0.0001*
Construction year	−0.018	0.002	<0.0001*
Lot size [Ln(acres + 0.1)]	0.454	0.025	<0.0001*
Distance from "water"[a] (km)	$2.700e - 05$	$1.672e - 05$	0.106
Distance from recreational land (km)	$1.305e - 05$	$2.051e - 05$	0.525
Distance from natural land (km)	$1.485e - 05$	$1.076e - 05$	0.168
Distance from vacant land (km)	$-1.110e - 04$	$1.902e - 05$	<0.0001*
Distance from agricultural land (km)	$-8.412e - 07$	$4.980e - 06$	0.866
Distance from transportation land (km)	$2.605e - 05$	$1.358e - 05$	0.055
Indicator of pool within 0.8 km	0.020	0.051	0.690
Indicator of residential location	0.184	0.057	0.001*

Source: Modified from *Landscape Urban Plann.*, 100, Baum, K.A., Kolodziej, E.Y., Tchakerian, M.D., Birt, A.G., Sherman, M., and Coulson, R.N., Spatial distribution of Africanized honey bees in an urban landscape, 153–162, Copyright (2011), with permission from Elsevier.

Note: Parameter values with a P-value <0.05 are indicated with an asterisk.

[a] Most lots classified as water only temporarily held water and were dry for part of the year; some were washes that only contained water during the monsoon season [78].

TABLE 6.4

Parameter Estimates, Standard Errors (SE), and Two-Tailed P-Values from the Poisson Model to Predict the Number of Times a Water Meter Box Was Occupied over the 12-Year Study Period

Parameter	Parameter Estimates	SE	P-Values
(Intercept)	31.71	5.69	<0.0001*
Percent neighbors with colonies	23.43	3.34	<0.0001*
Construction year	−0.018	0.003	<0.0001*
Lot size [Ln(acres +0.1)]	0.503	0.039	<0.0001*
Distance from "water"[a] (km)	2.529e − 05	2.523e − 05	0.316
Distance from recreational land (km)	4.816e − 06	2.260e−05	0.831
Distance from natural land (km)	1.909e−05	1.686e−05	0.258
Distance from vacant land (km)	−1.329e − 04	2.398e − 05	<0.0001*
Distance from agricultural land (km)	2.087e − 06	7.540e − 06	0.782
Distance from transportation land (km)	2.990e − 05	2.065e − 05	0.148
Indicator of pool within 0.8 km	0.054	0.082	0.508
Indicator of residential location	0.153	0.009	<0.0001*

Source: Modified from *Landscape Urban Plann.*, 100, Baum, K.A., Kolodziej, E.Y., Tchakerian, M.D., Birt, A.G., Sherman, M., and Coulson, R.N., Spatial distribution of Africanized honey bees in an urban landscape, 153–162, Copyright (2011), with permission from Elsevier.

Note: Parameter values with a P-value <0.05 are indicated with an asterisk.

[a] Most lots classified as water only temporarily held water and were dry for part of the year; some were washes that only contained water during the monsoon season [78].

different land use types could have contributed to their not being important predictors of water meter box occupancy or multiple occupancies, with recreational, natural, and agricultural lands tending to be distributed around the periphery of the greater Tucson metropolitan area. Resource availability within these different land use types also could have contributed to the observed patterns.

Overall, 2.46% of nearby water meter boxes (within 0.8 km of any given water meter box) were occupied, and these occupations were concentrated in south central Tucson (Figure 6.4). High probabilities of occupancy and of multiple occupancies were concentrated in south central Tucson, with additional high probability occupancies occurring in several suburbs around the periphery of the greater Tucson metropolitan area (Figures 6.5 and 6.6). These patterns suggest that the feral honey bee population is either denser in this area or that more colonies are located in water meter boxes in this area. We expect that cavity sources are abundant in the greater Tucson metropolitan area [1,6] and that the population is denser in this area. A tendency to reuse previously used cavities could contribute to this pattern as well, which could be driven by a preference for previously used cavities [20], a tendency to form aggregations (see previous discussion), or by similar nest site selection criteria and/or decisions [20–23].

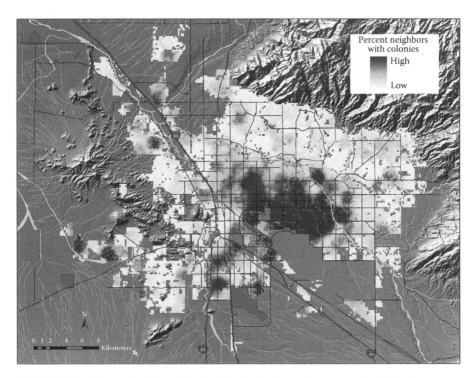

FIGURE 6.4 (See color insert.) Spatial distribution of percent of neighboring water meter boxes with colonies (within 0.8 km) for every water meter box in the water meter box dataset. (Modified and reprinted from *Landscape Urban Plann.*, 100, Baum, K.A., Kolodziej, E.Y., Tchakerian, M.D., Birt, A.G., Sherman, M., and Coulson, R.N., Spatial distribution of Africanized honey bees in an urban landscape, 153–162, Copyright [2011], with permission from Elsevier.)

6.5 CONCLUSIONS, ADDITIONAL CONSIDERATIONS, AND FUTURE DIRECTIONS

Our research on feral honey bees in urban landscapes has focused on two unique datasets developed from nontraditional sources. An analysis of the pest removal dataset revealed that colony and swarm removals were not spatiotemporally clustered after the initial arrival of Africanized honey bees in the study area. However, that pattern shifted in subsequent years, with positive spatiotemporal clustering in 1999 and 2000 and negative spatiotemporal clustering in 2001. We identified winter rainfall as a good predictor of honey bee abundance in urban areas in the desert Southwest and a potential indicator of high human–honey bee interactions. An analysis of the water meter dataset found that water meter boxes were more likely to be occupied when associated with residential locations, older structures, smaller lots, closer proximity to vacant land, and if nearby water meter boxes were occupied. Water meter box occupancies were concentrated in south central Tucson, suggesting this area is highly suitable for honey bees. Taken together, the results from these two

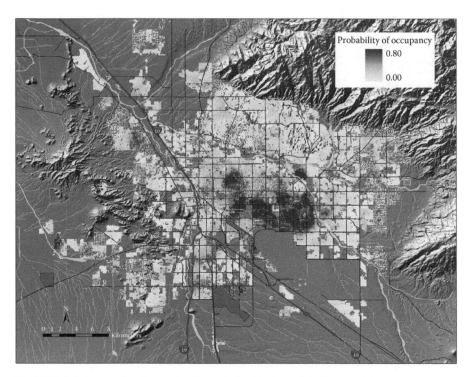

FIGURE 6.5 (See color insert.) Spatial distribution of Africanized honey bee colony occupancy probabilities predicted by the logistic regression model for the water meter box dataset. Dark shaded areas show parts of the greater Tucson metropolitan area where water meter boxes are predicted to have high occupancy rates, whereas light shaded areas are predicted to have low occupancy rates. (Reprinted from *Landscape Urban Plann.*, 100, Baum, K.A., Kolodziej, E.Y., Tchakerian, M.D., Birt, A.G., Sherman, M., and Coulson, R.N., Spatial distribution of Africanized honey bees in an urban landscape, 153–162, Copyright [2011], with permission from Elsevier.)

datasets suggest that control efforts should be focused in areas where colonies and/or swarms are found and that source colonies likely exist in south central Tucson. They indicate that control efforts should be increased following wet winters, when swarming activity is likely to be high. The results also highlight the importance of the "capture" and removal of colonies in water meter boxes, which likely contributes significantly to the control of Africanized honey bees in the study area.

For both datasets, we acknowledge that socioeconomic factors could have influenced the observed patterns [1,6]. The tolerance of stinging insects and perceptions about Africanized honey bees (aka "killer bees") could influence the likelihood of a homeowner or business owner choosing to remove a colony himself/herself or choosing to call a pest control company, as well as the ability to pay for honey bee removal services. Resource availability (*e.g.*, pollen, nectar, water, nest sites, etc.) can vary with social and economic variables as well, with land use, family income, and housing age being predictors of plant diversity in the Central Arizona–Phoenix region [76,77]. These factors could also influence detectability, based on

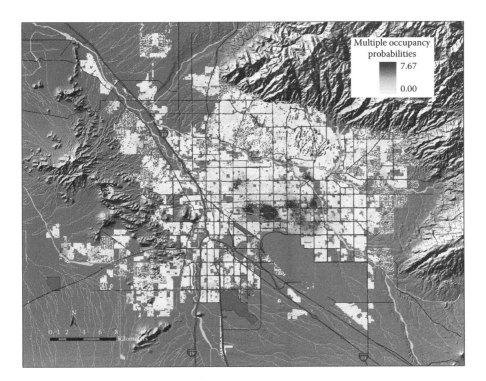

FIGURE 6.6 (See color insert.)   Spatial distribution of Africanized honey bee colony multiple occupancy probabilities predicted by the Poisson model for the water meter box dataset. Dark shaded areas show parts of the greater Tucson metropolitan area where water meter boxes are predicted to have high multiple occupancy rates, whereas light shaded areas are predicted to have low multiple occupancy rates. (Reprinted from *Landscape Urban Plann.*, 100, Baum, K.A., Kolodziej, E.Y., Tchakerian, M.D., Birt, A.G., Sherman, M., and Coulson, R.N., Spatial distribution of Africanized honey bees in an urban landscape, 153–162, Copyright [2011], with permission from Elsevier.)

the locations of honey bee resources and human behaviors and/or activities that can vary with socioeconomic considerations, such as the frequency of recreational activities focused in areas with resources shared with honey bees. Census data, such as population density, housing density, median household income, median house value, education level, and unemployment rate, could be used to provide insight into these factors.

It is also important to note that detection is important for the control of Africanized honey bees. For example, redesigning water meter boxes to restrict honey bee access could lead to an increase in the size of the Africanized honey bee population in the greater Tucson metropolitan area, especially if cavities are not limiting (which is likely the case up to a certain point) and fewer colonies are subsequently detected and removed. The implementation of smart meter programs also could influence these patterns if water meters are remotely monitored and water meter boxes are not checked on a regular basis.

Understanding the spatial ecology of feral honey bees can provide important insights into the biology and ecology of feral honey bees. Although each of our datasets is accompanied by certain constraints and/or limitations, examining complementary types of spatial data with the appropriate spatial techniques can expand our understanding of how feral honey bees respond to spatial and temporal variability in urban environments. Furthermore, our research identifies the importance of non-traditional sources of data for feral honey bees, which has important implications for urban ecological research in general, as well as developing strategies to control Africanized honey bees in urban areas.

REFERENCES

1. K.A. Baum, E.Y. Kolodziej, M.D. Tchakerian, A.G. Birt, M. Sherman, and R.N. Coulson, *Spatial distribution of Africanized honey bees in an urban landscape*, Landscape Urban Plann. 100 (2011), pp. 153–162.
2. K.A. Baum, W.L. Rubink, and R.N. Coulson, *Trapping of feral honey bee workers (Hymenoptera: Apidae) in a coastal prairie landscape: Effects of season and vegetation type*, Can. Entomol. 138 (2006), pp. 228–234.
3. K.A. Baum, W.L. Rubink, R.N. Coulson, and V.M. Bryant, *Pollen selection by feral honey bee (Hymenoptera: Apidae) colonies in a coastal prairie landscape*, Environ. Entomol. 33 (2004), pp. 727–739.
4. K.A. Baum, W.L. Rubink, R.N. Coulson, and V.M. Bryant, *Diurnal patterns of pollen collection by feral honey bee colonies in southern Texas, USA*, Palynology 35 (2011), pp. 85–93.
5. K.A. Baum, W.L. Rubink, M.A. Pinto, and R.N. Coulson, *Spatial and temporal distribution and nest site characteristics of feral honey bee (Hymenoptera: Apidae) colonies in a coastal prairie landscape*, Environ. Entomol. 34 (2005), pp. 610–618.
6. K.A. Baum, M.D. Tchakerian, S.C. Thoenes, and R.N. Coulson, *Africanized honey bees in urban environments: A spatio-temporal analysis*, Landscape Urban Plann. 85 (2008), pp. 123–132.
7. R.N. Coulson, M.A. Pinto, M.D. Tchakerian, K.A. Baum, W.L. Rubink, and J.S. Johnston, *Feral honey bees in pine forest landscapes of east Texas*, For. Ecol. Manage. 215 (2005), pp. 91–102.
8. W.L. Rubink, K.D. Murray, K.A. Baum, and M.A. Pinto, *Long term preservation of DNA from honey bees* (Apis mellifera) *collected in aerial pitfall traps*, Tex. J. Sci. 55 (2003), pp. 159–168.
9. M.A. Pinto, J.S. Johnston, W.L. Rubink, R.N. Coulson, J.C. Patton, and W.S. Sheppard, *Identification of Africanized honey bee (Hymenoptera: Apidae) mitochondrial DNA: Validation of a rapid polymerase chain reaction-based assay*, Ann. Entomol. Soc. Am. 96 (2003), pp. 679–684.
10. M.A. Pinto, W.L. Rubink, R.N. Coulson, J.C. Patton, and J.S. Johnston, *Temporal pattern of Africanization in a feral honeybee population from Texas inferred from mitochondrial DNA*, Evolution 58 (2004), pp. 1047–1055.
11. M.A. Pinto, W.L. Rubink, J.C. Patton, R.N. Coulson, and J.S. Johnston, *Africanization in the United States: Replacement of feral European honeybees* (Apis mellifera L.) *by an African hybrid swarm*, Genetics 170 (2005), pp. 1653–1665.
12. M.A. Pinto, W.S. Sheppard, J.S. Johnston, W.L. Rubink, R.N. Coulson, N.M. Schiff, I. Kandemir, and J.C. Patton, *Honey bees (Hymenoptera: Apidae) of African origin exist in non-Africanized areas of the southern United States: Evidence from mitochondrial DNA*, Ann. Entomol. Soc. Am. 100 (2007), pp. 289–295.

13. T.D. Seeley and R.A. Morse, *Nest of honey bee* (Apis mellifera *L.*), Insect. Soc. 23 (1976), pp. 495–512.

14. B.R. Johnson and N. Baker, *Adaptive spatial biases in nectar deposition in the nests of honey bees*, Insect. Soc. 54 (2007), pp. 351–355.

15. S. Camazine, *Self-organizing pattern formation on the combs of honey bee colonies*, Behav. Ecol. Sociobiol. 28 (1991), pp. 61–76.

16. B.R. Johnson, *Spatial effects, sampling errors, and task specialization in the honey bee*, Insect. Soc. 57 (2010), pp. 239–248.

17. M. Gil and R.J. De Marco, *Decoding information in the honeybee dance: Revisiting the tactile hypothesis*, Anim. Behav. 80 (2010), pp. 887–894.

18. T.D. Seeley, A.S. Mikheyev, and G.J. Pagano, *Dancing bees tune both duration and rate of waggle-run production in relation to nectar-source profitability*, J. Comp. Physiol. A 186 (2000), pp. 813–819.

19. K. von Frisch, *The Dance Language and Orientation of Bees*, Harvard University Press, Cambridge, MA, 1967.

20. T.D. Seeley and R.A. Morse, *Nest site selection by the honey bee,* Apis mellifera, Insect. Soc. 25 (1978), pp. 323–337.

21. P.K. Visscher, R.A. Morse, and T.D. Seeley, *Honey bees choosing a home prefer previously occupied cavities*, Insect. Soc. 32 (1985), pp. 217–220.

22. P.K. Visscher, *Group decision making in nest-site selection among social insects*, Annu. Rev. Entomol. 52 (2007), pp. 255–275.

23. J.O. Schmidt and R. Hurley, *Selection of nest cavities by Africanized and European honey bees*, Apidologie 26 (1995), pp. 467–475.

24. S. Taber, III, *A population of feral honey bee colonies*, Am. Bee J. 118 (1979), pp. 842–847.

25. L.C. McNally and S.S. Schneider, *Spatial distribution and nesting biology of colonies of the African honey bee* Apis mellifera scutellata *(Hymenoptera: Apidae) in Botswana, Africa*, Environ. Entomol. 25 (1996), pp. 643–652.

26. S. Schneider and R. Blyther, *The habitat and nesting biology of the African honey bee* Apis mellifera scutellata *in the Okavango River Delta, Botswana, Africa*, Insect. Soc. 35 (1988), pp. 167–181.

27. G.M. Loper, D. Sammataro, J. Finley, and J. Cole, *Feral honey bees in southern Arizona 10 years after varroa infestation*, Am. Bee J. 146 (2006), pp. 521–524.

28. B.P. Oldroyd, S.H. Lawler, and R.H. Crozier, *Do feral honey-bees* (Apis mellifera) *and regent parrots* (Polytelis anthopeplus) *compete for nest sites*, Aust. J. Ecol. 19 (1994), pp. 444–450.

29. P.K. Visscher and T.D. Seeley, *Foraging strategy of honeybee colonies in a temperate deciduous forest*, Ecology 63 (1982), pp. 1790–1801.

30. R.A. Morse, S. Camazine, M. Ferracane, P. Minacci, R. Nowogrodzki, F.L.W. Ratnieks, J. Spielholz, and B.A. Underwood, *The population density of feral colonies of honey bees (Hymenoptera, Apidae) in a city in upstate New York*, J. Econ. Entomol. 83 (1990), pp. 81–83.

31. M. de Mello, E.A. da Silva, and D. Natal, *Africanized bees in a metropolitan area of Brazil: Shelters and climatic influences*, Rev. Saude Publ. 37 (2003), pp. 237–241.

32. F.L.W. Ratnieks, M.A. Piery, and I. Cuadriello, *The natural nest and nest density of the Africanized honey bee (Hymenoptera, Apidae) near Tapachula, Chiapas, Mexico*, Can. Entomol. 123 (1991), pp. 353–359.

33. B. Oldroyd, A. Smolenski, S. Lawler, A. Estoup, and R. Crozier, *Colony aggregations in* Apis mellifera *L.*, Apidologie 26 (1995), pp. 119–130.

34. B.P. Oldroyd, E.G. Thexton, S.H. Lawler, and R.H. Crozier, *Population demography of Australian feral bees* (Apis mellifera), Oecologia 111 (1997), pp. 381–387.

35. A. Avitabile, D.P. Stafstrom, and K.J. Donovan, *Natural nest sites of honeybee colonies in trees in Connecticut, USA*, J. Apicult. Res. 17 (1978), pp. 222–226.

36. M.M. Boreham and D.W. Roubik, *Population changes and control of Africanized honey bees (Hymenoptera: Apidae) in the Panama Canal area*, Bull. Entomol. Soc. Am. 33 (1987), pp. 34–39.

37. A.M. Wenner, *Bee-lining and ecological research on Santa Cruz Island*, Am. Bee J. 129 (1989), pp. 808–809.

38. P. Gambino, K. Hoelmer, and H.V. Daly, *Nest sites of feral honey bees in California, USA*, Apidologie 21 (1990), pp. 35–45.

39. S.S. Schneider, *Nest characteristics and recruitment behavior of absconding colonies of the African honey bee,* Apis mellifera scutellata, *in Africa*, J. Insect Behav. 3 (1990), pp. 225–240.

40. W. Wattanachaiyingcharoen, S. Wongsiri, and B.P. Oldroyd, *Aggregations of unrelated* Apis florea *colonies*, Apidologie 39 (2008), pp. 531–536.

41. D. vanEngelsdorp, J.D. Evans, C. Saegerman, C. Mullin, E. Haubruge, B.K. Nguyen, M. Frazier, J. Frazier, D. Cox-Foster, Y. Chen, R. Underwood, D.R. Tarpy, and J.S. Pettis, *Colony collapse disorder: A descriptive study*, PLoS One 4 (2009), p. e6481.

42. J.J. Bromenshenk, C.B. Henderson, C.H. Wick, M.F. Stanford, A.W. Zulich, R.E. Jabbour, S.V. Deshpande, P.E. McCubbin, R.A. Seccomb, P.M. Welch, T. Williams, D.R. Firth, E. Skowronski, M.M. Lehmann, S.L. Bilimoria, J. Gress, K.W. Wanner, and R.A. Cramer, Jr., *Iridovirus and microsporidian linked to honey bee colony decline*, PLoS One 5 (2010), p. e13181.

43. K.E. Clarke, T.E. Rinderer, P. Franck, J.G. Quezada-Euan, and B.P. Oldroyd, *The Africanization of honeybees* (Apis mellifera L.) *of the Yucatan: A study of a massive hybridization event across time*, Evolution 56 (2002), pp. 1462–1474.

44. R.F.A. Moritz, F. Bernhard Kraus, P. Kryger, and R.M. Crewe, *The size of wild honeybee populations* (Apis mellifera) *and its implications for the conservation of honeybees*, J. Insect Conserv. 11 (2007), pp. 391–397.

45. R. Jaffe, V. Dietemann, M.H. Allsopp, C. Costa, R.M. Crewe, R. Dall'olio, P. de la Rua, M.A.A. El-Niweiri, I. Fries, N. Kezic, M.S. Meusel, R.J. Paxton, T. Shaibi, E. Stolle, and R.F.A. Moritz, *Estimating the density of honeybee colonies across their natural range to fill the gap in pollinator decline censuses*, Conserv. Biol. 24 (2010), pp. 583–593.

46. M.A. Stevenson, H. Benard, P. Bolger, and R.S. Morris, *Spatial epidemiology of the Asian honey bee mite* (Varroa destructor) *in the North Island of New Zealand*, Prev. Vet. Med. 71 (2005), pp. 241–252.

47. L. Belloy, A. Imdorf, I. Fries, E. Forsgren, H. Berthoud, R. Kuhn, and J.-D. Charriere, *Spatial distribution of* Melissococcus plutonius *in adult honey bees collected from apiaries and colonies with and without symptoms of European foulbrood*, Apidologie 38 (2007), pp. 136–140.

48. J. Paar, B.P. Oldroyd, E. Huettinger, and G. Kastberger, *Genetic structure of an* Apis dorsata *population: The significance of migration and colony aggregation*, J. Hered. 95 (2004), pp. 119–126.

49. B.A. Underwood, *Seasonal nesting cycle and migration patterns of the Himalayan honey bee* Apis laboriosa, Natl. Geogr. Res. 6 (1990), pp. 276–290.

50. B.P. Oldroyd, K.E. Osborne, and M. Mardan, *Colony relatedness in aggregations of* Apis dorsata *Fabricius (Hymenoptera, Apidae)*, Insect. Soc. 47 (2000), pp. 94–95.

51. T.E. Rinderer, B.P. Oldroyd, L.I. de Guzman, W. Wattanachaiyingchareon, and S. Wongsiri, *Spatial distribution of the dwarf honey bees in an agroecosystem in southeastern Thailand*, Apidologie 33 (2002), pp. 539–544.

52. T.D. Seeley, R.H. Seeley, and P. Akratanakul, *Colony defense strategies of the honeybees in Thailand*, Ecol. Monogr. 52 (1982), pp. 43–63.

53. J.O. Schmidt, *Dispersal distance and direction of reproductive European honey bee swarms (Hymenoptera, Apidae)*, J. Kans. Entomol. Soc. 68 (1995), pp. 320–325.

54. S.S. Schneider, *Swarm movement patterns inferred from waggle dance activity of the neotropical African honey bee in Costa Rica*, Apidologie 26 (1995), pp. 395–406.

55. T.D. Seeley and R.A. Morse, *Dispersal behavior of honey bee swarms*, Psyche 84 (1977), pp. 199–209.

56. E.R. Jaycox and S.G. Parise, *Homesite selection by Italian honey bee swarms,* Apis mellifera ligustica *(Hymenoptera: Apidae)*, J. Kans. Entomol. Soc. 53 (1980), pp. 171–178.

57. E.R. Jaycox and S.G. Parise, *Homesite selection by swarms of black-bodied honey bees,* Apis mellifera caucasica *and* Apis mellifera carnica *(Hymenoptera, Apidae)*, J. Kans. Entomol. Soc. 54 (1981), pp. 697–703.

58. J.O. Schmidt and S.C. Thoenes, *The efficiency of swarm traps: What percent of swarms are captured and at what distance from the hive*, Am. Bee J. 130 (1990), pp. 811–812.

59. M.L. Winston, *The Biology of the Honey Bee*, Harvard University Press, Cambridge, MA, 1987.

60. N. Koeniger, G. Koeniger, and H. Pechhacker, *The nearer the better? Drones* (Apis mellifera) *prefer nearer drone congregation areas*, Insect. Soc. 52 (2005), pp. 31–35.

61. B.P. Oldroyd and S. Wongsiri, *Asian Honey Bees: Biology, Conservation and Human Interactions*, Harvard University Press, Cambridge, MA, 2006.

62. M.L. Winston, *The biology and management of Africanized honey bees*, Annu. Rev. Entomol. 37 (1992), pp. 173–193.

63. S.S. Schneider, G.D. Hoffman, and D.R. Smith, *The African honey bee: Factors contributing to a successful biological invasion*, Annu. Rev. Entomol. 49 (2004), pp. 351–376.

64. M.L. Winston, O.R. Taylor, and G.W. Otis, *Some differences between temperate European and tropical African and South American honey bees*, Bee World 64 (1983), pp. 12–21.

65. J.F. Harrison, J.H. Fewell, K.E. Anderson, and G.M. Loper, *Environmental physiology of the invasion of the Americas by Africanized honeybees*, Integr. Comp. Biol. 46 (2006), pp. 1110–1122.

66. M.J. Rabe, S.S. Rosenstock, and D.I. Nielsen, *Feral Africanized honey bees* (Apis mellifera) *in Sonoran Desert habitats of southwestern Arizona*, Southwest. Natl. 50 (2005), pp. 307–311.

67. M.-J. Fortin and M. Dale, *Spatial Analysis: A Guide for Ecologists*, Cambridge University Press, Cambridge, MA, 2005.

68. E.J. Dietz, *Permutation tests for association between 2 distance matrices*, Syst. Zool. 32 (1983), pp. 21–26.

69. M.A. Dimmit, *Plant ecology of the Sonoran Desert region*, in *A Natural History of the Sonoran Desert*, S.J. Phillips and P.W. Comus, eds., Arizona-Sonoran Desert Museum Press/University of California Press, Tucson, AZ/Berkeley, CA, 2000, pp. 129–151.

70. R.J. O'Neal and G.D. Waller, *On the pollen harvest by the honey bee* (Apis mellifera *L.*) *near Tucson, Arizona (1976–1981)*, Desert Plant. 6 (1984), pp. 81–109.

71. T.E. Rinderer, B.P. Oldroyd, and W.S. Sheppard, *Africanized bees in the United States*, Sci. Am. 269 (1993), pp. 84–90.

72. S.S. Schneider and L.C. McNally, *Factors influencing seasonal absconding in colonies of the African honey bee,* Apis mellifera scutellata, Insect. Soc. 39 (1992), pp. 403–423.

73. S.S. Schneider, T. Deeby, D.C. Gilley, and G. DeGrandi-Hoffman, *Seasonal nest usurpation of European colonies by African swarms in Arizona, USA*, Insect. Soc. 51 (2004), pp. 359–364.

74. M.L. Winston, G.W. Otis, and O.R. Taylor, *Absconding behavior of the Africanized honeybee in South America*, J. Apic. Res. 18 (1979), pp. 85–94.

75. M. Beekman, D.J.T. Sumpter, N. Seraphides, and F.L.W. Ratnieks, *Comparing foraging behaviour of small and large honey bee colonies by decoding waggle dances made by foragers*, Funct. Ecol. 18 (2004), pp. 829–835.

76. C.A. Martin, P.S. Warren, and A.P. Kinzig, *Neighborhood socioeconomic status is a useful predictor of perennial landscape vegetation in residential neighborhoods and embedded small parks of Phoenix, AZ*, Landscape Urban Plann. 69 (2004), pp. 355–368.

77. D. Hope, C. Gries, W.X. Zhu, W.F. Fagan, C.L. Redman, N.B. Grimm, A.L. Nelson, C. Martin, and A. Kinzig, *Socioeconomics drive urban plant diversity*, Proc. Natl. Acad. Sci. USA 100 (2003), pp. 8788–8792.

78. W.W. Shaw, L.K. Harris, and M. Livingston, *Vegetative characteristics of urban land covers in metropolitan Tucson*, Urban Ecosyst. 2 (1998), pp. 65–73.

FIGURE 1.6 Bee with an RFID tag glued to the back side of her thorax. (Reproduced with kind permission of Axel Decourtye, ACTA.)

| Day 1 | Day 4 | Day 7 | Day 10 | Day 12 | Day 14 |

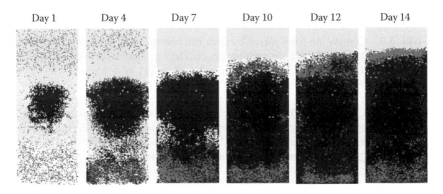

FIGURE 3.2 Agent-based model of pattern formation during a period of high nectar intake. Honey cells are yellow, pollen cells are red, and brood cells are black. Initially, pollen is scattered throughout the nest, but once open brood is present, pollen accumulates at the bottom of the nest. Nectar shows an immediate bias toward being unloaded at the top of the nest due to the upward movement of nectar receivers. By day 7, the complete pattern is almost present with the exception of the pollen band, which forms by day 14. (From Johnson, B.R., *Proc. Roy. Soc. B Biol. Sci.*, 276, 255, 2009.)

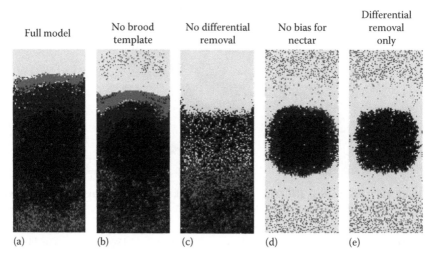

Full model   No brood template   No differential removal   No bias for nectar   Differential removal only

(a)   (b)   (c)   (d)   (e)

FIGURE 3.3 Contrasting roles played by each mechanism underlying pattern formation. The pattern is shown at 14 days in each image. (a) The full model. (b) Without the queen-based template for unloading pollen near open brood, pollen is scattered throughout the honey zone. (c) Without differential removal of food from near the brood (the SO mechanism), the pattern is far from complete in that pollen does not localize and the sections of the nest are not distinct. (d) Without the recipe's effect of upward bias in unloading of nectar, the pattern is not vertical (honey on top brood on the bottom). (e) The original self-organization model of Camazine (differential removal of food from near brood alone) does not lead to complete pattern formation. (From Johnson, B.R., *Proc. Roy. Soc. B Biol. Sci.*, 276, 255, 2009.)

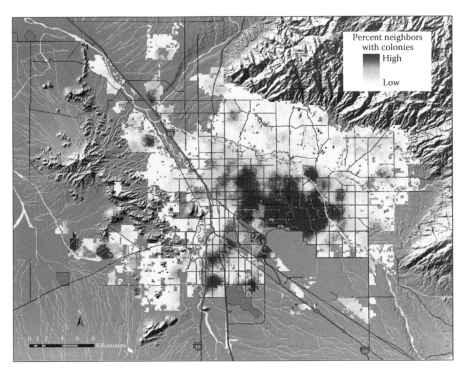

FIGURE 6.4 Spatial distribution of percent of neighboring water meter boxes with colonies (within 0.8 km) for every water meter box in the water meter box dataset. (Modified and reprinted from *Landscape Urban Plann.*, 100, Baum, K.A., Kolodziej, E.Y., Tchakerian, M.D., Birt, A.G., Sherman, M., and Coulson, R.N., Spatial distribution of Africanized honey bees in an urban landscape, 153–162, Copyright [2011], with permission from Elsevier.)

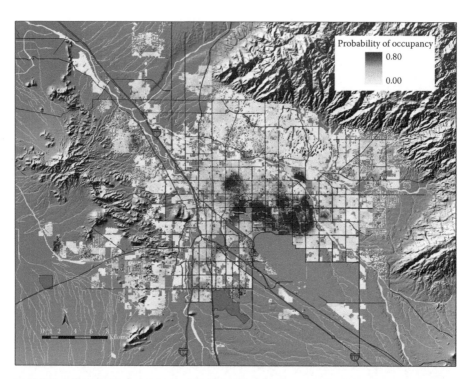

FIGURE 6.5 Spatial distribution of Africanized honey bee colony occupancy probabilities predicted by the logistic regression model for the water meter box dataset. Dark shaded areas show parts of the greater Tucson metropolitan area where water meter boxes are predicted to have high occupancy rates, whereas light shaded areas are predicted to have low occupancy rates. (Reprinted from *Landscape Urban Plann.*, 100, Baum, K.A., Kolodziej, E.Y., Tchakerian, M.D., Birt, A.G., Sherman, M., and Coulson, R.N., Spatial distribution of Africanized honey bees in an urban landscape, 153–162, Copyright [2011], with permission from Elsevier.)

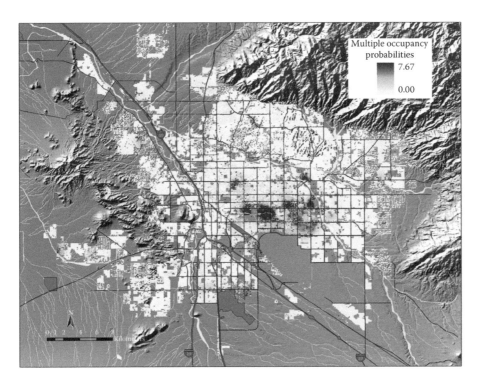

FIGURE 6.6 Spatial distribution of Africanized honey bee colony multiple occupancy probabilities predicted by the Poisson model for the water meter box dataset. Dark shaded areas show parts of the greater Tucson metropolitan area where water meter boxes are predicted to have high multiple occupancy rates, whereas light shaded areas are predicted to have low multiple occupancy rates. (Reprinted from *Landscape Urban Plann.*, 100, Baum, K.A., Kolodziej, E.Y., Tchakerian, M.D., Birt, A.G., Sherman, M., and Coulson, R.N., Spatial distribution of Africanized honey bees in an urban landscape, 153–162, Copyright [2011], with permission from Elsevier.)

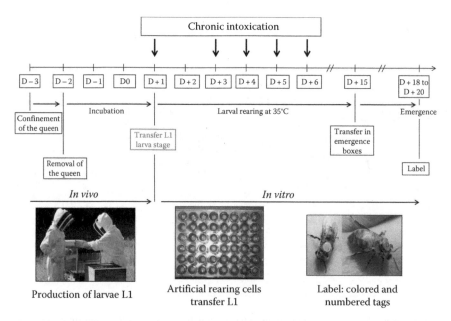

FIGURE 9.7 Principles of the larval test developed by Aupinel and coworkers. (From Aupinel, P. et al., *Bull. Insectol.,* 58, 107, 2005; Aupinel, P. et al. *Pest Manag. Sci.* 63, 1090, 2007; Aupinel, P. et al., *Julius-Kühn-Archiv.,* 423, 96, 2009.)

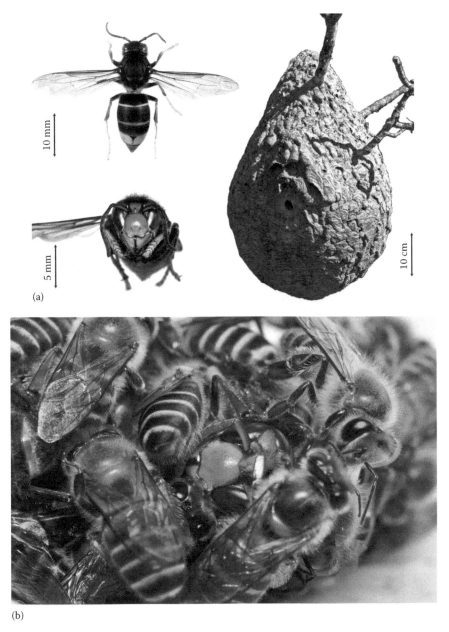

(a)

(b)

FIGURE 11.1 (a) Adult worker and nest of *Vespa velutina nigrithorax*. (b) Heat-balling behavior of *Apis cerana* on *Vespa velutina auraria* (Nepal).

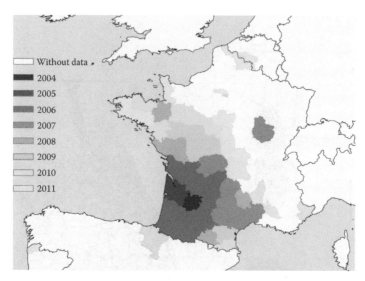

FIGURE 11.2 Annual distribution of *Vespa velutina* in Europe.

FIGURE 11.3 Predicted potential invasion risk of *V. v. nigrithorax* in Europe, based on ensemble forecast models using eight climatic data from WorldClim. The suitability probability is increasing from dark blue to red. (From Villemant, C. et al., *Biol. Conserv.*, 144, 2142, 2011.)

FIGURE 11.4 Predicted potential invasion risk of *V. v. nigrithorax* in the world, based on ensemble forecast models using eight climatic data from WorldClim. The suitability probability is increasing from dark blue to red. (From Villemant, C. et al., *Biol. Conserv.*, 144, 2142, 2011.)

7 QSAR Modeling of Pesticide Toxicity to Bees

James Devillers

CONTENTS

ABSTRACT

Quantitative structure–activity relationship (QSAR) models are increasingly used in toxicology, ecotoxicology, and pharmacology for predicting the activity of organic molecules from their physicochemical properties and/or their structural characteristics. The available congeneric and noncongeneric QSAR models used to estimate the acute toxicity of pesticides to honey bees were critically analyzed. From the available data, new models were also computed from linear and nonlinear methods for comparison purposes. Generally, these models provided better prediction results than those published in the literature. The advantages and limitations of these modeling tools are discussed.

KEYWORDS

QSAR model, Acute toxicity, Artificial neural network, Support vector machine

7.1   INTRODUCTION

The history of modern agriculture starts after 1945. Many countries sought to reach self-sufficiency in their food production to avoid the supply problems experienced during the Second World War. Farms became larger, their activities more mechanized, and the tendency for specialization in one or more crops became more prominent [1,2]. This system of intensive agriculture has less use for crop rotation than the earlier systems, although it is well known that crop rotation is quite effective in controlling numerous pests. Indeed, the downside of this intensive agriculture was the increasing use of fertilizers to enhance productivity and that of pesticides to control pests. The extensive use of pesticides has induced resistances in target organisms, contamination of aquatic and terrestrial ecosystems, and adverse effects on nontarget species. Among this category of organisms, the honey bee occupies a key place due to her economic and ecological importance. Pesticides act in two ways to reduce bee populations. First, many pesticides necessary in crop production are highly toxic to honey bees. In this category, we principally find insecticides. Second, the use of herbicides reduces the acreages of attractive plants for the bees to forage on. Pesticide damage to bee colonies takes many forms. Bees can be poisoned directly when they feed on nectar or pollen contaminated with certain pesticides. They can also be poisoned when they fly through a cloud of insecticide dust or spray or walk on the treated parts of a plant. The other members of the colony can also be contaminated directly or indirectly by pesticides. Thus, if the pesticide is brought back to the hive by foragers, the inhabitants constituting the colony will be contaminated [2].

Obviously, different laboratory tests have been developed for assessing the adverse effects of pesticides on honey bees [3]. However, they are expensive, time consuming, and require trained personnel. Surprisingly, while in aquatic toxicology the quantitative structure–activity relationship (QSAR) methodology is widely used to overcome these problems [4,5], the number of QSAR models derived on the honey bee (*Apis mellifera*) is very scarce. In this context, after having recalled the main principles of QSAR modeling, the different models available on bees will be reviewed. Strengths and weaknesses of these models will also be discussed, especially in the frame of hazard assessment of xenobiotics to honey bees.

7.2 WHAT IS A QSAR MODEL?

QSAR modeling is rooted on the fact that the biological activity of molecules directly depends on their intrinsic structure. For example, it is well known that the toxicity of chlorophenols against aquatic organisms depends only on the number and position of the chlorine atoms on the phenol nucleus [6–9]. The mathematical formulation of such a type of relationship leads to the design of a QSAR model. Thus, for deriving a QSAR model, three different types of elements are required (Figure 7.1). First, measured activities for a set of molecules have to be available for the endpoint of interest. Second, chemicals must be described by means of their physicochemical properties, topological indices, and/or structural features. Last, a statistical method must be used for linking the first two elements [10,11]. Specific constraints exist for each ingredient and their strict respect allows to obtain models that will be usable to make predictions with a satisfying level of confidence [5,12]. Indeed, the raison d'être of a QSAR model is to predict the activity of untested chemicals. This can be done in lieu of a laboratory test or to rationalize its design by optimizing the range of concentrations or doses to test from the modeling results. Due to the flexibility offered by such prediction tools, QSAR models are also used to better understand the mechanism of action of chemicals. Indeed, it is easier to test the influence of the

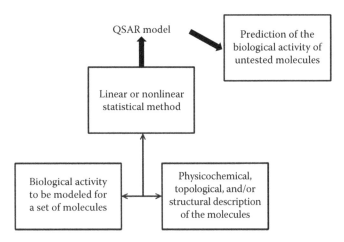

FIGURE 7.1 Principle of QSAR model design and functioning.

addition or deletion of an atom or a functional group on the studied activity. QSAR modeling also finds applications in the design of safer chemicals. There are different ways of categorizing the different types of QSAR models irrespective of the considered endpoint. One of the most convenient ways consists in considering the congeneric versus the noncongeneric models. In the former category, the model is computed from a set of molecules presenting the same mechanism of action and/or rather similar structures. In the second category, the learning set includes structurally diverse chemicals encompassing different mechanisms of action. Crudely speaking, congeneric QSAR models are focused on mechanistic interpretations, while noncongeneric QSAR models are primarily screening tools, having generally a large domain of application [12–14]. This classification will be used to discuss the different QSAR models available for the honey bee.

7.3 CONGENERIC QSAR MODELS ON BEES

The very first QSAR model derived for *Apis mellifera* was presented in 1990 by Davide Calamari during the Fourth International Workshop on QSAR in Environmental Toxicology [15] at Veldhoven, the Netherlands. In this pioneering work, in a first step, bees were tested for their oral toxicity for 14 organophosphorus insecticides that are inhibitors of acetylcholinesterase (AChE). The mortality was recorded after 24 h of exposure to calculate LD_{50} values (*i.e.*, doses of chemicals at which 50% of the test organisms die) expressed in μmol/bee.

The best QSAR model included six parameters as shown in the following equation:

$$\log 1/LD_{50} = 1.14 \log P - 0.28 \log P^2 + 0.28\,^1\chi - 0.76\,^2\chi_{ox}^v - 1.09\,\gamma_3$$

$$+ 0.096\,\gamma_3^2 + 12.29 \tag{7.1}$$

$$n = 14, \quad r^2 = 0.908, \quad s = 0.28$$

where
   n is the number of chemicals of the training set;
   r^2 is the coefficient of determination; and
   s is the standard error of estimate.

The molecular descriptors involved in Equation 7.1 encode different physicochemical and structural information. The 1-octanol/water partition coefficient (log P) characterizes the hydrophobicity of the molecules [16,17]. The molecular connectivity indices, $^1\chi$ and $^2\chi^v_{ox}$, provide topological information on the molecules. $^1\chi$ is a simple molecular connectivity index of order 1, while $^2\chi^v_{ox}$ is a valence connectivity index of order 2 [18]. Moreover, $^2\chi^v_{ox}$ was calculated on the oxygen metabolite. This explains the subscript in the topological index [15]. γ_3 is an electronic descriptor calculated by Vighi et al. [15].

Equation 7.1 does not satisfy the current principles of good modeling practices in QSAR [12,19]. Indeed, the model is overparameterized, and it was not tested on an external test set (*i.e.*, not used to compute the model). However, more than 20 years ago, we were at the beginning of the environmental QSAR and it was rather commonplace to do so. Note that it was also usual to work on his/her own experimental data to derive QSARs. Anyway, this study was definitively a visionary work, paving the way for models that aim to predict the toxicity of pesticides to honey bees. It is noteworthy that Vighi et al. [15] derived an additional QSAR regression model with the same set of descriptors by including the herbicide glyphosate in the previous set of 14 organophosphorus insecticides. The statistical results were broadly the same.

Because carbamates present the same mechanism of action as the organophosphorus insecticides, Dulin et al. [20] derived different QSAR models from 45 toxicity data for bees retrieved from various sources. The data consisted of 30 contact LD_{50} values for organophosphorus insecticides and 15 contact LD_{50} values for carbamates, all being measured after 48 h of exposure and expressed in μmol/bee. This database was randomly split into a training set of 39 insecticides for computing the QSAR models and an external test set of 6 insecticides for estimating the predictive power of the models. All the chemicals were described by means of more than 2,500 topological and physicochemical descriptors [21,22]. The QSARs were built from genetic function approximation (GFA) [23]. Briefly, GFA generates a population of QSAR models that are scored according to their performances. In this work, the performance of each model was estimated from the calculation of r^2 and root mean square error (RMSE) and by using leave-one-out and Y-randomization procedures. The GFA simulation conditions were 1,000 generations and 500 populations. The total number of model equations in the evolving population was 500. The functional form of the model was linear and the maximum number of terms in each model was fixed to five [20]. The 10 best QSAR models were kept and averaged to constitute a consensus model ($r^2 = 0.805$, RMSE $= 0.350$). The consensus model obtained good predictions on the external test set of three carbamates and three organophosphorus insecticides with r^2 and RMSE values of 0.850 and 0.218, respectively (Tables 7.1 and 7.2). Three descriptors were frequently found in the

TABLE 7.1

Observed [20] and Calculated Contact Toxicity Values (log $1/LD_{50}$, μmol/Bee) Obtained with the Consensus (C10), Multiple Regression (MRA), Three-Layer Perceptron (TLP), and Support Vector Machine (SVM) Models

Nb.	Insecticide	Obs.	C10	MRA	TLP	SVM
1	Acephate	2.19	2.03	1.90	2.14	2.02
2	Alanycarb[a]	2.77	3.00	2.75	2.88	2.87
3	Aldicarb	2.83	3.14	3.28	3.11	2.90
4	Aminocarb	3.24	2.90	3.19	3.18	3.41
5	Azinphos-methyl	2.88	2.77	2.49	2.52	2.70
6	Bendiocarb	2.72	2.86	2.91	3.00	3.15
7	Benfuracarb[a]	3.41	3.57	3.40	3.16	3.08
8	Bufencarb	3.14	3.09	3.18	3.07	2.93
9	Carbaryl	3.16	2.95	3.15	3.13	3.02
10	Carbofuran	3.79	3.11	3.11	3.09	3.60
11	Chlorpyrifos	3.78	3.01	2.78	3.05	3.07
12	Diazinon	2.92	2.81	2.80	3.07	3.10
13	Dichlorvos	2.65	2.38	2.43	2.90	2.47
14	Dimethoate	3.29	3.35	3.21	3.15	3.47
15	EPN	3.13	3.24	3.05	3.07	3.31
16	Ethion	1.27	1.35	1.31	1.11	1.27
17	Ethoprofos	1.64	1.39	1.36	0.81	1.82
18	Famphur	2.90	2.88	2.87	3.25	3.07
19	Fenamiphos	3.04	2.87	2.92	2.96	3.13
20	Fenitrothion	4.19	3.91	3.82	3.59	4.02
21	Fensulfothion	2.97	2.80	2.74	3.24	2.80
22	Fenthion	2.96	3.39	3.44	3.13	3.15
23	Fonofos	1.88	2.42	2.38	2.49	2.53
24	Formetanate	1.20	1.87	1.94	1.77	1.38
25	Fosthiazate[a]	3.04	3.19	3.21	3.45	2.87
26	Malathion	3.22	2.85	2.47	2.85	3.04
27	Methamidophos	2.02	2.08	2.65	1.99	2.20
28	Methidathion	3.37	3.22	3.19	3.11	3.19
29	Methiocarb	2.99	3.37	3.67	3.19	2.80
30	Methomyl	3.01	3.24	3.47	3.45	3.19
31	Naled	2.90	2.75	2.66	2.95	3.08
32	Oxamyl	2.67	2.21[b]	2.26	2.45	2.48
33	Parathion	3.23	3.41[b]	3.29	3.37	3.40
34	Phenthoate	3.02	3.48[b]	3.31	3.13	3.21
35	Phosalone	1.93	2.30[b]	1.94	2.00	2.11
36	Phosmet	3.16	2.65[b]	2.25	2.42	2.98
37	Phosphamidon	2.31	2.27[b]	2.20	2.27	2.38

(continued)

TABLE 7.1 (continued)

Observed [20] and Calculated Contact Toxicity Values (log $1/LD_{50}$, µmol/Bee) Obtained with the Consensus (C10), Multiple Regression (MRA), Three-Layer Perceptron (TLP), and Support Vector Machine (SVM) Models

Nb.	Insecticide	Obs.	C10	MRA	TLP	SVM
38	Profenofos	3.60	3.36	3.07	3.37	3.42
39	Propoxur	2.19	2.58	2.65	2.81	2.87
40	Pyrimicarb	1.11	1.30	1.67	1.43	1.28
41	Quinalphos[a]	3.24	3.18	3.02	3.04	3.30
42	Temephos[a]	2.48	1.90	1.95	2.31	2.53
43	Tetrachlorvinphos	2.43	3.01	3.02	3.00	2.79
44	Thiodicarb[a]	2.06	2.01	1.99	2.06	1.92
45	Trichlorfon	0.64	0.93	1.50	0.88	0.81

[a] Test set chemicals.

[b] Activity reconstituted from the information found in the supplementary material of Dulin et al. [20] and inspection of the original papers and database cited therein.

TABLE 7.2

Distribution of Residuals (Absolute Values), Differences between the Experimental and the Calculated Contact Toxicity Values (log $1/LD_{50}$, µmol/Bee) Obtained with the Consensus (C10), Multiple Regression (MRA), Three-Layer Perceptron (TLP), and Support Vector Machine (SVM) Models

Range	C10	MRA	TLP	SVM
<0.25	20(5[a])	17(5)	20(4)	34(5)
0.25–0.50[b]	13	10	10(2)	2(1)
0.50–0.75[b]	5(1)	8(1)	7	3
0.75–1.00[b]	1	3	2	0
≥1.00	0	1	0	0

[a] Including number of residuals from the external test set.

[b] Excluded value.

individual models. Dipole_X_DMol3 that encodes the polarity of the molecules is found in the 10 individual models with always a positive sign. GATS6e is also included in all the individual models with a positive contribution. This parameter belongs to the category of the autocorrelation descriptors [24] that are topological indices encoding physicochemical information [25]. Here, the property of concern is the Sanderson electronegativity, and the calculation is made by considering a distance of 6. CIC2 is the complementary information content of the molecular graph calculated from the second-order neighborhood of the vertices. It is particularly suited to account for branching and complexity information in a chemical structure [26]. CIC2 contributed negatively in eight equations. In the two other models, it was replaced by SIC2, the structural information content for the second-order neighborhood of vertices in the hydrogen-filled graph [26]. This topological parameter, which positively contributes in the two equations, also encodes branching and complexity information.

The importance of Dipole_X_DMol3, GATS6e, and CIC2 to explain the toxicity of carbamate and organophosphorus insecticides to honey bees was confirmed through the use of a honey bee AChE homology model [27] (ID: Q4LEZ2_APIME), which was designed from the 2.7 Å resolution crystal structure (PDB: 1DX4) of *Drosophila melanogaster* AChE (DmAChE) complex with tacrine derivative as a template (sequence identity = 61%).

Despite satisfying statistics, it is noteworthy that the consensus model proposed by Dulin et al. [20] is rather complex, averaging 10 equations and a total of 16 molecular descriptors. Consequently, using the same training and test sets, an attempt was made to see whether it was possible to obtain simpler QSAR models showing acceptable statistics, only from the three main molecular descriptors, namely, Dipole_X_DMol3, GATS6e, and CIC2. A multiple regression analysis performed from these three descriptors led to r^2 values of 0.65 and 0.87 for the training set and external test set, respectively. The prediction performances for the test set are comparable to those obtained with the consensus model of Dulin et al. [20] (Table 7.2). Conversely, those obtained for the learning set are lower, with some larger residuals (difference between the observed and calculated log $1/LD_{50}$ values) than with the consensus model [20]. Thus, with the consensus model, one insecticide presents a residual ≥0.75 (in absolute value) versus four with the multiple regression model, one being ≥1 (absolute value). Because the artificial neural networks (ANNs) [28] have demonstrated their interest in environmental toxicology [29,30], a three-layer perceptron (TLP) [31] was experienced on the reduced set of three descriptors. The TLP is perhaps the most popular supervised ANN in use in QSAR (see, *e.g.*, [32–39]). Because its functioning has been widely described in the literature (see, *e.g.*, [40]), only the basic principles are recalled here. A TLP includes one input layer with a number of neurons corresponding to the number of selected molecular descriptors, one output layer of one neuron corresponding to the modeled activity, and one hidden layer, between the two aforementioned layers, with an adjustable number of neurons for distributing the information (Figure 7.2). Too many hidden neurons often lead to overfitting, and hence, to avoid problems, their number has to be limited. This is done by a trial-and-error procedure and also by the use of pruning algorithms. The neurons of each layer are connected in the forward direction

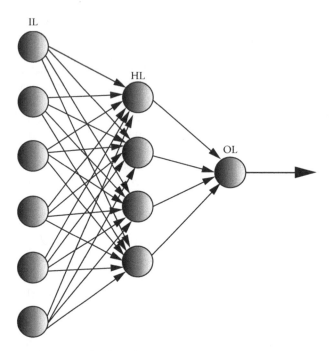

FIGURE 7.2 Architecture of a three-layer perceptron, which includes an input layer (IL), a hidden layer (HL), and an output layer (OL).

(*i.e.*, input to output) and are activated by means of activation functions. Each connection is associated with a weight. The weights are adjusted during the learning process, which aims to minimize an error computed from the target and calculated outputs. Numerous learning algorithms are available, among which the backpropagation, the Levenberg–Marquardt, and Broyden–Fletcher–Goldfarb–Shanno algorithms were tested alone or in combination [40]. The calculations were performed with the data mining module of the Statistica™ software (StatSoft, Fr).

In this study, the maximum number of hidden neurons was fixed to three to have a TLP with the smallest possible configuration. Among all the different architectures, learning algorithms, and parameter values tested, the best results were obtained with a 3/3/1 TLP with a bias connected to the hidden and output layers and using the BFGS (Broyden–Fletcher–Goldfarb–Shanno) second-order training algorithm. The hidden and output activation functions were a hyperbolic tangent activation function (tanh) and a logistic activation function, respectively. The convergence was obtained after only 47 cycles. With such a configuration, the r^2 values for the training and test sets were equal to 0.77 and 0.79, respectively. The calculated log $1/LD_{50}$ values with this TLP model are listed in Table 7.1 and the distribution of the residuals is given in Table 7.2. Despite the presence of outliers, inspection of Tables 7.1 and 7.2 shows that the predictive performances of the TLP are not so bad, especially for the test set chemicals that always have residuals <0.5 (in absolute value), while it is not the case for temephos (#42), which presents a residual value of 0.58 with the consensus model [20] (Table 7.1). Note that for this test set insecticide, a residual value of 0.17

was obtained with the TLP (Table 7.1). It is noteworthy that the addition of a fourth descriptor as input neuron (*e.g.*, LUMO_Energy_DMol3) allowed to significantly outperform the prediction performances of the consensus model [20] on both the training and test sets.

Support vector machines (SVMs) [41,42] have also gained prominence in SAR and QSAR due to their effective performances in comparison with the classical linear methods commonly used in structure–activity modeling (see, *e.g.*, [43–47]). SVMs [41,42] are rooted on two key ideas. The first one is the notion of maximum margin. The margin is the distance between the boundary of separation and the closest samples. These are called support vectors. The separation boundary is chosen as the one that maximizes the margin. In order to deal with cases where the data are not linearly separable, the second key idea of SVMs is to transform the representation space of the input data into a space of greater dimension (possibly infinite), in which it is likely that there exists a linear separation. This is achieved through a kernel function. The following kernel functions are commonly tested: linear, polynomial, radial basis function (RBF), and sigmoid. To construct an optimal hyperplane, SVM uses an iterative training algorithm that minimizes an error function. According to the form of the error function, the SVM will be a classifier or a regression tool. Two types of support vector regressions (SVRs) can be computed, the so-called SVR type 1 (ε-SVR) and type 2 (v-SVR) with their own error functions. Both were tested in the present study. The advantages of an SVM are the presence of a global minimum solution resulting from the minimization of a convex programming problem, relatively fast training speed, and sparseness in solution representation. The calculations were performed with the data mining module of the Statistica software (StatSoft, Fr).

In our study, SVR type 1, with a RBF as kernel, provided the best prediction results with the adjustable parameters C, ε, and γ set to 20, 0.1, and 0.60, respectively. In that case, the r^2 values for the training and test sets were both equal to 0.89. The SVR model using only the three main descriptors outperforms the consensus model of Dulin et al. [20] as well as the TLP model (Tables 7.1 and 7.2).

Our trials confirm that the Dipole_X_DMol3, GATS6e, and CIC2 descriptors allow encoding of the main structural characteristics of the studied carbamate and organophosphorus insecticides because rather satisfying prediction results can be obtained only from them. The sine qua non to obtain such results is to use a nonlinear method in place of a classical linear approach. Even if the interpretation of a TLP or an SVM is less straightforward than a simple multiple regression analysis, their use will simplify the modeling of these toxicity data by using a limited number of descriptors.

7.4   NONCONGENERIC QSAR MODELS ON BEES

One hundred pesticide toxicity values on honey bees (LC_{50} or LD_{50} in µmol/bee) obtained under well-defined and controlled laboratory conditions (Table 7.3) were used by Devillers et al. [34,35] for deriving a QSAR model based on a TLP. The different families of pesticides were described by means of autocorrelation descriptors [24,25]. From the fragmental constants of Rekker and Mannhold [48], an autocorrelation vector H representing lipophilicity was derived. Second, an autocorrelation

TABLE 7.3
Observed (Obs.) [34,35], Calculated (Cal.), and Residual (Res.) Acute Toxicity Values (log $1/LD_{50}$, μmol/Bee) Obtained with the TLP Model

Nb.[a]	Pesticide	Obs.	Cal.	Res.
1	TEPP	5.16	5.16	0.00
2	Bioethanomethrin	4.02	3.45	0.57
3	Resmethrin	3.74	3.29	0.45
4	Pay-off (flucythrinate)	3.76	3.70	0.06
5	Deltamethrin	3.88	3.69	0.19
6	Chlorpyrifos	3.50	2.68	0.82
7	Parathion-methyl	3.38	3.55	−0.17
8	Dieldrin	3.46	3.12	0.34
9	Carbofuran	3.17	2.80	0.37
10	Permethrin	3.39	3.42	−0.03
11	Parathion	3.22	2.87	0.35
12	Fenitrothion	3.20	2.96	0.24
13	Dimethoate	3.08	2.86	0.22
14	Methidathion	3.11	2.70	0.41
15	EPN	3.13	2.03	1.10
16	Etrimfos	3.04	3.07	−0.03
17	Aldicarb	2.84	3.15	−0.31
18	Mexacarbate	2.87	2.51	0.36
19	Dicrotophos	2.89	3.15	−0.26
20	Mevinphos	2.87	2.80	0.07
21	Fenthion	2.94	2.50	0.44
22	Fensulfothion	2.96	2.98	−0.02
23	Aldrin	3.02	3.03	−0.01
24	Monocrotophos	2.80	2.78	0.02
25	Diazinon	2.91	2.70	0.21
26	Methiocarb	2.78	2.61	0.17
27	Fenvalerate	3.01	3.75	−0.74
28	Famphur	2.90	3.16	−0.26
29	Azinphos-methyl	2.87	2.47	0.40
30	Bendiocarb	2.72	2.58	0.14
31	Naled	2.89	2.84	0.05
32	Dichlorvos	2.64	2.69	−0.05
33	Heptachlor	2.85	2.25	0.60
34	Isofenphos	2.76	2.71	0.05
35	Carbosulfan	2.75	3.43	−0.68
36	Malathion	2.66	2.74	−0.08
37	Azinphos-ethyl	2.56	2.61	−0.05
38	Aminocarb	2.27	2.48	−0.21
39	Phosmet	2.45	2.27	0.18
40	Acephate	2.18	2.76	−0.58

TABLE 7.3 (continued)
Observed (Obs.) [34,35], Calculated (Cal.), and Residual (Res.) Acute Toxicity Values (log $1/LD_{50}$, µmol/Bee) Obtained with the TLP Model

Nb.[a]	Pesticide	Obs.	Cal.	Res.
41	Methomyl	2.10	2.06	0.04
42	Propoxur	2.19	2.08	0.11
43	Methamidophos	2.01	2.02	−0.01
44	Stirofos (tetrachlorvinphos)	2.42	3.22	−0.80
45	Fenamiphos	2.33	1.89	0.44
46	Phosphamidon	2.32	2.27	0.05
47	Carbaryl	2.12	1.89	0.23
48	Pyrazophos	2.30	2.06	0.24
49	Temephos	2.52	2.29	0.23
50	Trichloronate	2.22	1.86	0.36
51	Crotoxyphos	2.13	2.16	−0.03
52	Oxydemeton-methyl	1.94	1.87	0.07
53	Profenofos	2.03	1.12	0.91
54	Terbufos	1.85	1.53	0.32
55	Ethoprophos	1.64	1.56	0.08
56	Ronnel	1.76	2.42	−0.66
57	Disulfoton	1.65	1.73	−0.08
58	DDT	1.76	2.35	−0.59
59	Ethiofencarb	1.52	1.54	−0.02
60	Thiodicarb	1.70	2.12	−0.42
61	Sulprofos	1.65	2.05	−0.40
62	Fonofos	1.45	0.72	0.73
63	Chlordane	1.67	2.15	−0.48
64	Phosalone	1.61	2.23	−0.62
65	Phorate	1.40	1.82	−0.42
66	Oxamyl	1.33	1.77	−0.44
67	Carbophenothion	1.42	2.13	−0.71
68	AC 303,630	3.31	2.87	0.44
69	Alanycarb	2.70	2.37	0.33
70	Chlorpyrifos-methyl	2.93	2.97	−0.04
71	Bensultap	1.22	1.13	0.09
72	Azamethiphos	3.51	3.86	−0.35
73	EPTC	1.24	1.23	0.01
74	Napropamide	0.35	0.49	−0.14
75	Dicloran	0.06	0.40	−0.34
76	Diethofencarb	1.13	1.01	0.12
77	Dithiopyr	0.70	0.47	0.23
78	Orbencarb	0.40	0.79	−0.39
79	Fenazaquin	1.57	2.50	−0.93

(continued)

TABLE 7.3 (continued)
Observed (Obs.) [34,35], Calculated (Cal.), and
Residual (Res.) Acute Toxicity Values (log 1/LD$_{50}$,
μmol/Bee) Obtained with the TLP Model

Nb.[a]	Pesticide	Obs.	Cal.	Res.
80	Quinalphos	3.63	3.62	0.01
81	Pyridaben	2.82	2.80	0.02
82	Mephosfolan	1.89	1.94	−0.05
83	Imazalil	0.87	1.59	−0.72
84	Metconazole	0.52	1.50	−0.98
85	Pyridaphenthion	3.63	3.25	0.38
86	Pyrifenox	0.70	0.90	−0.20
87	Quizalofop	0.84	1.09	−0.25
88	Tebutam	0.37	0.40	−0.03
89	Thiometon	2.64	2.15	0.49
90	Cypermethrin	4.08	3.66	0.42
91	Diafenthiuron	2.26	2.79	−0.53
92	Tralomethrin	3.74	3.74	0.00
93	Propargite	1.37	1.82	−0.45
94	Diclomezine	1.41	1.35	0.06
95	Phenthoate	3.43	2.91	0.52
96	Piperophos	1.07	1.25	−0.18
97	Silafluofen	2.91	2.76	0.15
98	Propachlor	−0.17	0.39	−0.56
99	Tralkoxydim	0.79	0.65	0.14
100	Propisachlor	0.45	1.01	−0.56

[a] Pesticides #1–89 were used as training set and #90–100 as external test set.

vector MR, encoding molar refractivity, was designed from the fragmental constants of Hansch and Leo [49] or directly from the Lorentz–Lorenz equation. Last, autocorrelation vectors encoding the H-bonding acceptor ability (HBA) and H-bonding donor ability (HBD) of the molecules were also calculated by means of Boolean contributions (*i.e.*, 0/1). The autocorrelation vectors were calculated by means of AUTOCOR™ 2.4 (CTIS, France) using SMILES (Simplified Molecular Input Line Entry System) notation as inputs. For the autocorrelation vectors H, MR, and HBA, a truncation of the vectors was performed in order to obtain five autocorrelation components (distances 0–4 in the graphs of the molecules). For the HBD autocorrelation vector, only the first component was selected. The database was randomly split into a training set of 89 chemicals and an external test set of 11 chemicals. However, an attempt was made to have toxicity data of the same origin in the learning set.

The calculations were performed with the ANN module of the Statistica software, version 6 (StatSoft, Fr). The backpropagation algorithm was used to train the ANN. Different series of runs were performed in order to determine the number of descriptors to use as input neurons to obtain good prediction results, especially on the external test set. The H, MR, and HBA autocorrelation vectors of five components were first introduced separately in the ANN. They were not able to give acceptable prediction results on both sets. Addition of HDB_0 did not increase the quality of modeling results. Conversely, the use of the 15 descriptors (*i.e.*, H_0–H_4, MR_0–MR_4, HBA_0–HBA_4) as inputs in the TLP allowed to obtain good prediction results on both sets with a hidden layer of three neurons in about 2,000 cycles. The feature selection process of the Statistica software, mainly based on genetic algorithm [50], was used to reduce the number of descriptors. The best simulation results were obtained with H_1–H_4, MR_0–MR_4, and HBA_2–HBA_4 as input neurons. A 12/3/1 ANN with a learning rate $\eta = 0.5$ and a momentum $\alpha = 0.9$ led to acceptable predictions in 2,492 cycles. Indeed, the r^2 and RMS (root mean square residual) values for the training set were equal to 0.82 and 0.43, respectively. The r^2 and RMS values for the external test set were equal to 0.94 and 0.39. The inspection of Table 7.3 shows the presence of significant outliers that are EPN (#15), profenofos (#53), fenazaquin (#79), and metconazole (#84) with residual values of 1.10, 0.91, −0.93, and −0.98, respectively.

It is noteworthy that runs were also performed by changing the composition of the training and test sets but an 89/11 ratio was always maintained. No significant changes were noted in the prediction performances. Attempts were also made to derive a model from a partial least squares (PLS) regression [51,52] analysis. Unfortunately, this did not yield satisfying results. Indeed, in all cases, very large outliers were obtained for numerous pesticides.

Amaury et al. [53] used a database of 105 structurally diverse pesticides to derive semiquantitative (three classes of activities) and quantitative (log $1/LD_{50}$, µmol/bee) structure–activity models, experiencing various types of molecular descriptors and different linear and nonlinear statistical methods. Seventeen molecules were selected to constitute the external test set. The remaining pesticides were used to derive the models. The most interesting modeling approach was the hybrid QSAR model combining an ANN and a PLS model. The full characteristics of the ANN, including seven descriptors, were not given but it showed r^2 values of 0.71 and 0.80 for the training set and external test, respectively. Twelve descriptors were used to derive the PLS model that presented r^2 values of 0.66 and 0.69 for the training set and external test set, respectively. The hybridization of the two models allowed to obtain an r^2 value of 0.74 for the training set and an r^2 value of 0.84 for the external test set. The 19 descriptors used in the two individual models and found in the hybrid model were topological indices and structural fragments. The former type of descriptors encodes shape and branching information while the latter type accounts for the reactivity of specific atoms or functional groups. This is in agreement with our results [34,35] because the autocorrelation components of zero and first orders encode the shape of the molecules while those of higher orders account for branching information. In addition, the autocorrelation vectors being computed from atomic contributions of specific properties, their components of different orders also account for the reactivity of specific structural features in the molecules [24,25].

The hybrid model was further refined by the deletion of outliers in the training and test sets. The new hybrid model derived from 80 pesticides described by 16 molecular descriptors showed an r^2 value equal to 0.77. An r^2 value equal to 0.83 was obtained on the new external test set of 15 pesticides [53].

Last, a noncongeneric model was derived by Toropov and Benfenati [54] from molecular descriptors directly obtained from the SMILES strings of 105 molecules tested on bees. Twenty molecules were randomly selected as a test set. The r^2 values obtained for the training and test sets were equal to 0.68 and 0.72, respectively.

While the models of Devillers et al. [34,35], Amaury et al. [53], and Toropov and Benfenati [54] present rather similar domains of application, the inspection of the r^2 values obtained on their training and test sets shows that the prediction performances of the former significantly outperform those of the two other models.

7.5   ADVANTAGES AND LIMITATIONS OF QSAR MODELS FOR PREDICTING THE TOXICITY OF CHEMICALS TO BEES

The raison d'être of the QSAR models is the prediction of the activity of organic molecules from their physicochemical properties and/or their structural characteristics coded by means of topological indices or by important constitutive structural fragments that can be atoms or functional groups. Thus, the models presented in this study may be used to predict the acute toxicity of pesticides to the honey bee. Depending on their domain of application, they can be used to fill data gaps rapidly and at reduced cost, to select the optimal range of concentrations prior to conducting a laboratory test, to detect chemicals that present specific behaviors, or to help in the understanding of their mechanism of action. It is worth noting that the homology models are particularly suited to consolidate mechanistic information provided by QSAR results [20]. They can also directly intervene in the search for structure–activity relationships [55].

However, the QSAR results cannot be directly used to fully assess the hazard of xenobiotics to honey bees. Indeed, they are derived from standardized laboratory test results performed on technical grade material of high purity while formulations with different characteristics are used in practice. These formulations behave differently on *Apis mellifera* than active molecules [56]. This is also true for non-*Apis* bees such as leafcutting bees (*Megachile rotundata*), alkali bees (*Nomia melanderi*), and bumblebees (*Bombus* spp.) [56]. Attempts have been made to derive QSAR models showing better practical realism [32,33,37,57] but no such models have yet been derived on bees.

Existing QSAR models on bees allow the prediction of acute toxicity effects while it is well known that sublethal effects can adversely impact a bee colony (see, *e.g.*, [58–62]). The lack of QSAR models on sublethal endpoints relies on the scarcity of biological data. Indeed, it is necessary to have sufficient results for different molecules to obtain a statistically significant QSAR. In the same way, because the existing QSAR models on bees were derived from pesticide data, they can only be used to predict the toxicity of this type of molecules. In other words, they cannot be employed to predict the acute toxicity of other man-made chemicals that can potentially contaminate honey bees and their hives such as solvents, polycyclic aromatic hydrocarbons, dioxins, and polychlorinated biphenyls [63,64]. Last, it is obvious that the available QSAR models on bees cannot be used to predict the toxicity of a mixture of chemicals.

The aforementioned weaknesses are logical because whatever the type of model, it is important to keep in mind that it can only be used for the goal for which it was built.

REFERENCES

1. K.A. Hassall, *The Biochemistry and Uses of Pesticides*, 2nd ed., John Wiley & Sons Ltd., Weinheim, Germany, 1990.
2. J. Devillers, *Acute toxicity of pesticides to honey bees*, in *Honey Bees: Estimating the Environmental Impact of Chemicals*, J. Devillers and M.H. Pham-Delègue, eds., Taylor & Francis, London, UK, 2002, pp. 56–66.
3. S. Cluzeau, *Risk assessment of plant protection products on honey bees. Regulatory aspects*, in *Honey Bees: Estimating the Environmental Impact of Chemicals*, J. Devillers and M.H. Pham-Delègue, eds., Taylor & Francis, London, UK, 2002, pp. 42–55.
4. Anonymous, *QSARs in the assessment of the environmental fate and effects of chemicals*, Tech. Rep. no. 74, ECETOC, Brussels, Belgium, 1998.
5. J. Devillers, *Application of QSARs in aquatic toxicology*, in *Computational Toxicology. Risk Assessment for Pharmaceutical and Environmental Chemicals*, S. Ekins, ed., John Wiley & Son Ltd., Hoboken, NJ, 2007, pp. 651–675.
6. J.M. Ribo and K.L.E. Kaiser, *Effects of selected chemicals to photoluminescent bacteria and their correlations with acute and sublethal effects on other organisms*, Chemosphere 12 (1983), pp. 1421–1442.
7. J. Devillers and P. Chambon, *Toxicité aiguë des chlorophénols sur Daphnia magna et Brachydanio rerio*, J. Fr. Hydrol. 17 (1986), pp. 111–120.
8. T. Kishino and K. Kobayashi, *Acute toxicity and structure-activity relationships of chlorophenols in fish*, Water Res. 30 (1996), pp. 387–392.
9. F. Briens, R. Bureau, S. Rault, and M. Robba, *Comparative molecular field analysis of chlorophenols. Application in ecotoxicology*, SAR QSAR Environ. Res. 2 (1994), pp. 147–157.
10. W. Karcher and J. Devillers, *Practical Applications of Quantitative Structure-Activity Relationships (QSAR) in Environmental Chemistry and Toxicology*, Kluwer Academic Publishers, Dordrecht, the Netherlands, 1990.
11. J. Devillers and W. Karcher, *Applied Multivariate Analysis in SAR and Environmental Studies*, Kluwer Academic Publishers, Dordrecht, the Netherlands, 1991.
12. J. Devillers, *Methods for building QSAR*, in *Computational Toxicology*, Vol. II, B. Reisfeld and A.N. Mayeno, eds., Humana Press, New York, 2012, pp. 3–27.
13. J. Devillers and D. Domine, *A noncongeneric model for predicting toxicity of organic molecules to Vibrio fischeri*, SAR QSAR Environ. Res. 10 (1999), pp. 61–70.
14. J. Devillers, *QSAR modeling of large heterogeneous sets of molecules*, SAR QSAR Environ. Res. 12 (2001), pp. 515–528.
15. M. Vighi, M.M. Garlanda, and D. Calamari, *QSARs for toxicity of organophosphorus pesticides to Daphnia and honeybees*, Sci. Total Environ. 109–110 (1991), pp. 605–622.
16. J. Devillers, D. Domine, C. Guillon, S. Bintein, and W. Karcher, *Prediction of partition coefficients (log Poct) using autocorrelation descriptors*, SAR QSAR Environ. Res. 7 (1997), pp. 151–172.

17. J. Devillers, D. Domine, C. Guillon, and W. Karcher, *Simulating lipophilicity of organic molecules with a back-propagation neural network*, J. Pharm. Sci. 87 (1998), pp. 1086–1090.

18. L.H. Hall and L.B. Kier, *Molecular connectivity chi indices for database analysis and structure-property modeling*, in *Topological Indices and Related Descriptors in QSAR and QSPR*, J. Devillers and A.T. Balaban, eds., Gordon and Breach, Amsterdam, the Netherlands, 1999, pp. 307–360.

19. Anonymous, *The principles for establishing the status of development and validation of (quantitative) structure-activity relationships (Q)SARs*, OECD document, ENV/JM/TG(2004)27.

20. F. Dulin, M.P. Halm-Lemeille, S. Lozano, A. Lepailleur, J. Sopkova-de Oliveira Santos, S. Rault, and R. Bureau, *Interpretation of honeybees contact toxicity associated to acetylcholinesterase inhibitors*, Ecotoxicol. Environ. Safety 79 (2012), pp. 13–21.

21. J. Devillers and A.T. Balaban, *Topological Indices and Related Descriptors in QSAR and QSPR*, Gordon & Breach, Amsterdam, the Netherlands, 1999.

22. R. Todeschini and V. Consonni, *Handbook of Molecular Descriptors*, John Wiley & Sons Ltd., Weinheim, Germany, 2000.

23. D. Rogers, *Some theory and examples of genetic function approximation with comparison to evolutionary techniques*, in *Genetic Algorithms in Molecular Modeling*, J. Devillers, ed., Academic Press, London, UK, 1996, pp. 87–107.

24. P. Broto and J. Devillers, *Autocorrelation of properties distributed on molecular graphs*, in *Practical Applications of Quantitative Structure-Activity Relationships (QSAR) in Environmental Chemistry and Toxicology*, W. Karcher and J. Devillers, eds., Kluwer Academic Publishers, Dordrecht, the Netherlands, 1990, pp. 105–127.

25. J. Devillers, *Autocorrelation descriptors for modeling (eco)toxicological endpoints*, in *Topological Indices and Related Descriptors in QSAR and QSPR*, J. Devillers and A.T. Balaban, eds., Gordon and Breach, Amsterdam, the Netherlands, 1999, pp. 595–612.

26. S.C. Basak, *Information theoretic indices of neighborhood complexity and their application*, in *Topological Indices and Related Descriptors in QSAR and QSPR*, J. Devillers and A.T. Balaban, eds., Gordon and Breach, Amsterdam, the Netherlands, 1999, pp. 563–593.

27. N. Marchand-Geneste and J. Devillers, *Comparative modeling review of the nuclear hormone receptor superfamily*, in *Endocrine Disruption Modeling*, J. Devillers, ed., CRC Press, Boca Raton, FL, 2009, pp. 97–142.

28. J. Devillers, *Neural Networks in QSAR and Drug Design*, Academic Press, London, UK, 1996.

29. J. Devillers, *Artificial neural network modeling in environmental toxicology*, in *Artificial Neural Networks: Methods and Protocols*, D. Livingstone, ed., Humana Press, Totowa, NJ, 2008, pp. 61–81.

30. J. Devillers, *Artificial neural network modeling of the environmental fate and ecotoxicity of chemicals*, in *Ecotoxicology Modeling*, J. Devillers, ed., Springer, New York, pp. 1–28.

31. R.C. Eberhart and R.W. Dobbins, *Neural Network PC Tools. A Practical Guide*, Academic Press, San Diego, CA, 1990.

32. J. Devillers, *Prediction of toxicity of organophosphorus insecticides against the midge, Chironomus riparius, via a QSAR neural network model integrating environmental variables*, Toxicol. Methods 10 (2000), pp. 69–79.

33. J. Devillers, *A general QSAR model for predicting the acute toxicity of pesticides to Lepomis macrochirus*, SAR QSAR Environ. Res. 11 (2001), pp. 397–417.

34. J. Devillers, M.H. Pham-Delègue, A. Decourtye, H. Budzinski, S. Cluzeau, and G. Maurin, *Structure-toxicity modeling of pesticides to honey bees*, SAR QSAR Environ. Res. 13 (2002), pp. 641–648.

35. J. Devillers, M.H. Pham-Delègue, A. Decourtye, H. Budzinski, S. Cluzeau, and G. Maurin, *Modeling the acute toxicity of pesticides to* Apis mellifera, Bull. Insect. 56 (2003) pp. 103–109.

36. D.V. Eldred, C.L. Weikel, P.C. Jurs, and K.L.E. Kaiser, *Prediction of fathead minnow acute toxicity of organic compounds from molecular structure*, Chem. Res. Toxicol. 12 (1999), pp. 670–678.

37. J. Devillers, *A QSAR model for predicting the acute toxicity of pesticides to gammarids*, in *Nature-Inspired Methods in Chemometrics: Genetic Algorithms and Artificial Neural Networks*, R. Leardi, ed., Elsevier, Amsterdam, the Netherlands, 2003, pp. 323–339.

38. J. Devillers, *Linear versus nonlinear QSAR modeling of the toxicity of phenol derivatives to* Tetrahymena pyriformis, SAR QSAR Environ. Res. 15 (2004), pp. 237–249.

39. J. Devillers, *A new strategy for using supervised artificial neural networks in QSAR*, SAR QSAR Environ. Res. 16 (2005), pp. 433–442.

40. J. Devillers, *Strengths and weaknesses of the backpropagation neural network in QSAR and QSPR studies*, in *Neural Networks in QSAR and Drug Design*, J. Devillers, ed., Academic Press, London, UK, 1996, pp. 1–46.

41. V.N. Vapnik, *The Nature of Statistical Learning Theory*, Springer, New York, 1995.

42. N. Cristianini and J. Shawe-Taylor, *An Introduction to Support Vector Machines and Other Kernel-Based Learning Methods*, Cambridge University Press, Cambridge, UK, 2000.

43. J.P. Doucet, F. Barbault, H.R. Xia, A. Panaye, and B.T. Fan, *Non linear SVM approaches to QSPR/QSAR studies and drug design*, Cur. Comput.-Aided Drug Des. 3 (2007), pp. 263–289.

44. Y. Wang, M. Zheng, J. Xiao, Y. Lu, F. Wang, J. Lu, X. Luo, W. Zhu, H. Jiang, and K. Chen, *Using support vector regression coupled with the genetic algorithm for predicting acute toxicity to the fathead minnow*, SAR QSAR Environ. Res. 21 (2010), pp. 559–570.

45. R. Darnag, A. Schmitzer, Y. Belmiloud, D. Villemin, A. Jarid, A. Chait, E. Mazouz, and D. Cherqaoui, *Quantitative structure-activity relationship studies of TIBO derivatives using support vector machines*, SAR QSAR Environ. Res. 21 (2010), pp. 231–246.

46. J. Devillers, J.P. Doucet, A. Doucet-Panaye, A. Decourtye, and P. Aupinel, *Linear and non-linear QSAR modelling of juvenile hormone esterase inhibitors*, SAR QSAR Environ. Res. 23 (2012), pp. 357–369.

47. J. Devillers, A. Doucet-Panaye, and J.P. Doucet, *Predicting highly potent juvenile hormone esterase inhibitors from 2D QSAR modeling*, in *Juvenile Hormones and Juvenoids. Modeling Biological Effects and Environmental Fate*, J. Devillers, ed., CRC Press, Boca Raton, FL, 2013, pp. 229–253.

48. R.F. Rekker and R. Mannhold, *Calculation of Drug Lipophilicity. The Hydrophobic Fragmental Constant Approach*, John Wiley & Sons Ltd., Weinheim, Germany, 1992.

49. C. Hansch and A. Leo, *Substituent Constants for Correlation Analysis in Chemistry and Biology*, John Wiley & Sons Ltd., New York, 1979.

50. J. Devillers, *Genetic Algorithms in Molecular Modeling*, Academic Press, London, UK, 1996.

51. J. Devillers, A. Chezeau, E. Thybaud, and R. Rahmani, *QSAR modeling of the adult and developmental toxicity of glycols, glycol ethers, and xylenes to* Hydra attenuata, SAR QSAR Environ. Res. 13 (2002), pp. 555–566.

52. J. Devillers, A. Chezeau, and E. Thybaud, *PLS-QSAR of the adult and developmental toxicity of chemicals to* Hydra attenuata, SAR QSAR Environ. Res. 13 (2002), pp. 705–712.

53. N. Amaury, E. Benfenati, E. Boriani, M. Casalegno, A. Chana, Q. Chaudhry, J.R. Chrétien, J. Cotterill, F. Lemke, N. Piclin, M. Pintore, C. Porcelli, N. Price, A. Roncaglioni, and A. Toropov, *Results of DEMETRA models*, in *Quantitative Structure-Activity Relationships (QSAR) for Pesticide Regulatory Purposes*, E. Benfenati, ed., Elsevier, Amsterdam, the Netherlands, 2007, pp. 201–281.

54. A.A. Toropov and E. Benfenati, *SMILES as an alternative to the graph in QSAR modelling of bee toxicity*, Comput. Biol. Chem. 31 (2007), pp. 57–60.

55. A. Rocher and N. Marchand-Geneste, *Homology modelling of the* Apis mellifera *nicotinic acetylcholine receptor (nAChR) and docking of imidacloprid and fipronil insecticides and their metabolites*, SAR QSAR Environ. Res. 19 (2008), pp. 245–261.

56. J. Devillers, M.H. Pham-Delègue, A. Decourtye, H. Budzinski, S. Cluzeau, and G. Maurin, *Comparative toxicity and hazards of pesticides to* Apis *and non-*Apis *bees. A chemometrical study*, SAR QSAR Environ. Res. 14 (2003), pp. 389–403.

57. J. Devillers and J. Flatin, *A general QSAR model for predicting the acute toxicity of pesticides to* Oncorhynchus mykiss, SAR QSAR Environ. Res. 11 (2000), pp. 25–43.

58. M.H. Pham-Delègue, A. Decourtye, L. Kaiser, and J. Devillers, *Behavioural methods to assess the effects of pesticides on honey bees*, Apidologie 33 (2002), pp. 425–432.

59. A. Decourtye, J. Devillers, S. Cluzeau, M. Charreton, and M.H. Pham-Delègue, *Effects of imidacloprid and deltamethrin on associative learning in honeybees under semi-field and laboratory conditions*, Ecotoxicol. Environ. Safety 57 (2004), pp. 410–419.

60. A. Decourtye, C. Armengaud, M. Renou, J. Devillers, S. Cluzeau, M. Gauthier, and M.H. Pham-Delègue, *Imidacloprid impairs memory and brain metabolism in the honeybee* (Apis mellifera L.), Pestic. Biochem. Physiol. 78 (2004), pp. 83–92.

61. A. Decourtye, J. Devillers, E. Genecque, K. Lemenach, H. Budzinski, S. Cluzeau, and M.H. Pham-Delègue, *Comparative sublethal toxicity of nine pesticides on olfactory learning performances of honeybee* Apis mellifera, Arch. Environ. Contam. Toxicol. 48 (2005), pp. 242–250.

62. A. Decourtye, S. Lefort, J. Devillers, M. Gauthier, P. Aupinel, and M. Tisseur, *Sublethal effects of fipronil on the ability of honeybees* (Apis mellifera L.) *to orientate in a complex maze*, Julius-Kühn-Archiv. 423 (2009), pp. 75–83.

63. G.C. Smith, J.J. Bromenshenk, D.C. Jones, and G.H. Alnasser, *Volatile and semi-volatile compounds in beehive atmospheres*, in *Honey Bees: Estimating the Environmental Impact of Chemicals*, J. Devillers and M.H. Pham-Delègue, eds., Taylor & Francis, London, UK, 2002, pp. 12–41.

64. J. Devillers and H. Budzinski, *Utilisation de l'abeille pour caractériser le niveau de contamination de l'environnement par les xénobiotiques*, Bull. Tech. Apic. 35 (2008), pp. 168–178.

8 Mathematical Models for the Comprehension of Chemical Contamination into the Hive

Paolo Tremolada and Marco Vighi

CONTENTS

ABSTRACT

Honey bees are very sensitive organisms, and their vulnerability toward chemical pollution is a priority environmental issue. In this chapter, the importance and the fundamental of a mathematical model able to predict the fate of chemicals in the hive are described. The "hive model" presented here is able to consider different contamination pathways: from inside the hive via pesticide treatments against bee pests or from outside by means of the eventual contamination present in nectar, pollen, resin, water, air, or vegetation. The input parameters of the model are the physical–chemical properties of the compound and the major characteristics of the hive ecosystem, from which it calculates contamination residues in bee products (honey, pollen, royal jelly, wax, and propolis) over time. The model can be applied to all nonionic chemicals with a measurable vapor pressure (volatile or semivolatile compounds). The model was validated with contamination data in bees, wax, and honey following tau-fluvalinate application in two experimental hives. The comparison between measured and modeled data over a period of 6 months was very good. Even if more experimental data should be provided for a better calibration of the model, it seems to correctly quantify the main contamination pathways from outside and within the hive.

KEYWORDS

Hive contamination, Pesticide, Modeling, Bee health, Bee activity

8.1 VULNERABILITY OF BEES TO CHEMICAL POLLUTION: BEES AS SENSITIVE SPECIES

Honey bees are considered a sensitive species indicator of environmental quality. The survival of honey bee colonies is related to the good individual performance (individual fitness) and to the efficacy of communication and cooperation between classes of bees (colony fitness). Moreover, honey bee breeding is certainly responsible for an artificial reduction of genetic diversity and for a lack of the natural selection benefits. These elements accentuate the vulnerability of honey bee colonies and their sensitivity to pollution. Since the late 1990s, widespread events of honey bee disappearance especially in the United States and in Europe (Spain, France, Italy and Germany) happened [1,2]. These events were considered alarming because it was not possible to relate them to a specific cause. This new pathology was called colony collapse disorder (CCD) [3,4]. Recrudescence of old and new pathologies [5,6], environmental stresses, including climate change, [7] and pesticide contamination [8,9] were considered as possible causes perhaps also in association among each other. The first two adversities are related to the genotypes of honey bee colonies, and they directly reflect the vulnerability of bees toward abiotic (climate change) and biotic (parasite) stresses, while the third one is directly related to human activity in terms of chemical contamination by xenobiotics.

Neonicotinoid insecticides were indicated as a possible cause of CCD events, and this possibility is a meaningful example of how xenobiotics can affect nontarget species in unsuspected ways (failure of risk assessment evaluation). Neonicotinoid insecticides act on the insect nicotinic (acetylcholine) receptor (nAChR) [10], and therefore they are toxic to bees too. Iwasa et al. [11] revealed that the nitro-substituted neonicotinoids were most toxic to honey bees with very low contact LD_{50} values even of 18 ng bee^{-1} as in the case of imidacloprid. Neonicotinoid insecticides are used in agriculture, mainly as a seed dressing in many crops such as corn and sunflower. Being systemic insecticides, application via seed treatment enables crop protection from soil invertebrates and sucking insects both during germination and later growth. Therefore, neonicotinoid insecticides affect pollinating insects as powder dispersion during the sowing procedures via the drill fan (pneumatic seed drills) [12–14] and through the contamination of pollen and nectar, produced by crops that have been grown from treated seeds [15].

This example demonstrates that chemical exposure may happen multiple ways, some of which are easily misunderstood. In fact, neonicotinoid insecticide exposure toward bees may happen during sowing operations, when pesticide-contaminated powders are dispersed into air, and later when the powders in air subsequently deposit on local soil and vegetation. Moreover, because of the systemic activity of these insecticides, exposure may happen during the flowering phase of plants grown from treated seeds, when contaminated pollen and nectar can be foraged by bees and other pollinator insects.

8.1.1 CHEMICAL CONTAMINATION SOURCES

Two main contamination sources toward bees can be distinguished: a direct contamination by the distribution of pesticides inside the hive to protect honey bees from parasites and an indirect contamination by all the chemicals that involuntarily come from outside.

The direct contamination is generally known and deals with few chemicals at high concentrations, as in the case of acaricides used against the mite *Varroa destructor*. Wallner [16], Bogdanov [17], and Johnson et al. [2] reviewed the problem of contamination of bee products following pesticide application within the hive. Fat-soluble and stable pesticides, such as bromopropylate, coumaphos, and τ-fluvalinate, are known to accumulate in wax, year by year, while fat-soluble and volatile compounds, such as essential oils, and unstable pesticides, such as amitraz, do not persist because they volatilize or degrade. On the other hand, water-soluble chemicals, such as formic acid, oxalic acid, and cymiazole, do not accumulate in wax, but they are found in honey, producing an unwanted taste.

The indirect contamination source is more complex because it deals with many chemicals generally at low concentrations. Barmaz et al. [18] developed a site-specific method for assessing risks to pollinators from pesticides used in the surroundings of a bee colony. They concluded that risk levels were very different depending on agronomic practices and crop type. The greatest risk is generally found at the beginning of the growing season for annual crops and later in June–July for permanent crops. As an example, the case of neonicotinoid insecticides highlighted how pesticides coming from outside can severely threaten honey bee survival. Another example of unsuspected contamination is the presence of

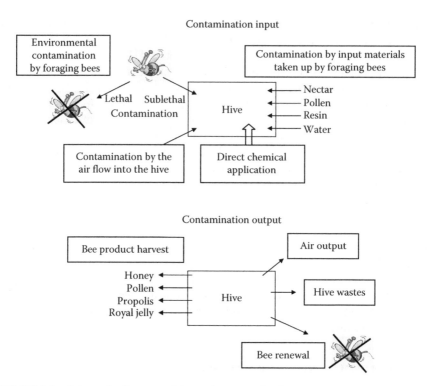

FIGURE 8.1 Schematic diagram of the main contamination sources and output ways for the hive ecosystem.

antibiotics in bee products, in the absence of direct applications to the hive. A possible explanation is the transfer of antibiotics from sewage of farms where these chemicals are extensively used for curing and preventing animal diseases. Antibiotic residues can be taken up by bees drinking sewage as water supply (bees take up sewage also in the presence of alternative cleaner water supply), and later bees can transfer antibiotics to honey bee products inside the hive. These examples highlight the complexity and the importance of the indirect contamination sources, which may affect hives in unsuspected ways. Figure 8.1 summarizes a scheme of the possible contamination sources and contamination output from the hive. Some of these pathways are well documented such as the contamination via direct pesticide application, but others were only recently considered such as those coming from pesticide-exposed pollen [19] or from environmental exposure of foraging bees [14,18,20]. The complexity of these multiple contamination sources and that of the system itself needs a careful description of the bee ecology in the frame of modeling applications.

8.1.2 Predicting Chemical Contamination with Mathematical Models

Mathematical models are useful tools for evaluating the environmental distribution and fate of chemicals [21]. Models forecast the fate of chemicals and also the resulting levels of exposure before their use. Moreover, they facilitate a greater comprehension of the distribution of chemicals [22]. Models based on fugacity [23,24] have been extensively applied to nonionized organic chemicals at various scales: global [25], regional [26], field [27], and microecosystem [22,28,29]. These models require three types of input data: the physicochemical properties of the compounds (molecular weight, water solubility, vapor pressure, and partition coefficients), the emission patterns (amounts emitted, space and time distribution of emissions), and the environmental characteristics of the ecosystem (temperatures, compartment volumes and compositions, and fluxes). The quantification of these aspects in a model system is able to furnish information about the fate of a specific chemical with an error (uncertainty of the prediction) related to the simplification and the generalization introduced in the model, to the goodness of the input data and to stochastic events not included in the model. A crucial point of models is the quantification of the mass fluxes and the exchange parameters between compartments. The former explains the transfer of a contaminant in different places of the same compartment (*e.g.*, the transfer of a pesticide by the air flux from outside to inside) and the second indicates the transfer from one compartment to another (*e.g.*, the transfer of a pesticide from stored honey into cell wax or vice versa). In fugacity models, these transport parameters are described as advection parameters (mass time^{-1}) and as a mass transfer coefficient (mass surface^{-1} time^{-1}). For applying them to the hive ecosystem, it is necessary to describe carefully and quantitatively bee activities and material fluxes.

8.2 QUANTIFICATION OF BEES' ACTIVITIES FOR MODELING PURPOSES

Productive honey bees and natural colonies present many substantial differences. Information reported here refers to typical productive honey bee colonies of *Apis mellifera ligustica* because the goal of this modeling application is the prevention of the contamination of bee products in relation to food safety. However, the same scheme can be applied to natural hives modifying the quantitative parameters when necessary. An extensive description of natural bee characteristics is reported by Seeley [30], while that of productive honey bees is reported by Chauvin [31], Grout [32], Snodgrass [33], Root [34], and Goodman and Fisher [35]. In addition, remaining information was taken from personal communications with bee experts.

8.2.1 Bee Biology: Development, Reproduction, and Renewal

Bee development passes through egg, larval, and adult stages. Egg deposition begins in January–February with around 1,000 eggs. Later, this number soars to a peak of 30,000–40,000 in May–June and finally declines gradually throughout the remainder of the summer, ceasing in September–October. Hatching happens 3 days after deposition, and the larval development completes in 21 days in the wax cells. Larval dimensions range from 1.5 mm (near the egg size) to the size of the cell in the final stage. Given a mean weight of 75 mg and an approximate volume of 90 mm^3 (1/4 of the cell volume), the density of larvae is about 0.83 mg mm^{-3}. Brood rearing begins in late winter with around 75 g of brood (nearly 1,000 cells containing brood). Later, brood rises to about 2.5 kg (30,000–40,000 larvae) in May–June and finally declines gradually until October. The total brood production is 15 kg year^{-1} (corresponding to 200,000 larvae).

Adult bees are specialized for different tasks depending on their sex and age. The majority of bees are sterile females (worker bees, ranging in number from 10,000 in winter to 50,000 or even 100,000 in June). Worker bees 0–2 days old clean the cells (age caste I, cell cleaner); bees 2–11 days old care for the brood living in the central part of the nest (age caste II, broodnest caste); bees 11–20 days old constitute a food-storage caste, whose tasks occur in the peripheral, food-storage region of the nest and that are able to produce wax (age caste III); and bees 20 days old or more become forager bees whose work is mostly outside the nest (age caste IV). Approximately 20% of forager bees spend a day or two serving as guards at the nest entrance (guard bees). The biological cycle of worker bees varies from 40 to 45 days in summer to 6 months or more in winter. The mean dimensions of a worker bee are 12–13 mm in length, 4 mm in width, and 130 mg in weight. Given a volume of approximately 160 mm^3, the density of a bee is about 0.8 mg mm^{-3}.

Male bees are intermediate in size between workers and the queen. They are much less numerous than worker bees (numbering between 1,000 and 5,000 depending on the period) and are important in reproductive activity with the queen, in hive thermoregulation, and in food exchange and transport within the hive. From a quantitative point of view, they have a limited impact consisting of around 10% of the sterile females.

8.2.2 Recruitment of Materials

Bees may explore about 300 ha of territory daily (with a foraging radius from 1 to more than 5 km) searching for food resources. Bees need only four resources for their subsistence: nectar, pollen, water, and resin. The average amounts collected yearly by a colony are 240, 30, 10, and 0.1 kg, respectively, for nectar, pollen, water, and resin. Nectar is the base material for honey production. Nectar and honey are 20%–40% and 80% sugar solutions, respectively. Honey has a density of 1.39–1.43 kg dm^{-3} [36]. Water is gathered for diluting honey for brood food and for evaporative cooling of the nest on hot days. Resins are the base materials for the production of propolis, which is composed of roughly 70% resin, 25% beeswax, and 5% volatile oils. Resins are taken from trees and carried home in pollen baskets. Propolis serves to plug unwanted openings and for hygienic purposes because of its antimicrobial and antifungal activity [37].

Another important material needed by bees is oxygen. Inside the hive, thermoregulation and other bee activities require a large energy supply and create great oxygen demand. The overall oxygen consumption of a hive has been quantified as 30 kg year^{-1} with the production of 52 kg year^{-1} and 34 kg year^{-1} of CO_2 and H_2O, respectively. The yearly amount of materials collected by bees is reported in Figure 8.2a.

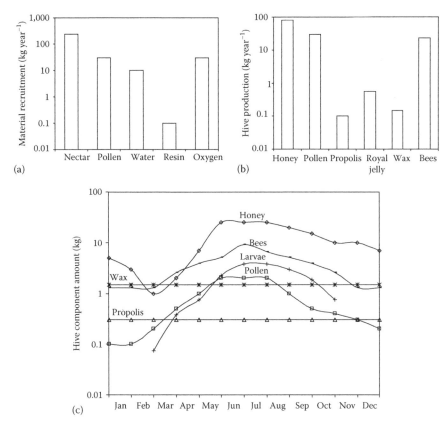

FIGURE 8.2 (a) Quantification of the material recruitment, (b) quantification of the hive products, and (c) year trends of the average amounts of hive compartments. All graphs are in logarithmic scale.

Nectar, pollen, and oxygen are the most abundant recruited materials, then water is highly requested when metabolic and nectar water are not sufficient. Resin is collected at a very low level.

8.2.3 Bee Products

Bees collect or produce six major products: honey, pollen, propolis, royal jelly, wax, and bees themselves (larval production). The typical honey reserve in summer is 15 kg, and the total amount produced in a year is about 80 kg, of which 60 kg year^{-1} is consumed for colony maintenance (25 kg in winter and 35 kg in summer) and 20 kg year^{-1} is typically removed by the beekeeper. Pollen is carried to the hive and stored by mixing the dry pollen grains with liquid (honey, nectar, or both). Pollen is generally composed of 10%–20% water, 10%–35% proteins, 15%–40% sugars, and 1%–10% lipids. The typical pollen reserve in summer is 2 kg, and the total amount foraged in a year is 30 kg. Honey and pollen are the primary food sources; honey furnishes energetic support and pollen supplies proteins, vitamins, fats, and minerals.

Daily food consumption depends on age and bee activity. The daily food intake of larvae ranges from a few milligrams in the initial days to tens of milligrams during the final stage. During the first 3 days, larvae are fed with about 3 mg day^{-1} of royal jelly. Later, the larval food changes in composition based on age. However, in order to define a mean value for the full larval period, we assume consumption of 12 mg day^{-1} of larval food composed of 50% honey and 50% pollen. An adult bee consumes food in proportion to its activity. During flight, about 6.77 mg of sugar is consumed per hour. Because a foraging bee is able to carry 10–20 mg and 40–60 mg of pollen and nectar, respectively, bees must make efficient choices of the most promising foraging areas in order to optimize the cost–benefit balance. We assume an average consumption of 10 mg day^{-1} of honey and 1 mg day^{-1} of pollen for adult bees. Propolis is used as an antiseptic agent and is a permanent material of the hive system present in variable amounts depending on the needs of the hive. We assume an average quantity of 0.3 kg with a renewal rate of 0.1 kg year^{-1}. Royal jelly is used for high-quality nutrition (queens and first-stage larvae) and is not stored in the hive but produced upon demand. We estimate an overall annual consumption of about 0.55 kg (10 and 540 g for the queen and larvae, respectively). These values are based on an average daily consumption of 20 and 1 mg for the queen and larvae, respectively. The queen (one individual) consumes this amount daily almost all year, while larvae (180,000 reared individuals) consume this amount for 3 days only. Royal jelly is generally composed of 57%–70% water, 6.4%–17% proteins, 6.8%–20% sugars, 1.3%–7.1% of lipophilic substances (phospholipids, sterols, phenols, fatty acids, glycerides, and waxes), and 0.75%–1.1% minerals and other minor constituents and has a mean density of 1.1 g cm^{-3} [38]. Wax is the structural material of the hive, and its quantity is nearly constant throughout the year. The average quantity of wax is 1.5 kg, most of it remaining for a long time. Bees further minimize the need of new wax by constantly recycling it. New wax is produced in abundance when a colony needs additional storage space for a large crop of honey. Bees continuously remove part of the wax (uncapping activity) and rebuild it (capping, fixing, and rebuilding activities). The amount of newly produced wax can be evaluated yearly based on the amount of cap wax (0.15 kg). The surface area of the wax cells (empty or full) can be evaluated by multiplying the internal surface area of one cell (360 mm^2) by the total number of cells (100,000). The resulting surface area of the wax cells is 36 m^2. The exchange surface of honey, pollen, and brood is equivalent to the proportion of the wax surface occupied by each component.

A colony is able to rear up to 200,000 bees annually (including 5,000–20,000 males and a few queens), consuming a total of 20–26 kg of pollen (130 mg of pollen are required to produce an adult bee) and 20–30 kg of honey (one honey cell is necessary to rear several larvae). A typical nest contains about 100,000 cells with 415 cells dm^{-2} on each side of the comb. Each hexagon-shaped cell typically has a perimeter of 18 mm and an area of 24 mm^2, and the wall-to-wall distance is 5 mm. Cell depth is variable; considering a depth of 15 mm and an area of 24 mm^2, the volume of a typical cell is about 360 mm^3. Root [34] reports cell volumes varying from 360 to 192 mm^3 depending on cell density (650–1,050 cells dm^{-2} for both sides of the comb). In summer, two- to three-fifths of the cells are filled with brood, two-fifths with honey, and less than one-fifth with pollen. A typical nest is arranged in

10 combs, each with external dimensions of 47×30 cm and wax foundation dimensions of 44×27 cm, giving a total surface area of 2.4 m². Cell walls and bases have only 0.073 ± 0.008 and 0.176 ± 0.028 mm of wax thickness, respectively. Based on these dimensions, rapid diffusion of pollutants is expected. The combs are composed of a total of about 1.4 dm³ (1.35 kg) of wax. With the cap wax (0.15 kg), the total amount of wax in a typical nest is about 1.5 kg. Wax is a complex blend of straight-chain monohydric alcohols esterified with carboxylic acids and hydroxycarboxylic acids mixed with various straight-chain alkanes. The density of wax is $0.958-0.970$ kg dm⁻³.

The year amounts of bee products are shown in Figure 8.2b. Honey and pollen are produced in high amounts together with bees, and much lower levels are necessary for wax and royal jelly. Production of propolis is very low.

8.2.4 ENVIRONMENTAL PARAMETERS WITHIN THE HIVE

Environmental conditions within the hive are quite peculiar. In the proximity of brood, bees maintain a constant temperature between 34.5°C and 35.5°C also in winter. Far from brood or in its absence, temperatures fluctuate. In winter, bees are able to maintain near 20°C around them using metabolic heat. The high temperatures in the brood nest accelerate the diffusion of contaminants within the hive. The intense exchange activity of bees for food supply and social communication further increases the distribution of any contaminants that may be present.

Another important parameter is the air flux into the hive for thermoregulation and for oxygen supply. The oxygen request (30 kg year⁻¹) corresponds to an air flux of 150 m³ year⁻¹. Considering that the air volume inside the hive is about 0.2 m³, the daily air flux (0.4 m³ day⁻¹) results in a twice-daily exchange of the full air volume.

8.2.5 YEAR CYCLE OF BEE ACTIVITY

An indicative quantification of the year cycle of the hive compartments is reported in Figure 8.2c. Throughout the year, some materials are almost permanent (wax and propolis), while others are highly variable such as food storages that reach their minimum in January–February. Larvae and adult bees start to increase from February (1,000 larvae and 10,000 adult bees) until June (30,000–40,000 larvae and 50,000–100,000 adult bees), and then they decline until October when brood rearing stops. These numbers can be transformed in mass and volume data by means of the mean weight and density of each stage. In February foraging activity starts again, and storages of pollen and honey rise slowly from 0.1 and 1 kg, respectively, in February until 2 and 25 kg in May–June when they reach their maximum. Royal jelly is almost absent as storage material, but it is produced upon request for the queen and for few-day-old larvae. From August, food storage slowly declines because of the reduction in environmental availability (daily flux into the hive), while the daily consumption is still high because of the high numbers of adult bees and larvae. From August, adult bees and larvae decrease, and brood rearing ceases in October. From this moment, only adult bees are present in the hive surviving as individuals from October to February and consuming mainly stored food (winter storage of honey) for themselves and for the next generation reared in February.

Throughout the year, wax and propolis can be considered substantially constant, while bees, larvae, honey, and pollen change, depending on the hive conditions and on the season. In order to simplify their variability, which is hive- and year-specific, a general parabolic trend depending on the season was chosen for each compartment. Therefore, the following second-order equations in function of the number of days (x) from January 1 were proposed for describing quantitatively the year trend of these compartments:

$$\text{Bees } (\text{kg}) = -0.0002 \cdot x^2 + 0.0752 \cdot x + 0.3 \tag{8.1}$$

$$\text{Larvae } (\text{kg}) = -0.0002 \cdot x^2 + 0.070 \cdot x - 2.6 \tag{8.2}$$

$$\text{Honey } (\text{kg}) = -0.007 \cdot x^2 + 0.27 \cdot x \tag{8.3}$$

$$\text{Pollen } (\text{kg}) = -0.00006 \cdot x^2 + 0.0214 \cdot x \tag{8.4}$$

These equations predict compartment dimensions with R^2 values of 0.71, 0.80, 0.70, and 0.68 for bees, larvae, honey, and pollen, respectively, referring to the 12 indicative data reported in Figure 8.2c. Larvae amount is 0 from November 1 to the middle of February and then follows predictions of Equation 8.2. By these equations and considering constant wax, propolis, and air compartments, the main hive components can be quantitatively described throughout the year on a day basis. The mass unit of Equations 8.1 through 8.4 can be transformed to a volumetric unit by means of their densities.

8.3 MATHEMATICAL EQUATIONS FOR PREDICTING THE FATE OF MOLECULES IN THE HIVE

8.3.1 Partition Parameters between Hive Compartments

In the hive, several compartments are quite persistent: first of all, the wax one is almost permanent, and only a small fraction is *ex novo* built or re-elaborated by bees, especially when stored food is deposited or consumed (capping and uncapping activities). Honey and pollen are also stored inside wax cells for months, and larvae itself stay within their cells for the time of their development (21 days). These quite long periods make possible a consistent transfer of contaminants between compartments by diffusion following the concentration gradient and the partition characteristics of the compound. The time to reach equilibrium by diffusion is generally long and depends on the media (viscosity), molecular characteristics, diffusion length (thickness), and temperature. In the case of several hive compartments, their thickness is very thin (*e.g.*, cell walls), the time for partitioning is quite long, and the temperature is high (near 35°C). These conditions facilitate consistent contaminant exchanges by diffusion, toward equilibrium conditions. Wallner [16] experimented over a 30-day period a consistent transfer of coumaphos residue from wax to honey. From these data and from others, reporting concentration measurements in honey and wax after a certain time from the pesticide application, experimental wax/honey partition coefficients (K_{wh}) were calculated supposing that near-equilibrium conditions were present. Tremolada et al. [28] reported mean log K_{wh} of 2.7 ± 0.66, 2.7 ± 0.50, 3.2 ± 0.95, and 2.8 ± 0.72 for coumaphos, malathion, fluvalinate, and bromopropylate, respectively. As wax is a hydrophobic material and honey a hydrophilic sugar solution, K_{wh} is comparable to K_{ow} (octanol/water partition coefficient), but a direct correlation was not attempted, because of the variability and the scarcity of the experimental data. Waiting for a more extensive experimental data set, these authors calculated K_{wh} by a theoretical approach, using fugacity. In this way, a partition coefficient (the mass ratio of a compound in two phases at equilibrium) is equal to the ratio between the capacity of the two compartments, intended as the amount of a compound that is retained by the unit of volume at the same fugacity (equilibrium condition). The unit of capacity is mol m^{-3} Pa^{-1} and its symbol is Z_i. Therefore, K_{wh} was calculated as Z_{wax}/Z_{honey} where the fugacity capacity of wax (Z_{wax}) was calculated throughout K_{oc} (organic carbon/water partition coefficient) and that of honey (Z_{wax}) throughout the capacity of water to which an additional capacity was added, for taking into account that honey has an extractable lipid residue of 0.02% on fresh weight:

$$Z_{wax} = \frac{K_{oc} \, ocf_{wax} \, \rho_{wax}}{H} \tag{8.5}$$

$$Z_{honey} = Z_{water} \cdot 0.9998 + Z_{wax} \cdot 0.0002 \tag{8.6}$$

where

Z_{water} is equal to $1/H$;

H is Henry's law constant in m^3 Pa mol^{-1}; and

K_{oc} is the organic carbon/water partition coefficient calculated by the relationship log K_{oc} = log K_{ow} − 0.21.

The organic carbon fraction of wax (ocf_{wax}) was approximated to 0.57, and the density of wax (ρ_{wax}) was fixed to 0.97 g cm^{-3}, according to literature indications. Predicted log K_{wh} values for coumaphos, malathion, fluvalinate, and bromopropylate were in the range of 2.7 and 3.2 as the experimental ones. Using these partition coefficients and their Z-values and adding to them the air compartment (Z_{air} = 1/ RT, where R = gas constant, 8.314 Pa m^3 mol^{-1} T^{-1}, and T = absolute temperature), a level I fugacity model was proposed [28]. Based on physical–chemical properties of pesticides and on the amounts that persist in the hive 1 month after application, predicted concentrations in wax and honey of coumaphos, malathion, fluvalinate, and bromopropylate were calculated. Predicted concentrations accorded quite well with measured data [28]. According to this work, this very simple "hive model" was further developed [29]. These authors took into account all the compartments that are present in the hive and the materials recruited and produced by bees. Nectar was considered more similar to water than honey because it is a more dilute sugar solution (20%–40% of sugar), and therefore it was approximated to water.

Bees and larvae were considered by their lipophilic and hydrophilic compositions, and their compartment capacity was derived by K_{ow} and by that of water:

$$Z_{bee/larvae} = Z_{water} \cdot 0.7 + \frac{K_{ow} \cdot lip_{bee/larvae}}{H} \tag{8.7}$$

The extractable lipid fraction of bees and larvae (lip_{bees} and lip_{larvae}) was evaluated in 0.02 (g lipid g^{-1} fresh weight of animal).

Pollen and royal jelly were considered similarly taking into account their typical compositions:

$$Z_{pollen} = Z_{water} \cdot 0.15 + \frac{K_{ow} \cdot lip_{pollen}}{H} \tag{8.8}$$

$$Z_{rj} = Z_{water} \cdot 0.65 + \frac{K_{ow} \cdot lip_{rj}}{H} \tag{8.9}$$

From literature data, the average lipid fraction of pollen and royal jelly (lip_{pollen} and lip_{rj}) was 0.05 and 0.04, respectively (g lipid g^{-1} fresh weight of product).

Propolis is mainly composed of resins (70%) and wax (25%); therefore it was considered analogous to wax using K_{oc}:

$$Z_{propolis} = \frac{K_{oc} \, ocf_{propolis} \, \rho_{propolis}}{H} \tag{8.10}$$

The organic carbon fraction of propolis ($ocf_{propolis}$) was approximated to 0.57 and its density ($\rho_{propolis}$) to 1 g cm^{-3}.

All these compartments were considered basically by their water and lipid content in order to represent their affinity for hydrophilic and hydrophobic compounds. At the moment, an indirect validation of these capacities is derived by their application to a case study of the fluvalinate distribution in the hive [29,39].

In order to simplify the model, several hive compartments (*e.g.*, honey, wax, propolis) can be considered together in a multicompartment: the "bee product" one. The larvae can also be considered a bee product because they are reared by bees within the wax cells (continuous contact with wax and adult bees) and because they are fed with bee products: royal jelly first and larvae food later. Because of this common origin, their contamination can be highly dependent on that of bees, and they can be grouped together in a bulk compartment (bee products), whose capacity (Z_{prod}) can be calculated by the sum of the capacities of the single components multiplied by their relative volume fraction (vf_i):

$$Z_{prod} = Z_{honey} \, vf_{honey} + Z_{wax} \, vf_{wax} + Z_{pollen} \, vf_{pollen} + Z_{propolis} \, vf_{propolis} + Z_{larvae} \, vf_{larvae}$$

$$(8.11)$$

Among bee products, pollen is peculiar because it is transported into the hive externally in pollen baskets and then it is stored in the wax cells nearly as it was originally (the contact with bees is limited), but its residence time in wax may be long, allowing contaminant exchanges with wax by diffusion. Therefore, it was decided to group it with the other bee products.

Royal jelly is not included in this bulk compartment because it is mainly produced on request and not stored in the hive: its volume fraction can be approximated to zero. Grouping all the bee products together allows reducing much of the number of compartments that may have different fugacity. By this solution, only three distinct bulk compartments were considered (air, adult bees, and bee products), having their own bulk capacities and fugacities (Z_{air}, Z_{bee}, Z_{prod}, and f_{air}, f_{bee}, f_{prod}, respectively).

Contamination sources can be different (cfr. 1.1): from the air as in the case of neonicotinoid insecticides or from bees as in the case of direct pesticide treatment into the hive. Several treatment modalities use a medium in order to deliver the pesticide into the hive (*e.g.*, Apistan strips carrying the active ingredient τ-fluvalinate). From this medium, the pesticide slowly migrates into the hive. The pesticide release depends on the medium's capacity to retain it inside, which depends on the affinity of the material to the pesticide characteristics. Because treatment media are often made of plastic materials or wood, they were modeled using K_{oc}:

$$Z_{treat} = \frac{K_{oc} \, ocf_{treat} \, \rho_{treat}}{H} \tag{8.12}$$

The organic carbon fraction of the treatment medium (ocf_{treat}) was approximated to 0.57 and its density (ρ_{treat}) to 0.9 g cm^{-3}.

8.3.2 Quantification of Contaminant Fluxes in and out of the Hive

Contaminants may enter the hive in four main ways as shown in Figure 8.3: by the air flux, if the contaminants are dispersed in this medium; by external food sources, if these materials are contaminated [19]; by foraging bees, if they come in contact with contaminated media [8,18]; and by chemical application within the hive [16]. These multiple contamination sources can be described as *D*-parameters within the framework of multicompartmental fugacity models [24]. These *D*-values are flux parameters with the unit of mol day^{-1} Pa^{-1}, and they refer to three different processes: mass fluxes (advection *D*-values), diffusion events (exchange *D*-values), or degradation processes (reaction *D*-values). Although *D*-values refer to these different processes, they have the same units, and therefore they are directly comparable for a given system (in this case, the hive ecosystem).

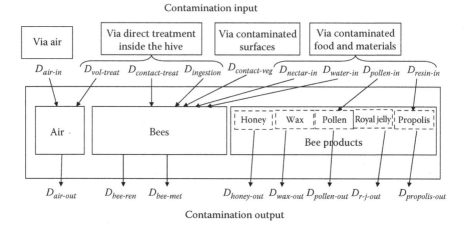

FIGURE 8.3 Modeling of the contaminant input and output for the hive ecosystem.

The advection of a compound by air (D_{air-in}) was calculated by the air flux into the hive for oxygen supply ($G_{air-in-out}$), which was quantified in 0.41 m³ day⁻¹. The outflow of air is clearly the same amount, and therefore the $D_{air-out}$ is equal to D_{air-in}:

$$D_{air-in} = D_{air-out} = G_{air-in-out} \, Z_{air} \tag{8.13}$$

The contaminant inflow by this way (mol day⁻¹) is defined by the product of the D_{air-in} by the fugacity of the air out of the hive ($f_{air-out}$ in Pa), which is a function of the contaminant concentration of the outside air ($C_{air-out}$ in mol m⁻³):

$$f_{air-out} = \frac{C_{air-out}}{Z_{air}} \tag{8.14}$$

The air inside the hive can be contaminated also by volatilization from the treatment medium in the case of a direct pesticide application into the hive. This advection flux ($D_{vol-treat}$) can be calculated by diffusion, using a double-layer resistance approach with two specific mass transport coefficients (MTC_{air} and MTC_{treat} for air and treatment-medium side, respectively). Volatilization depends also on the diffusion area (A_{treat}):

$$D_{vol-treat} = \frac{A_{treat}}{\left(1/(MTC_{air} \, Z_{air})\right) + \left(1/(MTC_{treat} \, Z_{treat})\right)} \tag{8.15}$$

MTC_{air} and MTC_{treat} were quantified in 1 and 0.0001 m day⁻¹ [29]. MTC_{air} was derived from already published data [24], while that of treatment medium was *ex novo* evaluated by the authors.

In the case of direct pesticide treatments into the hive, the contamination derives also from the contact of bees with the treatment medium or from the ingestion of the treatment solution (in the case of liquid medium, *e.g.*, Perizim).

The transfer of a pesticide by contact between the treatment surface and bees was evaluated by the following equation using the specific MTC (MTC_{treat} and $MTC_{bees\text{-}kitin}$). Both of them were considered having very low values (0.0001 m day^{-1}) [29]:

$$D_{contact\text{-}treat} = \frac{A_{treat}}{\left(1/(MTC_{treat}\,Z_{treat})\right)+\left(1/(MTC_{bees\text{-}kitin}\,Z_{bees})\right)} \qquad (8.16)$$

The contact area between bees and the treatment surface (A_{treat}) is roughly approximated to that of the treatment medium, but often this one is placed in between frames where a limited space is available, causing a lot of inadvertent contacts and making bee work more difficult. For this reason, bees are probably induced to reduce the strip interference, for example, by gluing them down to one frame with wax, to make for easier passage. The potential capacity of the treatment medium to release the pesticide by contact is probably constant during the treatment, but its efficacy probably declines over time because of the reduced actual contact between bees and the treatment area. This would also reduce pesticide transfer from the treatment medium to the bee compartment. Mathematically, this behavior was modeled by an exponential relationship of the contact area between the treatment medium and bees versus time (t) ($A_{treat} = 0.012 \times e^{-1.26 \times t}$). This equation was justified by experimental results of bee contamination after treatments with a pesticide acting by contact (*e.g.*, Apistan strips). In this case, experimental concentration showed an initial contamination peak in bees during the first day after treatment followed by a slow decrease even when the treatment lasted for 1 month [39].

Another direct contamination pathway is the ingestion of the treatment solution when the pesticide application is performed by a solution/suspension. In this case, the input flow (mol day^{-1}) was quantified by the product of the ingestion capacity of bees ($G_{ingestion}$ in m^3 day^{-1}) and the pesticide concentration in the treatment solution (C_{treat} in mol m^{-3}). This concentration becomes zero when the treatment ends. The ingestion capacity of bees can be calculated by multiplying the ingestion capacity of each bee (50 mg as in the case of a foraging bee with nectar) by the number of bees taking care of the hive (10,000), assuming that the density of the pesticide suspension is about 1.0 g cm^{-3}. The ingestion time of a single bee is rapid, but all bees cannot ingest the treatment solution simultaneously. Therefore, the overall ingestion time is much longer, on the order of many hours. In the model, 24 h are assumed as the time required to fill the overall ingestion capacity of bees, giving a value of 0.5 dm^3 day^{-1} as the overall ingestion capacity ($G_{ingestion}$).

The contamination can also be caused by the contact of foraging bees with contaminated vegetation (*e.g.*, during pesticide spraying to agricultural crops). This advection flux ($D_{contact\text{-}veg}$) can be modeled by diffusion between the vegetation surface and that of bees:

$$D_{contact\text{-}veg} = \frac{A_{bee}}{\left(1/(MTC_{veg}\,Z_{treat})\right)+\left(1/(MTC_{bees\text{-}kitin}\,Z_{bees})\right)} \qquad (8.17)$$

A crucial parameter is the area of bees in contact with the vegetation. Considering that a single bee comes in contact with the plant surfaces during its foraging activity, Barmaz et al. [20] proposed a representative contact area of 5 cm^2 bee^{-1} for each foraging flight, and supposing five flights in a day, this surface became 25 cm^2 bee^{-1} day^{-1}. Multiplying this value for the number of foraging bees (1/3 of the total), the daily contact area between bees and vegetation was derived (33 m^2). Barmaz et al. [20] suggested, as worst case, that the contamination from the vegetation surface can be calculated multiplying the contact area by the contaminant amount deposited onto the vegetation (mol m^{-2}). In this case, the whole amount present on the vegetation surface is considered to be transferred to bees. Even if bees are protected by a chitin skeleton, which limited the contaminant uptake, they are characterized also by a thick hair cover around the body and by a wide tracheal system, which has a high capacity of gas exchanges. Respiratory surfaces can also efficiently take up gas- and particulate-phase contaminants, so that the contaminant deposited on the plant surfaces can be efficiently transferred to the bee body and absorbed via the respiratory system. Alternatively, the amount of contaminant taken up by bees from the vegetation can be calculated on the basis of a diffusion process by Equation 8.17, in which the transferred amount depends on the two MTC values (MTC_{veg} and $MTC_{bees\text{-}kitin}$) that were set to 0.001 and 0.0001 m day^{-1}, respectively, and in this case, the amount of contaminant taken up by bees is much more limited. Further experimental trials are necessary to clarify this contamination source.

Another input pathway is the advection of a compound inside the nest via food and materials foraged by bees. These advection parameters are defined by the following equations in which the G-values are the flux of materials in m^3 day^{-1}:

$$D_{nectar\text{-}in} = G_{nectar\text{-}in}\, Z_{nectar} \tag{8.18}$$

$$D_{pollen\text{-}in} = G_{pollen\text{-}in}\, Z_{pollen} \tag{8.19}$$

$$D_{water\text{-}in} = G_{water\text{-}in}\, Z_{water} \tag{8.20}$$

$$D_{resin\text{-}in} = G_{resin\text{-}in}\, Z_{propolis} \tag{8.21}$$

Fluxes of nectar, pollen, water, and resin were quantified as 1.33, 0.167, 0.0555, and 0.000555 dm^3 day^{-1}, respectively, and the amount of contaminant entering in the hive with these materials is defined by the product of these D-values by their fugacity ($f_{nectar}, f_{pollen}, f_{water}$, and f_{resin}). These fugacities are derived from the concentrations of the contaminant in each material as shown in Equation 8.14.

The output pathways that are able to reduce the hive contamination (Figure 8.3) are those regarding the air compartment ($D_{air\text{-}out}$) as previously defined, the bee compartment (bee renewal and bee metabolism), and the bee product compartment, in the case of collection of materials by the beekeeper.

When adult bees end their life cycle, they usually die far from the hive or, if they die inside, they are thrown out by worker bees. In this way, bee renewal causes the outflow of contaminants that were contained in the bodies of dead bees.

This advection flux is quantified by $D_{bee\text{-}ren}$ obtained by the daily bee renewal $(G_{bee\text{-}ren})$ of 0.142 dm^3 day^{-1} (each bee has an individual volume of 160 mm^3 and a life cycle of 45 days):

$$D_{bee\text{-}ren} = G_{bee\text{-}ren}\, Z_{bees} \tag{8.22}$$

The loss of contaminants by bee metabolism is quantified by $D_{bee\text{-}met}$, which is based on the biotransformation rate constant in bees $(k_{bee\text{-}met}$ in day$^{-1})$, the volume of bees $(V_{bees}$ in m$^3)$, and the bee capacity $(Z_{bees}$ in mol m^{-3} Pa$^{-1})$:

$$D_{bee\text{-}met} = k_{bee\text{-}met}\, V_{bees}\, Z_{bees} \tag{8.23}$$

Contaminant removal by the material collected by the beekeeper is quantified by the terms: $D_{honey\text{-}out}$, $D_{wax\text{-}out}$, $D_{pollen\text{-}out}$, $D_{rj\text{-}out}$, and $D_{propolis\text{-}out}$ quantified by the corresponding $G_{i\text{-}out}$-values (m^3 day^{-1}) as in Equations 8.18 through 8.21. The type and amount of product withdrawals depend on the rearing practices and on the beekeeper choice; the model allows for evaluation of the contaminant output for the different alternatives in quantity and the type of materials taken out.

8.3.3 QUANTIFICATION OF INTERCOMPARTMENT EXCHANGES WITHIN THE HIVE

Once a pollutant reaches the hive, it can be dispersed by contact if the contamination is on the bee surface or deposited in storage cells if the contamination is within foraged materials. Wax can be contaminated directly by contact or indirectly through the cell contents. Air is in contact with bees and wax and can act as a redistribution medium inside the hive, especially for volatile compounds. Once a pollutant is stored in honey and wax, it is unlikely to degrade because these compartments are microbiologically stable. The pollutant can be recirculated when these products are used by the bees for their nutrition or for that of larvae.

Bees and air act as redistribution media with different efficacy depending on the compound properties: for example, for a low-volatile compound, contained mainly in wax, honey, or pollen, bees are the main redistribution pathway. Bees take up the contaminant when they feed on honey and pollen or when they work on wax, and they give it back when they feed the larvae or when they deposit newly secreted wax.

Considering the three compartments (air, bees, and bee products), all the exchange fluxes among them can be described by a specific D-value, as shown in Figure 8.4. When a given process determines the transfer of a certain amount of a contaminant from one compartment to another, this process determines an output flow for the first (giving) compartment and an input flow for the second (receiving) compartment. For example, D_{vol}, which represents the pesticide volatilization from wax to air, determines an output flow from wax and an input flow to the air compartment.

For the bee compartment, the major input flows within the hive come from the consumption of contaminated honey and pollen stored in the hive ($D_{honey\text{-}cons}$ and $D_{pollen\text{-}cons}$) for their feeding and for the food preparation for the larvae and the queen. In addition, bees inherit a contaminant residue from the larvae from which they are

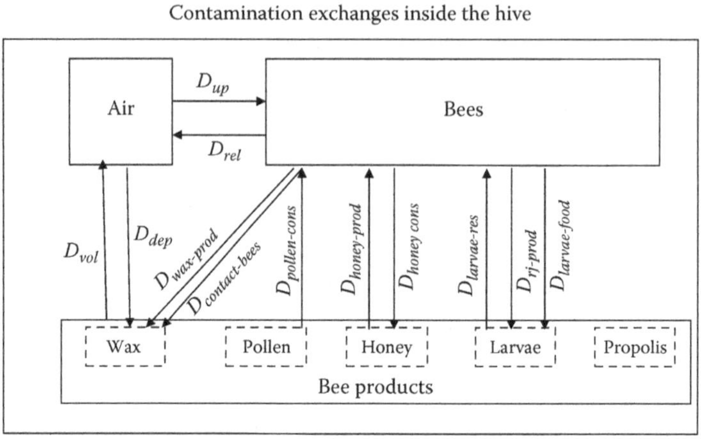

Contamination exchanges inside the hive

FIGURE 8.4 Modeling of contaminant exchanges within the hive ecosystem.

formed ($D_{larvae-res}$) and uptake it from the air through the respiratory surfaces (D_{up}). These D-values were defined as follows:

$$D_{honey-cons} = G_{honey-cons} \, Z_{honey} \tag{8.24}$$

$$D_{pollen-cons} = G_{pollen-cons} \, Z_{pollen} \tag{8.25}$$

$$D_{larvae-res} = G_{larvae-res} \, Z_{larvae} \tag{8.26}$$

$$D_{up} = D_{rel} = \frac{A_{bees}}{\left(1/(MTC_{air} Z_{air})\right) + \left(1/(MTC_{bees-resp} \, Z_{bees})\right)} \tag{8.27}$$

Summer honey and pollen consumption ($G_{honey-cons}$ and $G_{pollen-cons}$) were estimated into 0.286 and 0.04 dm³ day⁻¹, respectively, basing on an adult bee number of 40,000 and on an individual need of 10 and 1 mg bee⁻¹ day⁻¹ of honey and pollen, respectively, normalized by their densities. Mean larvae flux ($G_{larvae-res}$) of 0.0675 dm³ day⁻¹ was calculated based on an individual volume of 90 mm³ larvae⁻¹, the whole larvae number produced in a year (180,000 larvae), and the production period from February to October (240 days).

The absorption and desorption from the respiratory surfaces of bees were calculated considering the respiratory surface of a single bee (9.1 cm² bee⁻¹), the mean bee number in summer (40,000 bees), and the air- and the bee-respiratory-side MTC of 1 and 0.01 m day⁻¹, respectively, according to Mackay [24]. The respiratory surfaces of bees were evaluated indirectly knowing that the oxygen consumption of bees during flight is 5 mL of oxygen per hour, which corresponds to 641 mL of oxygen per kilogram of body weight per minute. It is known that the tracheal system and air sacs of bees are characterized by a high exchange area [33], but quantitatively this datum is not available. In the absence of specific data, the respiratory surface of bees was estimated by a proportion with the exchange surface able to furnish oxygen to man. During intense sporting activity, a man consumes 94 mL of oxygen per kilogram of body weight per minute using an oxygen exchange surface of 50–90 m²; even if this extrapolation is highly approximated, a respiratory surface of 9.1 cm² bee⁻¹ was, proportionally, calculated.

The bee compartment releases the accumulated contaminant by the production of wax, honey, larvae food, and royal jelly ($D_{wax\text{-}prod}$, $D_{honey\text{-}prod}$, $D_{larvae\text{-}food}$, and $D_{r\text{-}j\text{-}prod}$) and by the release of the compound into the air through the respiratory surfaces (D_{rel} as defined in Equation 8.27). The release of contaminants via the respiratory surface happens in the same way that absorption happens, but the net amount exchanged depends on the relative contamination levels of air and bees and thus on their fugacities.

The other contaminant released from bees through their products was quantified as follows:

$$D_{wax\text{-}prod} = G_{wax\text{-}prod} Z_{wax} \tag{8.28}$$

$$D_{honey\text{-}prod} = G_{honey\text{-}prod} Z_{honey} \tag{8.29}$$

$$D_{larvae\text{-}food} = G_{larvae\text{-}food} Z_{larvae\text{-}food} \tag{8.30}$$

$$D_{rj\text{-}prod} = G_{rj\text{-}prod} Z_{rj} \tag{8.31}$$

The amounts of honey, wax, larvae food, and royal jelly produced daily by bees were evaluated as 0.238, 0.000648, 0.178, and 0.00231 dm³ day⁻¹, based on a total honey production of 80 kg year⁻¹, a cap wax production of 0.15 kg year⁻¹, a larvae food production of 25 kg year⁻¹ of honey and 25 kg year⁻¹ of pollen, and a royal jelly production of 0.61 kg year⁻¹. All these year productions were divided by 8 months of activity (240 day year⁻¹).

For the "bee products" compartment, the major input flows come from contact with bees ($D_{contact\text{-}bees}$), deposition of contaminated wax, honey, larvae food, and royal jelly ($D_{wax\text{-}prod}$, $D_{honey\text{-}prod}$, $D_{larvae\text{-}food}$, and $D_{r\text{-}j\text{-}prod}$), and deposition from the air (D_{dep}). The major output flow results from the production of new bees from larvae ($D_{larvae\text{-}res}$), from the consumption of honey and pollen ($D_{honey\text{-}cons}$, $D_{pollen\text{-}cons}$, and $D_{larvae\text{-}food}$), and from the volatilization of the compound to the air compartment (D_{vol}). The majority of these input and output flows were already defined for the bee compartment. Only $D_{contact\text{-}bees}$, D_{vol}, and D_{dep} were not previously defined, and they were quantified as diffusion processes starting from the overall wax surface, which is in contact with air and with bees, and from their relative MTC values:

$$D_{contact\text{-}bees} = \frac{A_{wax\text{-}bees}}{\left(1/(MTC_{bees\text{-}kitin} Z_{bees})\right) + \left(1/(MTC_{wax} Z_{wax})\right)} \tag{8.32}$$

$$D_{dep} = D_{vol} = \frac{A_{wax}}{\left(1/(MTC_{air}\,Z_{air})\right) + \left(1/(MTC_{wax}\,Z_{wax})\right)} \tag{8.33}$$

The wax surface in contact with air (A_{wax}) was geometrically calculated from the honeycomb surface (2.4 m^2), while that in contact with bees ($A_{wax\text{-}bees}$) was reduced to one-third because not all the wax surface is in contact with bees, even if the intense in-hive activities require a continuous wax–bee contact. Bee-chitin-, wax-, and air-side MTCs were set to 0.0001, 0.001, and 1 m day^{-1}, respectively.

The wax surface was approximated as that of the 10 combs (2.4 m^2) rather than that of the wax cells (36 m^2) because most of the cells are capped or filled with food or larvae. Therefore, the surface of wax in contact with air was approximated to the plain surface of combs, as if all the cells were capped and as if this surface was plain. These approximations were considered acceptable.

For the air compartment, the major input flows derive from the compound volatilization from wax (D_{vol}) and from the release from bees to air through the respiratory surfaces (D_{rel}). The major output flow results from the deposition from air to wax (D_{dep}) and from the uptake by the respiratory surfaces of bees (D_{up}).

8.3.4 Non-steady-State Mass Balance of Input and Output Flows in the Hive

The overall balance of input and output flows among the hive compartments, which were reduced to three (bees, "bee products," and air), was described by a system of three non-steady-state mass balance equations (one for each compartment) using fugacity (f, units of Pa) as the unknown time-dependent variable (t=time, units of day). Non-steady-state conditions allow evaluating the evolution of the contamination within the hive after a possible contamination event. The system of ordinary differential equations describing the contamination in the three compartments is as follows:

$$\frac{df_{bee}}{dt} = \Big(D_{contact\text{-}treat}\, f_{treat} + G_{ingestion}\, C_{in} + D_{contact\text{-}veg}\, f_{veg} + D_{nectar\text{-}in}\, f_{nectar} + D_{water\text{-}in}\, f_{water}$$

$$+ D_{honey\text{-}cons}\, f_{prod} + D_{pollen\text{-}cons}\, f_{prod} + D_{larvae\text{-}res}\, f_{prod} + D_{up}\, f_{air} - \big(D_{wax\text{-}prod}$$

$$+ D_{honey\text{-}prod} + D_{larvae\text{-}food} + D_{rj\text{-}prod} + D_{rel} + D_{bee\text{-}ren} + D_{bee\text{-}met} \big)\, f_{bee} \Big) / V_{bee}\, Z_{bee}$$

$$\tag{8.34}$$

$$\frac{df_{prod}}{dt} = \Big(D_{pollen\text{-}in}\, f_{pollen} + D_{resin\text{-}in}\, f_{resin} + D_{contact\text{-}bees}\, f_{bee} + D_{wax\text{-}prod}\, f_{bee}$$

$$+ D_{honey\text{-}prod}\, f_{bee} + D_{larvae\text{-}food}\, f_{bee} + D_{r\text{-}j\text{-}prod}\, f_{bee} + D_{dep}\, f_{air}$$

$$- \big(D_{larvae\text{-}res} + D_{honey\text{-}cons} + D_{pollen\text{-}cons} + D_{vol} \big)\, f_{prod} \Big) / V_{prod}\, Z_{prod} \tag{8.35}$$

$$\frac{df_{air}}{dt} = \frac{\left(D_{air\text{-}in}f_{out} + D_{vol\text{-}treat}f_{treat} + D_{vol}f_{prod} + D_{rel}f_{bees} - \left(D_{air\text{-}out} + D_{dep} + D_{up}\right)f_{air}\right)}{V_{air}Z_{air}}$$

(8.36)

where

f_i is the fugacity with the units of Pa;

D_i represents the various D-values (advection, exchange, and degradation) with the units of mol day^{-1} Pa^{-1};

V_i represents the compartment volumes with the units of m^{-3};

Z_i is the compartment capacities with the units of mol m^{-3} Pa^{-1};

f_{bee}, f_{prod}, and f_{air} are the fugacity of the compound in bees, bee products, and air, respectively; and

f_{treat} is the fugacity of the contaminant in the treatment medium determining a flux to bees by contact and to air by volatilization.

This term is considered for pesticide treatments into the hive acting by contact (*e.g.*, Apistan), while for treatments acting by ingestion (*e.g.*, Perizin), the pesticide input is obtained by multiplying the ingestion capacity of bees ($G_{ingestion}$, units of m^3 day^{-1}) by the pesticide concentration in the treatment solution (C_{in}, units of mol m^{-3}). f_{nectar} and f_{water} in Equation 8.34 and f_{pollen} and f_{resin} in Equation 8.35 are the fugacities of contaminated nectar, water, pollen, and resin, respectively. In Equation 8.36, the term f_{out} is the fugacity of the air outside the hive, if it is contaminated. These fugacity values (Pa) can be calculated by the ratio between the concentration of the compound in each medium (mol m^{-3}) and its capacity (mol m^{-3} Pa^{-1}) as shown in Equation 8.14.

The differential equation system exposed earlier calculates day to day the fugacities in air, bees, and "bee products" as a result of input flows toward the considered compartment (D-values with positive algebraic signs) and output flows from the same compartment (D-values with negative algebraic signs). Output flows can produce an input flow toward another compartment, taking the contaminant inside the hive, or, if they are addressed outside, they constitute a contaminant loss.

The contamination from outside can be derived from many pathways, each represented by a specific D-value multiplied by its fugacity, which represents the intensity of the contamination. The input flow is addressed to the specific compartment that initially receives the contaminant. For example, $D_{nectar\text{-}in}$ and $D_{water\text{-}in}$, which represent the eventual pesticide inputs from nectar and water, are considered as input flows only for the bee compartment, because nectar and water are effectively ingested by bees and not simply carried in external structures such as pollen baskets. These external structures are used by bees to carry pollen and resin, and the contaminants present in these materials are deposited directly into the "bee products" compartment. For this reason, $D_{pollen\text{-}in}$ and $D_{resin\text{-}in}$ are considered as input flows for the "bee products" compartment.

8.3.5   APPLICATION OF THE "HIVE MODEL": A CASE STUDY
WITH THE INSECTICIDE/ACARICIDE τ-FLUVALINATE

The equation system reported earlier was applied to a specific contamination trial [29,39]. The pesticide used was τ-fluvalinate, a pyrethroid insecticide/acaricide commonly used against the bee pest *Varroa destructor.*

Following common treatment methodologies, the pesticide was administered by two plastic strips hung between combs for 1 month. From the strips, the compound passes to adult bees principally by contact, and from them the pesticide is able to reach the other hive compartments, including larvae too, in order to protect both adult bees and larvae. If bees are contaminated with fluvalinate, they tend to produce contaminated products or to contaminate products by contact during handling. Once fluvalinate residues are transferred to bee products (wax, honey, pollen, larvae, and propolis), the products themselves become secondary sources of contamination for bees. All of these exchanges can be mathematically expressed by the differential system of Equations 8.34 through 8.36, and using the physical–chemical properties and treatment conditions of τ-fluvalinate, this equation system was able to quantify the pesticide fugacity in bees, "bee products," and air, and from them, the concentrations in the different hive compartments as a function of time. Figure 8.5

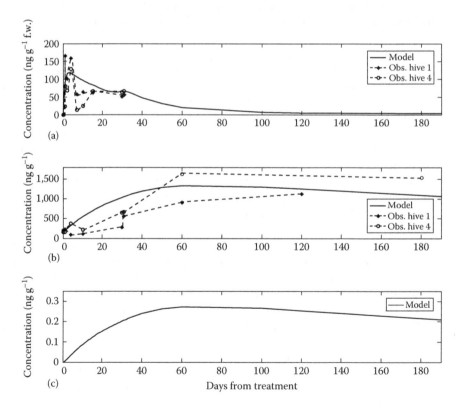

FIGURE 8.5 Comparison of predicted and measured concentrations of τ-fluvalinate in (a) bees, (b) wax, and (c) honey.

reports fluvalinate concentrations in bee wax and honey predicted by the model (continuous lines), starting just before the treatment (Time 0) up to 180 days after it. Experimental data measured in the two experimental hives (hives 1 and 4) were also reported (dotted lines).

Measured concentrations in experimental hives gave consistent results between them, and the two data sets drew quite clear time trends. Bees were not contaminated before treatment; thereafter, their contamination level rose to a peak and later decreased to a nearly constant value (50 ng g^{-1} f.w.) until the end of the treatment (30 days). In contrast, measured concentrations in wax gradually increased after treatment, reaching a plateau slightly above 1,000 ng g^{-1} at 60 days after treatment. Measured contamination levels in honey were always below the limit of quantification of 2.5 ng g^{-1}. Predicted concentrations closely follow these trends, showing predicted levels near or intermediate between measured ones. Predicted concentrations for honey were always below the limit of quantification, according to measurements.

The model predicts a rapid increase in concentration in bees just after treatment, as suggested by the measured data, which shows an initial high variability. This behavior can be explained considering that only a fraction of the bees working in the nest can simultaneously visit the treatment strips. Later, the general contamination of all hive compartments (secondary sources) and the reduced visitation of bees to the treatment strips (primary source) tend to make the concentrations in bees more uniform (as shown by measurements). Predicted data show a concentration peak in bees during the first few days of treatment and then slowly decrease because of the reduced contact between strips and bees. After 30 days, when the treatment strips are removed, concentrations in bees tend to decrease more rapidly because the primary pesticide source (the strips) is no longer present and secondary sources (contaminated hive products) remain. Predicted concentrations in wax and honey follow a slowly rising trend until about 60 days, according to the measured data. After removal of the strips, the hive products still receive the pesticide through contact with bees until their fugacity becomes equal to that of bees. Thereafter, concentrations in the hive products begin to slowly decline because, in the absence of new pesticide input, the various loss processes begin to reduce the overall contamination. Concentrations decrease more rapidly in bees than in bee products (wax and honey) because of the persistence of this pesticide in these compartments. Biotransformation and advection out of the hive by dead bees are the primary processes by which the pesticide is lost from the hive system. The slow decline of the contamination in wax is consistent with the observation of the wax contamination before the treatment, presumably derived from previous treatments at least 1 year before.

8.4   CONCLUSIONS AND FUTURE PERSPECTIVES

The presented "hive model" considers different contamination pathways: from outside, such as the contamination coming from nectar, pollen, resin, water, air, and vegetation, or from inside via pesticide treatments against bee pests. It includes almost all the contamination threats toward bees, but, at the moment, it has been applied successfully to only a single case. However, as any initial scientific effort, it can be very useful in indicating the most urgent research need. Analyzing the input data of the model, it uses the physical–chemical properties of the compound and the major characteristics of the hive ecosystem, from which it calculates contamination residues in bee products (honey, pollen, royal jelly, wax, and propolis) over time. While the physical–chemical properties of a compound are quite well known, the hive characteristics, important for contamination modeling, are much less known. Even among some well-known hive parameters, there is a certain level of uncertainty. Of course, the food consumption of adult bees depends on food availability, bee age, and primary activities: bees that are preparing royal jelly require more pollen than those that are producing new wax. However, for modeling purposes referring to an average value of food intake is sufficient, and this average value may represent the material fluxes into the hive with a sufficient detail. For this reason, the hive parameters proposed in this chapter can be considered to be sufficiently supported by the present knowledge, even if it is clear that they may be improved and corrected. On the other hand, many more uncertainties exist in the evaluation of the contaminant behavior inside the hive. For example, the compartment capacities (Z-values) and the exchange parameters (D-values) described in this chapter have, at the moment, only been proposed and indirectly validated with a single case study. It is clear that much more experimental work is needed for better describing them and for supporting these parameters with experimental data. For example, the capacity of chemicals to cross the chitin barrier or the diffusion velocity in wax is unknown. All these values were quantified by a specific MTC parameter defined by similarity from others already proposed. Specific experiments of volatilization from wax, honey, and pollen or bee contamination from air should be performed with more experimental data and evidence to quantify these exchange parameters. A first step could be the experimental evaluation of the wax–honey partitioning, starting from contaminated honey or wax and lasting through the chemical diffusion process until equilibrium is reached.

In the future, more experimental works could produce more experimentally based parameters, which can improve the present attempt of modeling the contamination of the hive ecosystem. A more validated "hive model" can be used in two ways: as a prognostic tool for assessing the possible contamination levels of bee products for voluntary contamination events (pesticide treatment into the hive) and as a diagnostic tool for understanding the contaminant fate of involuntary contamination events, such as those coming from outside. The "hive model" can be applied in order to test the contamination levels of different pesticides depending on their application modalities or to give unsuspected indications of severe contamination threat from outside.

REFERENCES

1. B. Hileman, *Why are the bees dying?* Chem. Eng. News 85 (2007), pp. 56–61.
2. R.M. Johnson, M.D. Ellis, C.A. Mullin, and M. Frazier, *Pesticides and honey bee toxicity—USA*, Apidologie 41 (2010), pp. 312–331.
3. D. vanEngelsdorp, R. Underwood, D. Caron, and J. Hayes, *An estimate of managed colony losses in the winter of 2006–2007: A report commissioned by the Apiary Inspectors of America*, Am. Bee J. 147 (2007), pp. 599–603.

4. B. Laszlo, *About the uncommon bee losses. Literature review*, Magy. Allatorvosok 130 (2008), pp. 551–557.
5. M. Ribiere, V. Olivier, P. Blanchard, F. Schurr, O. Celle, P. Drajnudel, J.P. Faucon, R. Thiery, and M.P. Chauzat, *The collapse of bee colonies: The CCD case ('Colony collapse disorder') and the IAPV virus (Israeli acute paralysis virus)*, Virologie 12 (2008), pp. 319–322.
6. M. Higes, R. Martín-Hernández, E. Garrido-Bailón, A.V. González-Porto, P. García-Palencia, A. Meana, M.J. del Nozal, R. Mayo, and J.L. Bernal, *Honeybee colony collapse due to* Nosema ceranae *in professional apiaries*, Environ. Microbiol. Rep. 1 (2009), pp. 110–113.
7. Y. Le Conte and M. Navajas, *Climate change: Impact on honey bee population and diseases*, Rev. Sci. Tech.-OIE 27 (2008), pp. 499–510.
8. A. Decourtye, J. Devillers, S. Cluzeau, M. Charreton, and M. Pham-Delègue, *Effects of imidacloprid and deltamethrin on associative earning in honeybees under semi-field and laboratory conditions*, Ecotox. Environ. Safe. 57 (2004), pp. 410–419.
9. R.M. Johnson, H.S. Pollock, and M.R. Berenbaum, *Synergistic interactions between in-hive miticides in* Apis mellifera, J. Econ. Entomol. 102 (2009), pp. 474–479.
10. S. Buckingham, B. Lapied, H. Corronc, and F. Sattelle, *Imidacloprid actions on insect neuronal acetylcholine receptors*, J. Exp. Biol. 200 (1997), pp. 2685–2692.
11. T. Iwasa, N. Motoyama, J.T. Ambrose, and R.M. Roe, *Mechanism for the differential toxicity of neonicotinoid insecticides in the honey bee,* Apis mellifera, Crop. Prot. 23 (2004), pp. 371–378.
12. M. Greatti, A.G. Sabatini, R. Barbattini, S. Rossi, and A. Stravisi, *Risk of environmental contamination by the active ingredient imidacloprid used for corn seed dressing. Preliminary results*, Bull. Insectol. 56 (2003), pp. 69–72.
13. M. Greatti, R. Barbattini, A. Stravisi, A.G. Sabatini, and S. Rossi, *Presence of the a.i. imidacloprid on vegetation near corn fields sown with Gaucho® dressed seeds*, Bull. Insectol. 59 (2006), pp. 99–103.
14. P. Tremolada, M. Mazzoleni, F. Saliu, M. Colombo, and M. Vighi, *Field trial for evaluating the effects on honeybees of corn sown using Cruiser® and Celest xl® treated seeds*, Bull. Environ. Contam. Toxicol. 85 (2010), pp. 229–234.
15. M.-P. Halm, A. Rortais, G. Arnold, J.N. Taséi, and S. Rault, *New risk assessment approach for systemic insecticides: The case of honey bees and imidacloprid (Gaucho)*, Environ. Sci. Technol. 40 (2006), pp. 2448–2454.
16. K. Wallner, *Varroacides and their residues in bee products*, Apidologie, 30, (1999) pp. 235–248.
17. S. Bogdanov, *Contaminants of bee products*, Apidologie 37 (2006), pp.1–18.
18. S. Barmaz, S.G. Potts, and M. Vighi, *A novel method for assessing risks to pollinators from plant protection products using honeybees as a model species*, Ecotoxicology 19 (2010), pp. 1347–1359.
19. S. Villa, M. Vighi, A. Finizio, and G. Bolchi Serini, *Risk assessment for honeybees from pesticide-exposed pollen*, Ecotoxicology 9 (2000), pp. 287–297.
20. S. Barmaz, *Plant protection product risk assessment: Distribution and experimental validation in terrestrial ecosystems*, PhD dissertation, University of Milano-Bicocca, Milan, Italy, 2009.
21. C.E. Cowan, D. Mackay, T.C.J. Feijtel, D. Van de Meent, A. Di Guardo, and Davies, *The Multimedia Fate Model: A Vital Tool for Predicting the Fate of Chemicals*, SETAC Press, Pensacola, FL, 1995.
22. P. Tremolada, M. Sugni, G. Gilioli, A. Barbaglio, F. Bonasoro, and M.D. Candia Carnevali, *A dynamic model for predicting chemical concentrations in water and biota during the planning phase of aquatic ecotoxicological tests*, Chemosphere 75 (2009), pp. 915–923.

23. D. Mackay and S. Paterson, *Finding fugacity feasible*, Environ. Sci. Technol. 15 (1979), pp. 1218–1223.

24. D. Mackay, *Multimedia Environmental Model, The Fugacity Model*, 2nd ed., Lewis Publishers, Boca Raton, FL, 2001.

25. F. Wania and D. Mackay, *A global distribution model for persistent organic chemicals*, Sci. Total. Environ. 160/161 (1995), pp. 211–232.

26. D. Mackay, S. Paterson, and W.Y. Shiu, *Generic models for evaluating the regional fate of chemicals*, Chemosphere 24 (1992), pp. 695–717.

27. A. Di Guardo, D. Calamari, G. Zanin, A. Consalter, and D. Mackay, *A fugacity model of pesticide runoff to surface water: Development and validation*, Chemosphere 28 (1994), pp. 511–531.

28. P. Tremolada, I. Bernardinelli, M. Colombo, M. Spreafico, and M. Vighi, *Coumaphos distribution in the hive ecosystem: Case study for modeling applications*, Ecotoxicology 13 (2004), pp. 589–601.

29. P. Tremolada, I. Bernardinelli, B. Rossaro, M. Colombo, and M. Vighi, *Predicting pesticide fate in the hive (part II): Development of a dynamic hive model*, Apidologie 42 (2011), pp. 439–456.

30. T.D. Seeley, *Honeybee Ecology. A Study of Adaptation in Social Life*, D. Thomas Publisher, Princeton University Press, Princeton, NJ, 1985.

31. R. Chauvin, *Traité de Biologie de l'Abeille*, Masson et C, Paris, France, 1968.

32. R.A. Grout, *The Hive and the Honey Bee*, Dadant & Sons, Hamilton, IL, 1973.

33. R.E. Snodgrass, *Anatomy of the Honey Bee*, Cornell University Press, London, UK, 1984.

34. A.I. Root, *The ABC and XYZ of Bee Culture*, A.I. Root Company, Medina, OH, 1990.

35. L.J. Goodman and R.C. Fisher, *The Behaviour and Physiology of Bees*, C.A.B. International, Wallingford, UK, 1991.

36. E. Crane, *Honey. A Comprehensive Survey*, William Heinemann Ltd, London, UK, 1976.

37. M.C. Marcucci, *Propolis: Chemical composition, biological properties and therapeutic activity*, Apidologie 26 (1995), pp. 83–99.

38. G. Lercker, M.F. Caboni, M.A. Vecchi, A.G. Sabatini, and A. Nanetti *Caratterizzazione dei principali costituenti della gelatina reale*, Apicoltura 8 (1992), pp. 11–21.

39. S. Bonzini, P. Tremolada, I. Bernardinelli, M. Colombo, and M. Vighi, *Predicting pesticide fate in the hive (part 1): Experimentally determined τ-fluvalinate residues in bees, honey and wax*, Apidologie 42 (2011), pp. 378–390.

9 Agent-Based Modeling of the Long-Term Effects of Pyriproxyfen on Honey Bee Population

James Devillers, Hugo Devillers,
Axel Decourtye, Julie Fourrier,
Pierrick Aupinel, and Dominique Fortini

CONTENTS

ABSTRACT

Honey bees are beneficial insects playing a key role in pollinating wild plants and crop plants. Unfortunately, during their foraging activity they can be exposed to pesticides and other xenobiotics. The members of the colony can also be poisoned indirectly by contaminated food and water brought back to the hive by foragers.

While the assessment of the potential adverse effects of a xenobiotic on individual bees can be easily done by using normalized laboratory tests, toxicity estimation at the population level suffers from technical difficulties and is costly. Fortunately, population modeling is increasingly used to overcome this problem of transition from the individual to the population level. In this context, an agent-based model, called SimBeePop, was designed for predicting the long-term effects of man-made chemicals on honey bee populations. After a detailed description of the structure of the model, its performances were estimated on pyriproxyfen, a juvenile hormone mimic. To do so, the physicochemical properties, environmental fate, use, and toxicity of this insecticide to honey bees were reviewed in order to elaborate realistic scenarios of contamination. Application of these scenarios in SimBeePop reveal that while pyriproxyfen is classified as moderately toxic or not toxic to bees under current regulations, it can induce significant adverse effects on bee populations at sublethal concentrations.

KEYWORDS

Apis mellifera, Agent-based model, Population dynamics, Pyriproxyfen, Endocrine disruptor

9.1   INTRODUCTION

A challenge in ecotoxicology is the translation of individual-level effects of exposure to chemicals, as measured, for instance, in laboratory tests, into population-level effects that are more ecologically relevant. Yet, such population risk assessments are seldom conducted, due primarily to the technical difficulties they entail. Broadly speaking, the more complex the life-history trait of the studied organism, the more difficult, time consuming, and costly the risk assessment of its population. Fortunately, population modeling is increasingly being considered to overcome this problem of transition from the individual to the population level. Population modeling can easily integrate birth, development stages, reproduction, death, immigration, emigration, and coaction to predict the future behavior of a population or to detect short- or long-term adverse effects on population structure or productivity. Because there exist *in silico* simulation tools, these models can be used to test various hypotheses and scenarios in order to predict the outcome of some effects or to identify the most relevant variables [1–6]. The literature is full of reviews and case studies stressing the strengths and weaknesses of the various types of population models that have found applications in ecotoxicology [7–11]. These models can include more or fewer biological details related to population structure, physiological and toxicological responses, spatiality, temporality, stochasticity, and so on. The choice of a model paradigm is problem dependent. It basically depends on the accuracy needed for the question asked. Each type of model serves to answer specific questions while having different data requirements and assumptions to reach a sufficient level of realism for practical use.

Thus, to study the endocrine disruption potential of xenobiotics on honey bee populations, we decided to use an agent-based model (ABM) [12] to account for the multifaceted aspects that endocrine disruption phenomena might take [13] on the different inhabitants of a hive.

Domestic bees (*Apis mellifera*) are less healthy and less abundant than they have been in the past. If this trend continues, it could have serious implications, since most plants rely on bees for their pollination. In the United States, the honey bee is responsible for pollinating $15 billion worth of agricultural crops each year. The California almond crop alone uses 1.3 million colonies of bees for pollination, approximately one-half of all the honey bees in the United States [14,15]. The causes of decline are multiple and include environmental and climatic changes, the action of pesticides, pathogens, and parasites, intensive migratory beekeeping, and maybe the effects of radiations [16–21]. The direct and indirect effects of endocrine-disrupting chemicals also cannot be neglected because it has been shown that xenobiotics could affect the endocrine system of insects [22] and crustaceans [23–25]. Because hormonal signaling regulates numerous processes in honey bees such as development, metamorphosis, reproduction [26,27], polyethism [28,29], and aggressiveness [30], we assumed that an ABM was the most suited modeling approach to encode these phenomena expressed at different *spatiotemporal* levels within a hive as well as their potential perturbations induced by xenobiotics. Indeed, conventional top-down modeling approaches, such as differential equations, approximate the average behavior of a system as a whole. They generally fail to correctly capture the different local relationships between the system components that can be the cause of high-level behaviors. On the other hand, ABMs are bottom-up modeling approaches that aggregate the different level components of the studied system to derive more realistic models [31,32]. Briefly, an ABM is constituted of a collection of autonomous decision-making entities called agents located within a defined environment. Each agent individually assesses its situation and makes decisions on the basis of a set of rules. Relationships between agents are specified, linking agents to other agents and/or components of the environment. These relationships can be simply reactive, the agents only performing tasks when triggered to do so by some external stimuli or, on the contrary, goal directed, the agents always having specific tasks to achieve. It is noteworthy that agents may be capable of evolving, allowing the emergence of unanticipated behaviors [12,33]. ABMs have been proposed to simulate the different activities of bees within and/or outside their hive [34–43].

In this context, this chapter presents an ABM that can be used to simulate the endocrine disruption potential of xenobiotics on bee populations. After a synthesized description of the main characteristics of the model, its potentialities in ecotoxicology modeling will be discussed from simulation results obtained with pyriproxyfen, 2-[1-methyl-2-(4-phenoxyphenoxy)ethoxy]pyridine (CAS RN 95737-68-1), an aromatic, nonterpenoid juvenile hormone agonist introduced by Sumitomo Chemical Co. Ltd in the 1990s [44]. Our motivation for selecting this insecticide was at least twofold. As it is mainly used in agriculture and in vector control, it is a potential contaminant for both terrestrial and aquatic ecosystems. As a result, it was expected to find enough environmental concentration data to run the model under realistic hypotheses of contaminations. Even more important, while pyriproxyfen is a juvenoid, on the basis of its oral and contact LD_{50} (lethal dose 50%) values, it is currently classified as moderately or not toxic to honey bees [45–47]. Pyriproxyfen was also found moderately toxic against the leafcutting bee (*Megachile rotundata*, Megachilidae), which is a solitary nesting bee mainly foraging on alfalfa plants (*Medicago sativa*), and against the alkali bee (*Nomia melanderi*, Halictidae), which is a solitary soil-dwelling bee [47]. Regarding *Bombus terrestris*, toxicity tests using micro-colonies showed that pyriproxyfen (Admiral, 10% EC) did not cause acute/chronic worker mortality by oral/contact exposure. Similarly, no effect on reproduction was reported when *B. terrestris* workers were exposed for 11 weeks to the maximum field-recommended concentration of pyriproxyfen (25 mg a.i. L^{-1}). In contrast, pollen exposure to pyriproxyfen (25 mg L^{-1}) resulted in a significantly higher number of removed third- and fourth-instar larvae, implying a lethal blockage of the development before metamorphosis [48,49].

Consequently, it was of primary interest to study the potential adverse effects of sublethal concentrations of pyriproxyfen on the main life-history traits of the honey bee to see whether this allocation has to be changed or not. To do so, an extensive bibliographical investigation was performed to compile all the data on the ecotoxicological effects of pyriproxyfen on honey bees. It was completed by original ecotoxicological data obtained in the frame of this study.

9.2 AGENT-BASED MODEL

9.2.1 MODEL DESCRIPTION AND PARAMETER SETTING

The process of designing an ABM starts with its conceptualization. Basic questions and goals, the elements of the system represented by the model environment and the different types of agents with their rules, and the measurable outcomes of interest have to be identified. It is crucial to establish whether the necessary simplification that has always to be made during the design process of a model will not affect the credibility and likelihood that the model provides realistic results [50,51]. The next step deals with data acquisition that is very often not a straightforward process due to the high number of parameter values to be set for designing most of the ABMs. This is particularly true for a population dynamics ABM for a honey bee colony. The difficulty relies on the scarcity of the biological data that are mostly obtained on different subspecies, under various environmental conditions, and very often without repetitions. Because it is never useful to reinvent the wheel, we always have to take advantage of the reflections, strategies, and tricks used by others who have made interesting population models on bees or models focused on a specific point of the bee life history. Thus, different models have been dissected with precision [42,52–55]. Among them, the most exploited was HoPoMo [54], which was first recomputed in R and its bugs corrected in order to be used as a benchmark simulator for testing the behavior of some parameters.

Our model, called SimBeeBop (version 1.04), was written in Java. It runs from 145 parameters including 113 parameters that characterize the structure and the properties of the agents and the environment, 27 parameters that allow modulations of different agent properties, such as the application of "perturbations," and 5 parameters for optimizing memory allocation. On the basis of a large literature review and the use of trial-and-error procedures for data adjustments, default values were proposed. Obviously, they can be changed by the users.

To be closer to reality, SimBeeBop considers a Dadant hive as the environment and all the inhabitants of the hive as agents. The Dadant hive, which is the most used in Europe, includes 10 frames (420×270 mm). The hive body is surmounted by one or more supers for storing honey and each comprises nine frames (420×135 mm). The frames in the hive are covered on both sides with hexagonal cells of different sizes and functions. The numbers of cells in the hive body and hive super were equal to 80,000 and 36,000, respectively [56]. Obviously, these figures can be changed. The hive super is only filled with honey. The hive body includes honey, pollen, and the brood. The spawning activity of the queen shows large fluctuations [57,58]. Such variations are partly due to the brood cells, which are not rapidly available or not cleaned enough. The abiotic factors, pollen availability, and the age of the queens are other influential factors. The egg-laying dynamics of the queen is rarely taken into account in the models inducing a significant bias. Very often, when the queen is taken into account, a laying rate of 1 egg min^{-1} is assigned [41,52]. Schmickl and Crailsheim [54], in HoPoMo, accounted for a variation in the egg-laying rate, but their model being parameterized along 1 year, they did not take into account the decrease in performance with age. Conversely, they integrated an individual variability through a factor of stochasticity. In SimBeePop, attempts were made to simulate at best the reality by accounting for the main factors influencing the egg-laying rate of the queen (Figure 9.1) during 2 years or more. Notice that the maximum queen's laying rate was equal to 2,000 eggs day^{-1} during the summer time [54]. The queen progressively starts to lay at $D+30$ and reaches a potential optimal laying activity from $D+60$ to $D+150$. Then, a decrease in laying activity is observed until $D+320$, when the queen stops to lay eggs. This laying cycle is common under our latitudes [58].

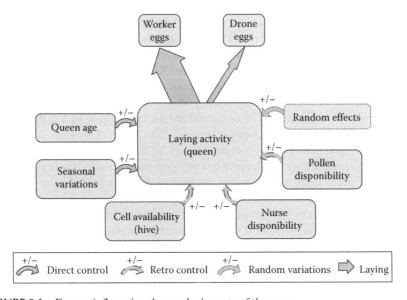

FIGURE 9.1 Factors influencing the egg-laying rate of the queen.

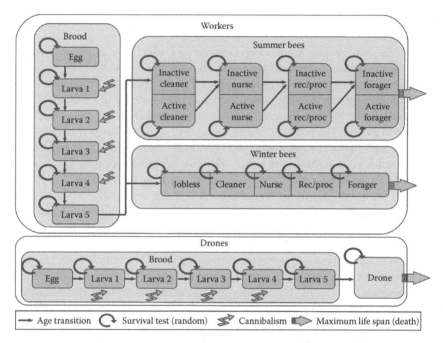

FIGURE 9.2   Different hive inhabitants.

It is noteworthy that it is not only the queen that is responsible for the amount of brood produced. During certain periods and critical situations in food resources, the workers are also critical to regulate brood rearing [58]. Indeed, cannibalism is a common mechanism of regulation observed in the first larval stages [59,60]. To our knowledge, only Schmickl and Crailsheim [54], in HoPoMo, accounted for the cannibalism phenomenon in the larvae except during their pupal stage. The same strategy was used in SimBeePop. The five larval stages (L1–L5) for the worker bees and the drones (Figure 9.2) were characterized from Rembold et al. [61], Michelette and Soares [62], and Lipinski et al. [63]. In colonies of *Apis mellifera*, drones typically compose 5%–10% of the adult population. However, the colony adjusts its investment in drones in accordance with biotic and abiotic factors [64]. In SimBeePop, 5% of the eggs laid by the queen are future drones.

Honey bee division of labor is characterized by temporal polyethism, in which young workers remain within the hive and perform various tasks whereas old workers perform more risky tasks outside the hive as foragers [65]. In the literature, there exists a considerable variability in the ages at which tasks are performed by workers. There also exists discrepancy in the number of different tasks performed by the bees within the hive [66]. From a modeling point of view, it is not necessary to consider all the types of tasks that could be observed in a bee population. Conversely, it is necessary to introduce a high plasticity between the selected castes. Thus, in SimBeePop, four castes were computed, namely, the cleaning bees, the nursing bees, the receiving/processing bees, and the foraging bees, with the age

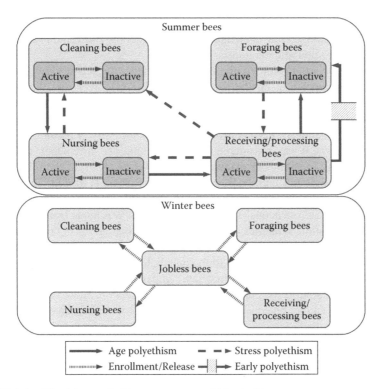

FIGURE 9.3   Plasticity applied between castes within the hive.

limits of 3, 12, and 21 days, respectively [67,68]. The plasticity applied between the castes in case of stresses appearing within and/or outside the hive is displayed in Figure 9.3.

To correctly estimate the long-term effects of xenobiotics on honey bees, especially when they act at sublethal concentrations on different bee targets, it is necessary to consider their population dynamics over 1 year. To do so, the winter bee population has to be included in the modeling process. Depending on external and internal factors, winter bees can start to appear in July, most frequently in August, but they mainly hatch in September [58,69]. The life duration expectancy for a winter bee varies from 150 to 200 days with a maximum of 243 days [70,71]. In SimBeePop, the queen starts to lay winter bee eggs in mid-August, and hence, the first winter bees emerge at the beginning of September (D + 250). Their maximum life duration is of 180 days. This means that it is possible to find winter bees in the hive till the first few days of May. This corresponds to reality [71]. As shown in Figure 9.3, the activity of winter bees is rather ubiquitous. Wintering bees play a key role at the beginning of the season to restart the colony development. To secure an optimal development of the colony during the active season, at least a population of 10,000 wintering bees is necessary at the beginning of the year [58]. This is the threshold selected by Pratt [53] as a starting value in his dynamic programming model, followed by a mid-season peak of 30,000, and back to 10,000 by the fall.

According to Imdorf and coworkers [58], a minimum population of 5,000–8,000 bees has to be guaranteed when wintering. In SimBeePop, a threshold value of 4,000 was selected for population collapse.

Foragers supply the colony with nectar and pollen collected at varying distances from the hive depending on the availability of the resources and the season. In SimBeePop, there are between 1% and 33% of the foragers that collect the pollen. This range depends on the needs of the hive. In fact, the recruitment of pollen foragers takes into account the needs of the hive and the necessity of maintaining a stock for holding 6 days. It was considered that nectar contained 40% sugar, which is the percentage found in sunflower plants [72]. The food needs of all the inhabitants of the hive were estimated mainly from the data of Rortais et al. [73] supplemented by some others coming from various studies [74–78]. Cosmetic adjustments were made due to the heterogeneous origin of the data. This was made from the use of the invaluable population dynamics data and corresponding curves obtained by Imdorf and coworkers [58] for broods and adults of different colonies. It is noteworthy that while water is absolutely necessary to the metabolism of the honey bee, to maintain the necessary humidity and an appropriate temperature within the hive [68,79,80], it was not considered in the current version of SimBeePop. However, this does not mean that water cannot be considered as a source of contamination even if it cannot be directly quantified.

9.2.2 SENSITIVITY ANALYSIS

Sensitivity analysis relies on the quantification of fluctuations of the outputs of the model when applying variations on the input parameter values. It allows assessment of the parameter's ability to influence the output of a model. The main goal of a sensitivity analysis is the identification of the most influential parameters of a system, as well as the assessment of how the response of a model depends on its input parameters. When designing an ABM, it is crucial to perform such an analysis to test the robustness, to possibly refine the parameter setting, and to better understand relationships that arise from the interactions between the agents. Due to space limitation, only some examples are given here.

Stochasticity is an invaluable asset of ABMs to provide realistic simulations. Generally, it consists in applying random variations on a given property of constitutive elements of the model (*i.e.*, an agent, a group of agents, the environment). In SimBeePop, the main factor of stochasticity deals with the laying activity of the queen. Briefly, at each step (day) a number of eggs ($N_{egg}(t)$) to lay is determined according to various factors (Figure 9.1). A random variation is then applied to the value of $N_{egg}(t)$. It is governed by a single parameter that indicates the amplitude of variations. Figure 9.4 provides model outputs for 2-year simulations considering different levels of stochasticity, from no stochasticity in Figure 9.4a to maximum stochasticity in Figure 9.4f (amplitude = 100% of $N_{egg}(t)$). Stochasticity catalyzes the emergence of phenomena from agent interactions. Too little (or no) stochasticity leads to a deterministic model (Figure 9.4a and b). Too much stochasticity leads to chaotic outputs (Figure 9.4e and f). In SimBeePop, a mid-value of stochasticity was selected corresponding to Figure 9.4c.

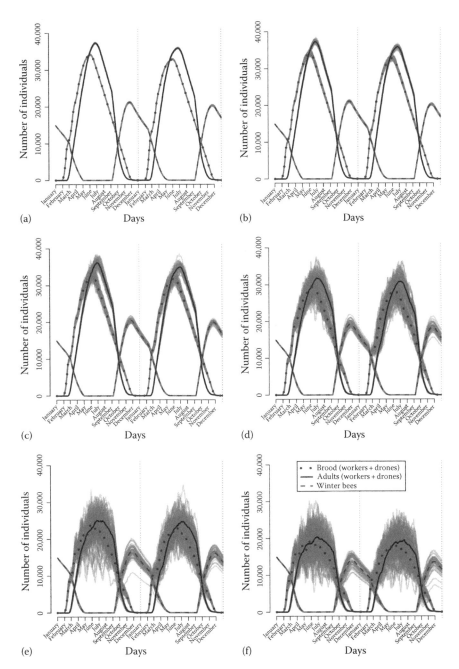

FIGURE 9.4 Sensitivity analysis (50 simulation trials) of the stochasticity applied to the number of eggs ($N_{egg}(t)$) laid by the queen: (a) no stochasticity, (b) 12.5% of $N_{egg}(t)$, (c) 25% of $N_{egg}(t)$, (d) 50% of $N_{egg}(t)$, (e) 75% of $N_{egg}(t)$, and (f) 100% of $N_{egg}(t)$.

Nursing requirement corresponds to the number of nurse bees that are necessary to feed the brood. Reference values for this parameter were proposed by Schmickl and Crailsheim [54] in their HoPoMo model, for the different larval stages. Figure 9.5 shows the outputs obtained for different values of nursing requirement, including the values from HoPoMo (Figure 9.5c), from no nursing requirement (Figure 9.5a) to a high nursing requirement (Figure 9.5f). When nursing demand is neutralized (Figure 9.5a), the feedback of nursing quality on egg laying and on cannibalism is null, explaining the different shapes of the curves obtained for the population dynamics. Increasing the demand in nurses activates polyethism interactions described in Figure 9.3, thus increasing cannibalism activity and lowering egg laying if necessary. If the nurse requirement is too high, bee population dynamics is drastically perturbed (Figure 9.5d through f), up to the collapse of the colony. In SimBeePop, values proposed by HoPoMo were kept as standard values (Figure 9.5c).

Cannibalism plays a crucial role in bee population dynamics, as it is a response to various stresses (*e.g.*, lack of nurses, lack of pollen). To our knowledge, HoPoMo [54] was the first model that includes cannibalism in the modeling process of a bee population. This aspect was characterized by parameters denoting the maximum proportions of larval population that can be eaten in case of maximum stress. In SimBeePop, cannibalism was implemented in a similar way. Different levels of cannibalism were investigated, including values from HoPoMo. Results are presented in Figure 9.6. When cannibalism is turned off (Figure 9.6a) or low (Figure 9.6b), output uncertainty seems to be higher than for mid-range cannibalism (Figure 9.6c and d). This is probably because, with a low level of cannibalism, stress response is less efficient. On the contrary, too high levels of cannibalism lead to important perturbations in population dynamics and possibly to its extinction (Figure 9.6f). In SimBeePop, the values from Figure 9.6c were selected as standard values for cannibalism.

The aforementioned three examples illustrate how sensitivity analyses helped us validate as well as find suitable values for the different parameters of SimBeePop and better understand the relationships between inputs and outputs.

9.3 PHYSICOCHEMICAL PROPERTIES AND ENVIRONMENTAL FATE OF PYRIPROXYFEN

At 25°C and pH=6, the solubility in water of pyriproxyfen (99.4%) is only equal to 0.367 ± 0.004 mg L^{-1}, while the value of its 1-octanol/water partition coefficient (log K_{ow}) is 5.37 at 25°C and pH=5.6. With a vapor pressure $<1.33 \times 10^{-5}$ Pa at 22.8°C, technical pyriproxyfen (100%) shows a low affinity for the air compartment [81]. Pyriproxyfen presents a strong propensity to adsorb onto sediments and soils. Adsorption coefficient (K_{oc}) values of 4,980, 11,000, 11,600, 12,600, 26,900, and 34,200 were found in lake sediment, clay loam, sand, sandy loam, silt loam, and silty clay loam soils, respectively [82]. Hydrolysis half-lives are ranged from about 148 to 605 days at pH=5, 241–1,293 days at pH=7, and 161–511 days at pH=9 [79]. Radiolabeled pyriproxyfen subjected to artificial sunlight, approximately equivalent to double the light intensity of natural midday sunlight at 43° N in July, in aqueous buffer solution at pH=7 and 25°C for 30 days, led to a rapid photodegradation with half-lives of 3.7 days for the [Pyr-^{14}C]-label and 6.4 days for the [Phe-^{14}C]-label

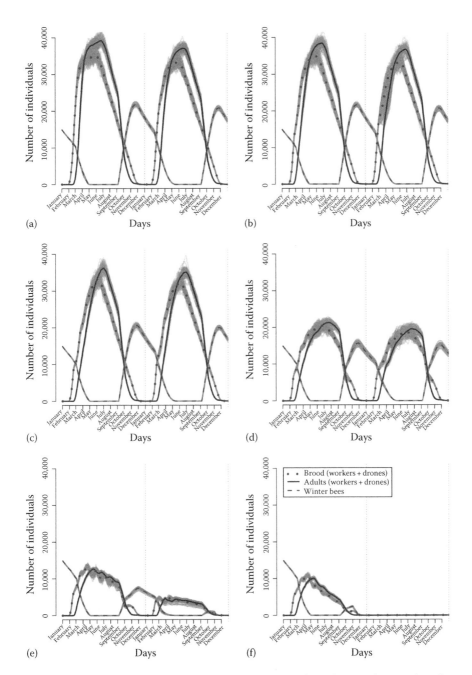

FIGURE 9.5 Sensitivity analysis (50 simulation trials) of nursing requirement based on values of reference from HoPoMo [54]: (a) no nurse is required, (b) half values of reference, (c) values of reference, (d) 1.5 times the values of reference, (e) twice the values of reference, and (f) 2.5 times the values of reference.

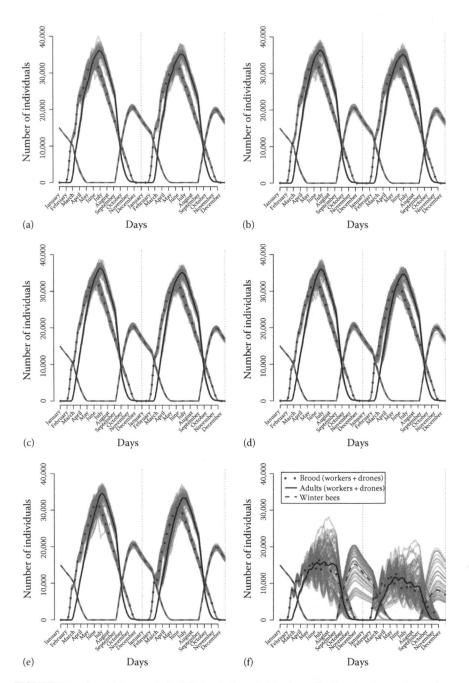

FIGURE 9.6 Sensitivity analysis (50 simulation trials) of cannibalism level based on values of reference from HoPoMo [54]: (a) no cannibalism, (b) half values of reference, (c) values of reference, (d) 1.5 times the values of reference, (e) twice the values of reference, and (f) 5 times the values of reference.

[81,82]. In natural sunlight and sandy loam soil, [Pyr-^{14}C]- and [Phe-^{14}C]-labeled pyriproxyfen degraded with half-lives of 10.3 and 12.5 weeks, respectively. In silty loam soil, the half-lives were equal to 21 and 18 weeks, respectively [82]. Field studies conducted in California on a loamy sand soil to evaluate the mobility and persistence of pyriproxyfen when applied to bare ground showed that 4-(4′-hydroxyphenoxy) phenyl (RS)-2-(2-pyridyloxy)-propyl ether (4′-OH-Pyr) and 2-(2-pyridyloxy)propionic acid (PYPAC), the major degradation metabolites of pyriproxyfen, were found to be stable in soil for up to a year. The greatest pyriproxyfen concentrations were found in the 0–6 in. soil layer. No pyriproxyfen residues were found above the level of quantitation below 12 in. The half-life of pyriproxyfen in the 0–6 in. soil-layer was 36 days. No 4′-OH-Pyr or PYPAC residues were found below 2 in. or greater than 120 days [82].

9.4   USE OF PYRIPROXYFEN AND RESIDUES FOUND IN ENVIRONMENTAL COMPARTMENTS

9.4.1   Agriculture

In agriculture, pyriproxyfen has been the most widely used insecticide to control the red scale (*Aonidiella aurantii*), a major pest in citrus groves worldwide [82,83]. Thus, for example, Cunha et al. [84] mentioned that Spain was the sixth major producer in the world with 6.3 million tons produced in 2009 from 318,385 ha. These authors sampled about 18,000 ha predominantly planted with oranges (71.2%), tangerines (26.1%), and lemons (2.7%). Interviews showed that farms between 1 and 10 ha area prevailed. Air-blast sprayers were the most common machines employed (58%) in foliar applications, followed by hand-held gun hydraulic sprayers (42%). Typically, operational characteristics of applications with air-blast sprayers in the citrus regions of Spain were spray volumes of 1,000–2,400 L ha^{-1} (25th–75th percentiles) on oranges and tangerines and 1,000 L ha^{-1} on lemons. [84]. Interestingly, Cunha et al. [84] tried to assess the risk of spray drift on the honey bee. Using a spray volume of 4,016 L ha^{-1} (90th percentile extrapolated from questionnaires), giving an applied dose of 0.301 kg a.s. ha^{-1} year^{-1}, a predicted drift at 3 m, and an LD$_{50}$ of 74 µg a.s. bee^{-1}, they concluded to an absence of risk to bees. However, it is noteworthy that the authors only considered an acute toxicity effect. In addition, the intensity of the spray drift depends on numerous conditions including the properties of the sprayed products, the characteristics of the culture, the machinery setup, and the meteorological conditions [85,86]. Thus, for example, to account for all the parameters possibly impacting the spray drifts, the technical document on Admiral® Pro [87], which is used to fight the San Jose scale (*Diaspidiotus perniciosus*), the white peach scale (*Pseudaulacaspis pentagona*), or the olive soft scale (*Saissetia oleae*), indicates that the insecticide (0.3 L ha^{-1}) must not be used if mulberry trees that are used as food for silkworms are located at less than 2 km of the applied dose. The number of available studies on the levels of residues found in citrus groves after the application of pyriproxyfen is rather limited. Orange groves in different sites in California, Florida, and Texas were treated at intervals of 20–22 days with pyriproxyfen using airblast sprayers [88]. Freezer storage periods ranging from 3 to 95 days were applied prior to analysis. Where 0.12 kg a.i. ha^{-1} was applied, residues ranging from 0.09 to 0.23 mg kg^{-1} were found in fruits. In some trials where measurements were made at t + 21, no significant decrease in the concentrations was noted. Where 0.25 kg a.i. ha^{-1} was applied, the residues of pyriproxyfen were 0.41 mg kg^{-1} in the fruit and 0.01 mg kg^{-1} in the edible portion, giving a factor of 0.024 for the "process" of peeling oranges [88]. Hattingh and Tate [89] and Hattingh [83] showed that pyriproxyfen residues on citrus leaves still induced 100% sterility in coccinellid *Chilocorus nigritus* after 20 weeks of field weathering. The same conclusions were obtained by Smith et al. [90]. More generally, the average (and maximum) concentrations in μg kg^{-1} of pyriproxyfen in strawberries, tomatoes, plums, green beans, almonds, apples, peaches, and cherries were found to be 0.2 (74), 0.2 (20), 0.08 (20), 0.06 (13), 0.04 (8.1), 0.03 (22), 0.02 (13), and 0.006 (2.4), respectively [91].

Pyriproxyfen has been also used for controlling the Homoptera *Bemisia tabaci* biotype B (*B. argentifolii*), a key pest of cotton and other vegetable crops worldwide. On cotton, Naranjo et al. [92] used aerial and ground applications of 60 g pyriproxyfen ha^{-1}. Pyriproxyfen is also used against the mulberry scale (*Pseudaulacaspis pentagona*), which is harmful to tea and difficult to control, because the insect produces 3–4 generations year^{-1}, parasitizing and sucking on the branches of tea plants. Isayama and Tsuda [93] showed that a high residual activity against mulberry scale continued 100–150 days after a treatment of pyriproxyfen at 90 ppm. Pyriproxyfen is one of several insecticides being used in California for the control of the red imported fire ant (*Solenopsis invicta*), a major agricultural, horticultural, and urban pest [94].

9.4.2 VECTOR CONTROL

Pyriproxyfen is also used in vector control. WHO [95] has assessed pyriproxyfen for use as a mosquito larvicide in drinking water in containers, especially for the control of dengue fever. The recommended dosage of pyriproxyfen in potable water in containers should not exceed 0.01 mg L^{-1} under the WHO Pesticides Evaluation Scheme [96,97]. For the control of *Aedes albopictus*, responsible for the Chikungunya epidemic on Réunion Island in 2005–2006, pyriproxyfen (Sumilarv® 0.5%, granule formulation) at the rate of 5–10 g a.s. ha^{-1} was recommended. Doses until 100 g ha^{-1} were suggested for highly contaminated zones. It was considered that a treatment remained active for 6 weeks, and two consecutive treatments were necessary to secure a maximum of efficacy in the control of the mosquitoes [46]. The same formulation applied at the rate of 0.01 mg a.i. L^{-1} to gem pits and river bed pools in a gem mining area in Sri Lanka inhibited the emergence of *Anopheles culicifacies* adults for 185 and 190 days, respectively, reducing significantly the incidence of malaria [98]. The granular formulation of 0.5% pyriproxyfen was evaluated for the inhibition of emergence of *Aedes togoi* in brackish water of rock pools near a coastal area in Pusan, South Korea [99]. Over 90% inhibition of emergence was observed 70 days after treatment at 0.1 and 0.05 mg L^{-1}. A concentration of 0.01 mg pyriproxyfen L^{-1} leads to about 60% inhibition of emergence 70 days posttreatment.

9.4.3   VETERINARY MEDICINE

Pyriproxyfen is marketed as a spot-on topical treatment for prophylactic flea control on cats and dogs, generally in combination with other biocides [100–102]. It has been also reported that pyriproxyfen kills adult fleas when applied in relatively high dosages via dog hair or when administered in the blood diet [101]. The residues that could be found in the environment resulting from this use are not documented and difficult to assess.

9.5   EFFECTS OF PYRIPROXYFEN ON HONEY BEES

In *Apis mellifera*, the process of cuticular melanization starts at the pupal stage. Bitondi et al. [103] showed that bees treated during their different developmental phases with pyriproxyfen showed important alterations in cuticular pigmentation and sclerotization.

Fifth-instar larvae at the feeding phase (LF5, 4–4.5 days after hatching) were topically treated with 1 μg pyriproxyfen dissolved in 1 μL acetone in their own comb cells. Controls were treated with 1 μL acetone or left untreated. Three areas of a comb containing larvae of the same age were delimited for juvenoid or acetone treatment and for controls, respectively. After treatment, the combs were returned to the hives. Interestingly, the authors noted that larvae treated with pyriproxyfen and subsequently reintroduced into the colonies were preferentially removed from their cells by the workers. However, this event was not quantified. At the beginning of the pupal period (about 8.5–9 days a.h.), pupae were removed from the combs and placed in an incubator at 34°C and 80% relative humidity in order to control their emergence [103].

Fifth-instar larvae at the spinning phase (LS5, 5–6 days a.h.), prepupal (PP, 7–8 days a.h) and pupal stages, white-eyed, unpigmented cuticle pupae (Pw, 8.5–9 days a.h.), pink-eyed, unpigmented cuticle pupae (Pp, 10 days a.h.), dark-pink-eyed, unpigmented cuticle pupae (Pdp, 11 days a.h.), and brown-eyed, unpigmented cuticle pupae (Pb, 12–13 days a.h.) were collected from the hives and treated with 1 μg pyriproxyfen. Pw pupae were also treated with other concentrations of pyriproxyfen (0.1 and 5 μg μL^{-1} acetone). Controls were treated with 1 μL acetone or left untreated. The bees were not returned to the hives but were maintained in an incubator at 34°C and 80% relative humidity [103].

When the treatment at 1 μg was applied during LF5, the pupal development was blocked, and pigmentation did not occur. Treatment of older larvae (LS5, PP, Pw) did not impair pigmentation, but, conversely, this process was accelerated, intensified, and abnormal. Juvenoid treatment during these developmental phases (LS5, PP, and Pw) induced earlier activity of phenoloxidase, an enzyme of the reaction chain leading to melanin synthesis. When pyriproxyfen at 1 μg was applied to Pp, Pdp, or Pb pupae, developmental periods during which the cuticle is no longer affected, the juvenile hormone analogue promoted an acceleration in pupal development and, hence, a precocious emergence. The authors [103] showed that treated pupae presented significantly higher enzymatic levels and a graded response in phenoloxidase activity after treatment with 0.1, 1, or 5 μg of pyriproxyfen. These results were consolidated in another study [104], in which they also showed that the topical application of pyriproxyfen (1 μg μL^{-1}) led to a delay of the pupal ecdysteroid peak by nearly 4 days. Santos et al. [105] demonstrated that the topical application of pyriproxyfen (1 μg μL^{-1}) to Pw bees accelerated the typical protein pattern from one phase to the next. The protein pattern observed 96 h after treatment was similar to that observed much later in controls (*i.e.*, 192 h). On the contrary, 20-hydroxyecdysone induced an arrest in the protein pattern development. A double hormonal treatment restored the normal temporal expression of epidermal proteins.

Pinto et al. [106] showed that pyriproxyfen affected the synthesis, secretion, and accumulation of vitellogenin in a dose-dependent fashion in young honey bee workers, Africanized *Apis mellifera*. Groups of 120 newly emerged bees were treated with 1 μL acetone containing different concentrations of pyriproxyfen (10, 5, 2.5, 1.25, 0.1, 0.01, or 0.001 μg). The control group received 1 μL acetone only. They were confined to small cages and placed in an incubator for 6 days. Because vitellogenin synthesis depends on pollen consumption, access to this resource as well as to water was guaranteed to all workers. Three hours after treatment (day 0) and every 24 h until the 6th day, hemolymph was extracted through a small superficial incision in the abdominal dorsal cuticle of the insect. Samples from 20 workers of the same age and cage cohort were pooled. Vitellogenin concentration in hemolymph was measured by rocket immunoelectrophoresis. In the controls, an increase in vitellogenin titer from day 4 to day 6 of adult life was observed. Pyriproxyfen inhibited this effect when topically applied at 10 μg to newly emerged workers. Consequently, the vitellogenin concentrations in 4–6-day-old treated bees were extremely low, comparable to those of 0–3-day-old bees.

The lack of vitellogenin accumulation in treated bees was reflected in the total protein content of the hemolymph. Protein dosage by the dye-binding assay showed that, by day 6, bees treated with 10 μg pyriproxyfen presented significantly lower protein titer than controls of the same age. The dose-dependent inhibitory effect of pyriproxyfen was evident in 6-day-old workers. After application of 1.25 to 10 μg of pyriproxyfen, protein titers decreased progressively. Treatment with doses of 0.1, 0.01, or 0.001 μg of pyriproxyfen did not affect the normal accumulation of protein in hemolymph [106]. The low vitellogenin titer in the hemolymph of pyriproxyfen-treated bees was not due to its enhanced sequestration by ovaries. None of the pyriproxyfen-treated bees had vitellogenic ovaries, as shown by ovary examination after dissecting bees under a stereomicroscope. In a complementary study, fat bodies attached to abdominal cuticles were dissected from confined 6-day-old workers treated with 0.01 or 10 μg pyriproxyfen or acetone (controls) at emergence. In each assay, two fat bodies were cultured *in vitro* (34°C for 24 h). Results showed that fat bodies obtained from bees treated with pyriproxyfen synthesized and secreted less V$_g$ *in vitro* than untreated controls [106].

The aim of the aforementioned studies is mainly to better understand the physiology and the developmental mechanisms of the honey bee rather than to estimate the adverse effects of pyriproxyfen on this beneficial insect. Pyriproxyfen is used in place of the juvenile hormone at concentrations that are very often higher than those belonging to the sublethal domain.

Machado Baptista and coworkers [107] showed that the direct spraying of pyriproxyfen (Cordial 100 EC—0.075) on bees led to a TL_{50} of 466 h. Utsumi et al. [108] at Sumitomo Chemical Co. Ltd. (Japan) tested the potential adverse effects of pyriproxyfen (10% EC) on honey bees under field conditions. Four colonies including 60,000–80,000 workers were used per group. Each group was placed within 0.5 ha located at 3 km from full-flowering *Phacelia tanacetifolia*. A treatment of 75 g a.s. ha^{-1} was applied. Both the eggs and larvae in the combs developed into healthy adult bees and no differences were found with the controls. The number of dead bees from the colony after treatment, obtained by continuous monitoring of the traps, was lower than that observed during the acclimation period of the hives and was the same as for the untreated colonies [108]. On the other hand, Yang et al. [109] stressed a significant impact on the honey bee brood. One-day-old larvae were reared in the laboratory and fed with 0, 0.1, 1, 10, and 100 ppm of pyriproxyfen in larval food. A dose–response effect was found in the larval development of the treated larvae. At 0.1 ppm, 58.9% of the larvae succeeded to pupation, and 52.6% of adult emergence was obtained. Among these adults, 2.8% showed deformed wings. At 1 ppm, only 15.5% of emergence was obtained and 13.1% of the bees presented deformed wings. At 10 and 100 ppm, all the honey bees died before emergence.

In the frame of a research program aiming at estimating the endocrine disruption potential of xenobiotics on the honey bee [110], the potential adverse effects of sublethal concentrations of pyriproxyfen on the larval developments of bees were thoroughly investigated by means of a larval test developed by Aupinel and coworkers [111–113]. The principles of this test are summarized in Figure 9.7. Briefly,

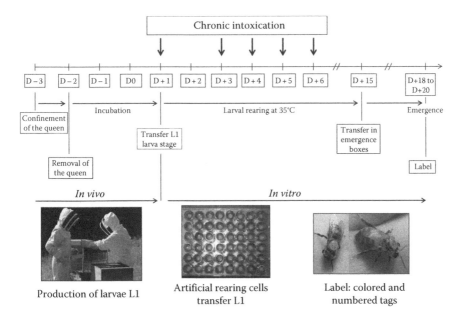

FIGURE 9.7 (See color insert.) Principles of the larval test developed by Aupinel and coworkers. (From Aupinel, P. et al., *Bull. Insectol.*, 58, 107, 2005; Aupinel, P. et al. *Pest Manag. Sci.* 63, 1090, 2007; Aupinel, P. et al., *Julius-Kühn-Archiv.*, 423, 96, 2009.)

first-instar larvae are produced at the apiary. Queens of experimental healthy colonies are confined on empty combs where they lay eggs. One day later, they are removed and the combs with eggs are returned to the colonies for incubation. Three days later, frames with first-instar larvae (L1) are removed. In the laboratory, larvae are transferred into culture plates with artificial rearing cells. They are reared in an incubator at 35°C. The contamination via the larval diet is made at $D+1$ and $D+3$ to $D+6$. During pupation stage ($D+15$), the plates are placed in boxes for adult emergence. This test is particularly interesting because it allows us to take into account many factors throughout the larval development of the bee. It offers the opportunity to assess the effects of a chronic exposure to a molecule during the larval stage at the individual and population level. Two different types of experiments were performed.

To estimate the delayed effects on nymphs and adults of a chronic exposure, attempts were made to measure the development of hypopharyngeal glands in the bees after emergence. Hypopharyngeal glands are the main organs responsible for the royal jelly secretion by the nurses [114,115]. Their size is age and food protein dependent and it is correlated to the amount of secretion as well as the weight of the head. Their development can be assessed by measuring the total protein contents of the glands from the Bradford method after their extraction [116]. Thus, the bees emerging from the larval test were reared in the laboratory until the age of 10 days. During this period, they were fed *ad libitum* with pollen and syrup. At $D+26$, the bees were sacrificed and the protein contents of their hypopharyngeal glands were measured by the Bradford method. Thus, we showed [110] that 305 ppb of pyriproxyfen, corresponding to a cumulative dose of 54 ng larva^{-1}, significantly reduced the protein contents in the hypopharyngeal glands. Consequently, we deduced that this concentration highly perturbed the activity of the nurses within the hive.

In another series of experiments, the larval test was performed with concentrations of 101 ppb (18 ng larva^{-1}) and 305 ppb (54 ng larva^{-1}) of pyriproxyfen. Water and acetone were used as controls. The cumulative larval ($D+7$), pupal ($D+15$), and adult mortalities at the end of the emergence period ($D+20$) were not significantly different. In contrast, exposed individuals presented shorter development time; these individuals emerged earlier than control groups. At emergence, bees showed malformations, mainly atrophied or damaged wings. The malformation rate was significantly different between controls and the highest contaminated dose. At 54 ng larva^{-1}, 20% of emerging bees were malformed versus less than 6% for the controls. In another series of experiments, 35% of malformations were obtained, but the controls included 16% of malformed. Conversely, no significant difference was observed between the controls and bees treated with pyriproxyfen at 101 ppb.

The normal emerging adults of each day were collected and labeled with colored and numbered tags. They were then released into a hive of observation (including glass sides with a grid) containing a receiving colony, which consisted of four frames (two frames of brood of all ages, one frame of honey and pollen, one empty frame for storing reserves). This colony was placed outside under natural semicontrolled conditions. Observations were performed every day over 16 days. Dead bees were counted, and behaviors were recorded from an ethogram with the scan technique from Kolmes, which consists of drawing a grid on the glass walls of the hive and recording the behavior of the first bee present in every square observed. Analysis was made by grouping behaviors into two categories, namely, the social (*e.g.*, brood care, congeners care) and nonsocial (inactivity, groom self) behaviors. Bees exposed to the two pyriproxyfen concentrations showed significantly fewer behaviors related to the social cohesion of the colony than the controls, especially those dealing with brood care. However, the most interesting behavior was the rejection of the emergent bees by their congeners following their introduction in the hive. This was recorded by counting the number of dead bees in front of the hive entrance. In a first trial, the concentrations of 101 and 305 ppb of pyriproxyfen led to about 38% and 80% of rejections, respectively. Less than 10% of rejection was recorded for the controls [117]. In another trial, the concentrations of 101 and 305 ppb of pyriproxyfen led to about 18% and 36% of rejection, respectively. Both controls only led to 6% of rejection [118].

Last, it is worth noting that while numerous pesticides can accumulate in hive products [119–122], pyriproxyfen was only found in wax at a maximum concentration of 8 ppb [119,120].

9.6 SIMULATION OF LONG-TERM EFFECTS OF PYRIPROXYFEN ON HONEY BEE POPULATIONS

Honey bees commonly forage within 1.5 km of their hive and exceptionally as far as 10–12 km, depending on their need for food and its availability. During their foraging flights, they visit numerous plants to gather nectar, pollen, honeydew, sap, and water [123]. Foragers also visit puddles, ponds, and other aquatic resources to collect the 10–40 L of water that are necessary annually for the colony [124,125]. Thus, the honey bees can be contaminated by pyriproxyfen used in agriculture and when it is applied as larvicide in vector control. The transfer of pyriproxyfen by *Aedes albopictus* and *Aedes aegypti* and subsequent inhibition of emergence of the mosquitoes have been recently demonstrated [126,127]. This kind of contamination is even more possible with the honey bee, due to the quantity of contaminated water that would be transferred within the hive and also because it is well known that a part of the water is brought back directly into the brood cells to guarantee the 90%–95% relative humidity necessary for the complete hatching of the eggs [128].

Our experiments and those by others show that sublethal concentrations of pyriproxyfen affect the development of honey bees and disturb the normal behavior of the colony. Inspection of the literature reveals that these effective concentrations are in the ranges of the residual concentrations found in the aquatic and terrestrial environments when pyriproxyfen is used in vector control and in agriculture. In addition, the persistence of pyriproxyfen in these media can be of few months.

Various scenarios were elaborated to fully estimate the long-term effects of sublethal concentrations of pyriproxyfen on a honey bee population. Thus, for example, Figure 9.8a shows the evolution of the different hive inhabitants during a 2-year period after a perturbation consisting in the rejection of 50% of the emergent bees by their congeners for 2 months from June 1. A linear decreasing effect was applied. In practice, this means that after 1 month, the rate of rejection was only 25%. This scenario is totally realistic. The inspection of Figure 9.8a, which is the mean of 50 simulation trials, shows that the normal dynamics is rather rapidly restored, the

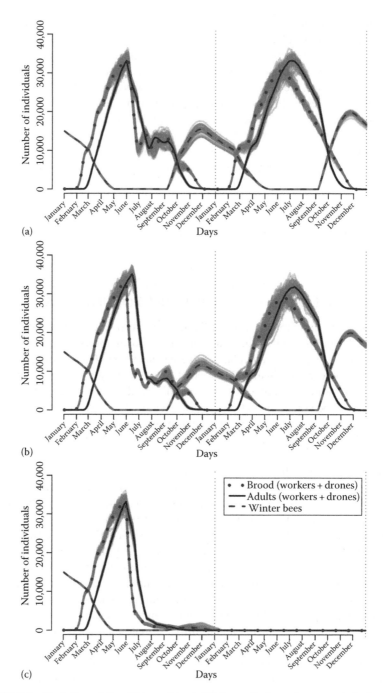

FIGURE 9.8 Evolution of the different hive inhabitants during a 2-year period after a perturbation consisting in (a) a rejection of 50% of the emergent bees by their congeners, (b) a significant impact on the efficacy of the nurses, and (c) both effects jointly during a 2-month period, starting on June 1 and with a linear decreasing effect.

population of the second year growing satisfactorily. In Figure 9.8b, the efficacy of the nurses was perturbed. The perturbation started on June 1 and was applied for 2 months with a linear decreasing effect. This scenario is also realistic because we showed that pyriproxyfen affected the hypopharyngeal glands of the nurses at sublethal concentrations. Figure 9.8b displays the result of 50 simulation trials. Again, the population was affected during the stress period but not enough to impact the dynamics of the population during the second year. Interestingly, Figure 9.8c shows the effects of both perturbations on the bee population dynamics. Fifty simulation trials were performed. When the two types of adverse effects are applied jointly, a definitive collapse is observed at the end of the first year. This clearly illustrates the interest of a population dynamics model, and especially an ABM, to study the long-term effects of ecotoxicological events observed at different levels of organizations in a complex system.

In another series of simulations, an attempt was made to test the effects of the date of application of the perturbation. Thus, in Figure 9.9, the evolution of the different hive inhabitants during a 2-year period after a perturbation consisting in a rejection of 30% of the emergent bees by their congeners and a significant impact on the efficacy of the nurses during a 2-month period with a linear decreasing effect was tested with the perturbation starting always on March 1 (a), April 1 (b), May 1 (c), June 1 (d), July 1 (e), or August 1 (f). Each time, 50 simulation trials were performed. Figure 9.9a shows that the perturbation is applied too early in the season, and, hence, most of the time, in the second year, the bee populations are highly perturbed. Such colonies will collapse sooner or later. In other trials, the colonies seem to be less affected in the second year, except as regards the number of individuals, which is lower than in a normal population. This difference of population dynamics, clearly shown from the ABM simulation results, corresponds to what is observed in practice. Indeed, it is well known that the different colonies within an apiary do not behave similarly when faced with an external stress. When the perturbation is applied on April 1 (Figure 9.9b) or May 1 (Figure 9.9c), many more populations present a normal development during the second year, except their size, but some of them remain perturbed. This phenomenon is even more accented when the perturbation is applied on June 1 (Figure 9.9d). Conversely, when the perturbation is applied on July 1 (Figure 9.9e), the populations will be strongly affected in the second year, leading very often to collapses. In all cases, with a maximum of about 20,000 adult bees, there is no doubt that a colony will not survive durably. When the perturbation is applied on August 1, a total collapse is observed at the end of the first year (Figure 9.9f). These different simulation results (Figure 9.9) are particularly interesting. They show that there exist critical periods during which the bee populations are more sensitive against stresses. Broadly speaking, they correspond to the beginning of the season when the colony starts its development. The strength of the bee colony must be enough to support a perturbation, otherwise its normal and durable development cannot be expected. The period during which the winter bees are produced corresponds to the second critical phase, which can lead to a total collapse of the colony. More generally, Figure 9.9 clearly illustrates the interest of an ABM, which tends to copy at best reality due to its inherent stochasticity.

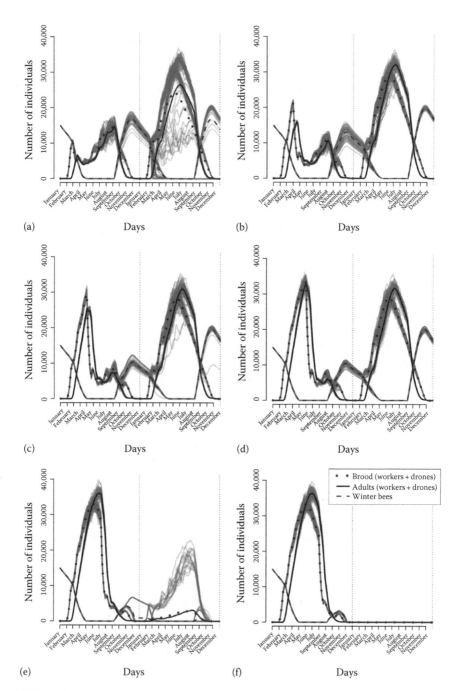

FIGURE 9.9 Evolution of the different hive inhabitants during a 2-year period after a perturbation consisting in a rejection of 30% of the emergent bees by their congeners and a significant impact on the efficacy of the nurses during a 2-month period with a linear decreasing effect, starting on (a) March 1, (b) April 1, (c) May 1, (d) June 1, (e) July 1, and (f) August 1.

9.7 CONCLUSIONS

Even if pyriproxyfen is currently classified as moderately or not toxic to honey bees on the basis of acute toxicity tests, recent experimental studies reveal that exposure to low doses of this juvenoid during larval stages can cause adverse effects at the colony level. This is clearly shown when our original ABM, called SimBeePop (version 1.04), is used to simulate the bee population dynamics during a 2-year period, following different realistic contamination scenarios.

From an ecotoxicological point of view, our study shows that the current regulatory tests are not suited to detect the potential adverse effects of chemicals on the endocrine system and development of honey bees. The forthcoming inclusion of the larval test in the battery of tests to perform on bees is welcomed to start to fill this gap in the regulation of insecticides.

From a modeling standpoint, our study also demonstrates the invaluable role of the population dynamics models to fully exploit the various results obtained with *in vivo* and *in vitro* experiments. Indeed, the complexity of the relationships between the adverse effects of a chemical observed at the individual level and the subsequent possible impacts at the population level makes necessary the use of modeling approaches to test various scenarios with the goal of better understanding and, if possible, simulating the toxicity of the chemicals.

ACKNOWLEDGMENT

The financial support from the French Ministry of Ecology, Sustainable Development, Transport and Housing (MEDDTL) is gratefully acknowledged (PNRPE program).

REFERENCES

1. D.H. Miller and G.T. Ankley, *Modeling impacts on populations: Fathead minnow* (Pimephales promelas*) exposure to the endocrine disruptor 17β-trenbolone as a case study*, Ecotoxicol. Environ. Safety 59 (2004), pp. 1–9.
2. C. Lopes, A.R.R. Péry, A. Chaumot, and S. Charles, *Ecotoxicology and population dynamics: Using DEBtox models in a Leslie modeling approach*, Ecol. Model. 188 (2005), pp. 30–40.
3. S. Charles, E. Billoir, C. Lopes, and A. Chaumot, *Matrix population models as relevant modeling tools in ecotoxicology*, in *Ecotoxicology Modeling*, J. Devillers, ed., Springer, New York, 2009, pp. 261–298.
4. S.A.L.M. Kooijman, J. Baas, D. Bontje, M. Broerse, C.A.M. van Gestel, and T. Jager, *Ecotoxicological applications of dynamic energy budget theory*, in *Ecotoxicology Modeling*, J. Devillers, ed., Springer, New York, 2009, pp. 237–259.
5. V.J. Kramer, M.A. Etterson, M. Hecker, C.A. Murphy, G. Roesijadi, D.J. Spade, J.A. Spromberg, M. Wang, and G.T. Ankley, *Adverse outcome pathways and ecological risk assessment: Bridging to population-level effects*, Environ. Toxicol. Chem. 30 (2011), pp. 64–75.
6. M. Wang and R. Luttik, *Population level risk assessment: Practical considerations for evaluation of population models from a risk assessor's perspective*, Environ. Sci. Eur. 24 (2012), pp. 1–11. Available at http://www.enveurope.com/content/24/1/3. Accessed on January 3, 2013.

7. J. Ares, *Time and space issues in ecotoxicology: Population models, landscape pattern analysis, and long-range environmental chemistry*, Environ. Toxicol. Chem. 22 (2003), pp. 945–957.

8. A.R. Brown, P.F. Robinson, A.M. Riddle, and G.H. Panter, *Population dynamics modeling. A tool for environmental risk assessment of endocrine disrupting chemicals*, in *Endocrine Disruption Modeling*, J. Devillers, ed., CRC Press, Boca Raton, FL, 2009, pp. 47–82.

9. V. Grimm, P. Thorbek, A. Schmolke, and P. Chapman, *State of the art of ecological modeling for pesticide risk assessment. A critical review*, in *Ecological Models for Regulatory Risk Assessments of Pesticides. Developing a Strategy for the Future*, P. Thorbek, V.E. Forbes, F. Heimbach, U. Hommen, H.H. Thulke, P.J. Van den Brink, J. Wogram, and V. Grimm, eds., CRC Press, Boca Raton, FL, 2010, pp. 77–87.

10. A. Schmolke, P. Thorbek, P. Chapman, and V. Grimm, *Ecological models and pesticide risk assessment: Current modeling practice*, Environ. Toxicol. Chem. 29 (2010), pp. 1006–1012.

11. N. Galic, U. Hommen, J.M. Baveco, and P.J. van der Brink, *Potential application of population models in the European ecological risk assessment of chemicals II: Review of models and their potential to address environmental protection aims,* Integ. Environ. Assess. Manag. 6 (2010), pp. 338–360.

12. E. Bonabeau, *Agent-based modeling: Methods and techniques for simulating human systems*, Proc. Nat. Acad. Sci. 99 (2002), pp. 7280–7287.

13. J. Devillers, *Endocrine Disruption Modeling*, CRC Press, Boca Raton, FL, 2009.

14. Anonymous, *Colony collapse disorder, action plan*, CCD Steering Committee, 2007. Available at http://www.ars.usda.gov/is/br/ccd/ccd_actionplan.pdf. Accessed on January 3, 2013.

15. Anonymous, *USDA's response to colony collapse disorder*, Audit Report 50099–0084-HY, United States Department of Agriculture, Office of Inspector General, Washington, DC, 2012.

16. E. Genersch, *Honey bee pathology: Current threats to honey bees and beekeeping*, Appl. Microbiol. Biotechnol. 87 (2010), pp. 87–97.

17. D. vanEngelsdorp and M.D. Meixner, *A historical review of managed honey bee populations in Europe and the United States and the factors that may affect them*, J. Invert. Pathol. 103 (2010), pp. S80–S95.

18. J.Y. Wu, C.M. Anelli, and W.S. Sheppard, *Sub-lethal effects of pesticide residues in brood comb on worker honey bee* (Apis mellifera) *development and longevity*, PLoS ONE 6 (2011), e14720. doi:10.1371/journal.pone.0014720.

19. N.R. Kumar, S. Sangwan, and P. Badotra, *Exposure to cell phone radiations produces biochemical changes in worker honey bees*, Toxicol. Int. 18 (2011), pp. 70–72.

20. S.S. Sainudeen, *Electromagnetic radiation (EMR) clashes with honey bees*, Int. J. Environ. Sci. 1 (2011), pp. 897–900.

21. C.H. Krupke, G.J. Hunt, B.D. Eitzer, G. Andino, and K. Given, *Multiple routes of pesticide exposure for honey bees living near agricultural fields*, PLoS ONE 7 (2012), e29268. doi:10.1371/journal.pone.0029268.

22. T. Soin and G. Smagghe, *Endocrine disruption in aquatic insects: A review*, Ecotoxicology 16 (2007), pp. 83–93.

23. K.N. Baer and K.D. Owens, *Evaluation of selected endocrine disrupting compounds on sex determination in* Daphnia magna *using reduced photoperiod and different feeding rates*, Bull. Environ. Contam. Toxicol. 62 (1999), pp. 214–221.

24. M.F.L. Lemos, C.A.M. van Gestel, and A.M.V.M. Soares, *Reproductive toxicity of the endocrine disrupters vinclozolin and bisphenol A in the terrestrial isopod* Porcellio scaber *(Latreille, 1804)*, Chemosphere 78 (2010), pp. 907–913.

25. E.M. Rodriguez, D.A. Medesani, and M. Fingerman, *Endocrine disruption in crustaceans due to pollutants: A review*, Comp. Biochem. Physiol. Part A: Molec. Integ. Physiol. 146 (2007), pp. 661–671.

26. K. Hartfelder, *Insect juvenile hormone: From "status quo" to high society*, Braz. J. Med. Biol. Res. 33 (2000), pp. 157–177.

27. A.R. Barchuk, R. Maleszka, and Z.L.P. Simões, Apis mellifera *ultraspiracle: cDNA sequence and rapid upregulation by juvenile hormone*, Insect Molec. Biol. 13 (2004), pp. 459–467.

28. Z.Y. Huang, G.E. Robinson, S.S. Tobe, K.J. Yagi, C. Strambi, A. Strambi, and B. Stay, *Hormonal regulation of behavioural development in the honey bee is based on changes in the rate of juvenile hormone biosynthesis*, J. Insect Physiol. 37 (1991), pp. 733–741.

29. Z.Y. Huang and G.E. Robinson, *Honeybee colony integration: Worker-worker interactions mediate hormonally regulated plasticity in division of labor*, Proc. Natl. Acad. Sci. USA 89 (1992), pp. 11726–11729.

30. A.N. Pearce, Z.H. Huang, and M.D. Breed, *Juvenile hormone and aggression in honey bees*, J. Insect Physiol. 47 (2001), pp. 1243–1247.

31. M. Kiran, S. Coakley, N. Walkinshaw, P. McMinn, and M. Holcombe, *Validation and discovery from computational biology models*, BioSystems 93 (2008), pp. 141–150.

32. L.H. Ren, Y.S. Ding, Y.Z. Shen, and X.F. Zhang, *Multi-agent-based bio-network for systems biology: Protein-protein interaction network as an example*, Amino Acids 35 (2008), pp. 565–572.

33. H. Devillers, J.R. Lobry, and F. Menu, *An agent-based model for predicting the prevalence of* Trypanosoma cruzi *I and II in their host and vector populations*, J. Theor. Biol. 255 (2008), pp. 307–315.

34. H. de Vries and J.C. Biesmeijer, *Modelling collective foraging by means of individual behaviour rules in honey-bees*, Behav. Ecol. Sociobiol. 44 (1998), 109–124.

35. H. de Vries and J.C. Biesmeijer, *Self-organization in collective honeybee foraging: Emergence of symmetry breaking, cross inhibition and equal harvest-rate distribution*, Behav. Ecol. Sociobiol. 51 (2002), pp. 557–569.

36. R. Thenius, T. Schmickl, and K. Crailsheim, *The "dance or work" problem: Why do not all honeybees dance with maximum intensity*, in *CEEMAS 2005*, M. Pechoucek, P. Petta, and L.Z. Varga, eds., Springer, Berlin, Germany, LNAI 3690, 2005, pp. 246–255.

37. M. Fehler, M. Kleinhenz, F. Klügt, F. Puppe, and J. Tautz, *Caps and gaps: A computer model for studies on brood incubation strategies in honeybees* (Apis mellifera carnica), Naturwissenschaften 94 (2007), pp. 675–680.

38. T. Schmickl and K. Crailsheim, *Analysing honeybees' division of labour in broodcare by a multi-agent model*, Artif Life 11 (2008), pp. 529–536.

39. T. Schmickl and K. Crailsheim, *An individual-based model of task selection in honeybee*, in, *SAB 2008*, M. Asada et al., eds., Springer, Berlin, Germany, LNAI 5040, 2008, pp. 382–392.

40. C. List, C. Elsholtz, and T.D. Seeley, *Independence and interdependence in collective decision making: An agent-based model of nest-site choice by honeybee swarms*, Phil. Trans. R. Soc. B 364 (2009), pp. 755–762.

41. B.R. Johnson, *Pattern formation on the combs of honeybees: Increasing fitness by coupling self-organization with templates*, Proc. R. Soc. 276 (2009), pp. 255–261.

42. B.R. Johnson, *Spatial effects, sampling errors, and task specialization in the honey bee*, Insect. Soc. 57 (2010), pp. 239–248.

43. B.R. Johnson and J.C. Nieh, *Modeling the adaptive role of negative signaling in honey bee intraspecific competition*, J. Insect. Behav. 23 (2010), pp. 459–471.

44. J. Devillers, *Juvenile hormones and juvenoids: A historical survey*, in *Juvenile Hormones and Juvenoids. Modeling Biological Effects and Environmental Fate*, J. Devillers, ed., CRC Press, Boca Raton, FL, 2013, pp. 1–14.

45. Pesticide Properties DataBase (PPDB), Available at http://sitem.herts.ac.uk/aeru/footprint/en/Reports/574.htm. Accessed on January 3, 2013.

46. Afsset, *La Lutte Antivectorielle dans le Cadre de l'Epidémie de Chikungunya sur l'Ile de la Réunion. Evaluation des Risques et de l'Efficacité des Produits Larvicides*, Afsset, Paris, France, 2007.

47. J. Devillers, M.H. Pham-Delègue, A. Decourtye, H. Budzinski, S. Cluzeau, and G. Maurin, *Comparative toxicity and hazards of pesticides to* Apis *and non-*Apis *bees. A chemometrical study*, SAR QSAR Environ. Res. 14 (2003), pp. 389–403.

48. V. Mommaerts, G. Sterk, and G. Smagghe, *Bumblebees can be used in combination with juvenile hormone analogues and ecdysone agonists*, Ecotoxicology 15 (2006), pp. 513–521.

49. V. Mommaerts and G. Smagghe, *Side-effects of pesticides on the pollinator* Bombus: *An overview*, in *Pesticides in the Modern World—Pests Control and Pesticides Exposure and Toxicity Assessment*, M. Stoytcheva, ed., InTech Publisher, Rijeka, Croatia, 2011, pp. 507–552.

50. C.J.E. Castle and A.T. Crooks, *Principles and Concepts of Agent-Based Modelling for Developing Geospatial Simulations*, Centre for Advanced Spatial Analysis, UCL Working Papers Series, paper 110, 2006.

51. J. Devillers, H. Devillers, A. Decourtye, and P. Aupinel, *Internet resources for agent-based modelling*, SAR QSAR Environ. Res. 21 (2010), pp. 337–350.

52. S. Camazine, *Self-organizing pattern formation on the combs of honey bee colonies*, Behav. Ecol. Sociobiol. 28 (1991), pp. 61–76.

53. S.C. Pratt, *Optimal timing of comb construction by honeybee* (Apis mellifera) *colonies: A dynamic programming model and experimental tests*, Behav. Ecol. Sociobiol. 46 (1999), pp. 30–42.

54. T. Schmickl and K. Crailsheim, *HoPoMo: A model of honeybee intracolonial population dynamics and resource management*, Ecol. Model. 204 (2007), pp. 219–245.

55. B.R. Johnson, *Division of labor in honeybees: Form, function, and proximate mechanisms*, Behav. Ecol. Sociobiol. 64 (2010), pp. 305–316.

56. F. Anchling, *Avril, c'est deux saisons dans un même mois*, Abeille Fr. 1995, pp. 913.

57. H. Fukuda, *The relationship between work efficiency and population size in a honeybee colony*, Res. Popul. Ecol. 25 (1983), pp. 249–263.

58. A. Imdorf, K. Ruoff, and P. Fluri, *Le Développement des Colonies chez l'Abeille Mellifère*, Station de recherche Agroscope Liebefeld-Posieux ALP, 2010.

59. J. Woke, *Cannibalism and brood-rearing efficiency in the honeybee*, J. Apic. Res. 16 (1977), pp. 84–94.

60. T. Schmickl and K. Crailsheim, *Cannibalism and early capping: Strategy of honeybee colonies in times of experimental pollen shortages*, J. Comp. Physiol. A 187 (2001), pp. 541–547.

61. H. Rembold, J.P. Kremer, and G.M. Ulrich, *Characterization of postembryonic developmental stages of the female castes of the honey bee*, Apis mellifera L, Apidologie 11 (1980), pp. 29–38.

62. E.R. Michelette and A.E.E. Soares, *Characterization of preimaginal developmental stages in Africanized honey bee workers* (Apis mellifera L.), Apidologie 24 (1993), pp. 431–440.

63. Z. Lipinski, K. Zoltowska, J. Wawrowska, and M. Zaleska, *The concentration of carbohydrates in the developmental stages of the* Apis mellifera carnica *drone brood*, J. Apic. Sci. 52 (2008), pp. 5–11.

64. K.E. Boes, *Honeybee colony drone production and maintenance in accordance with environmental factors: An interplay of queen and worker decisions*, Insect. Soc. 57 (2010), pp. 1–9.

65. S.N. Beshers, Z.Y. Huang, Y. Oono, and G.E. Robinson, *Social inhibition and regulation of temporal polyethism in honey bees*, J. Theor. Biol. 213 (2001), pp. 461–479.

66. M.L. Winston, *The Biology of the Honey Bee*, Harvard University Press, Cambridge, UK, 1987.

67. M.H. Pham-Delègue, *Les Abeilles*, Editions de la Martinière, 1998.

68. J. Devillers, *The ecological importance of honey bees and their relevance to ecotoxicology*, in *Honey Bees: Estimating the Environmental Impact of Chemicals*, J. Devillers and M.H. Pham-Delègue, eds., Taylor & Francis, London, UK, 2002, pp. 1–11.

69. R. Merz, L. Gerig, H. Wille, and R. Leuthold, *Das Problem der Kurz- und Langlebigkeit bei der Ein- und Auswinterung im Bienenvolk* (Apis mellifica *L.*): *eine Verhaltensstudie*, Rev. Suisse Zool. 86 (1979), pp. 663–671.

70. K.H. Nickel and L. Armbruster, *Vom Lebenslauf der Arbeitsbienen besonders auch bei Nosemaschäden*, Arch. Bienenkunde 18 (1937), pp. 257–287.

71. P. Fluri, *Quel Age Atteignent les Ouvrières?*, Centre Suisse de Recherches Apicoles, 1990.

72. M.H. Pham-Delègue and A. Bonjean, *La pollinisation du tournesol en production de semence hybride*, Bull. Tech. Agric. 10 (1983), pp. 211–218.

73. A. Rortais, G. Arnold, M.P. Halm, and F. Touffet-Briens, *Modes of honeybees exposure to systemic insecticides: Estimated amounts of contaminated pollen and nectar consumed by different categories of bees*, Apidologie 36 (2005), pp. 71–83.

74. I. Keller, P. Fluri, and A. Imdorf, *Pollen nutrition and colony development in honey bees. Part II*, Bee World 86 (2005), pp. 27–34.

75. S.E.R. Hoover, H.A. Higo, and M.L. Winston, *Worker honey bee ovary development: Seasonal variation and the influence of larval and adult nutrition*, J. Comp. Physiol. B 176 (2006), pp. 55–63.

76. N. Hrassnigg and K. Crailsheim, *Differences in drone and worker physiology in honeybees* (Apis mellifera *L.*), Apidologie 36 (2005), pp. 255–277.

77. D. Babendreier, N. Kalberer, J. Romeis, P. Fluri, and F. Bigler, *Pollen consumption in honey bee larvae: A step forward in the risk assessment of transgenic plants*, Apidologie 35 (2004), pp. 293–300.

78. R. Brodschneider and K. Crailsheim, *Nutrition and health in honey bees*, Apidologie 41 (2010), pp. 278–294.

79. T.D. Seeley, *Honeybee Ecology*, Princeton University Press, Princeton, NJ, 1985.

80. E.E. Southwick and G. Heldmaier, *Temperature control in honey bee colonies*, BioScience, 37 (1987), pp. 395–399.

81. WHO, *WHO Specifications and Evaluations for Public Health Pesticides: Pyriproxyfen, 4-Phenoxyphenyl (RS)-2-(2-pyridyloxy)propyl ether*, 2006, WHO Press, World Health Organization, Geneva, Switzerland.

82. J.J. Sullivan and K.S. Goh, *Environmental fate and properties of pyriproxyfen*, J. Pestic. Sci. 33 (2008), pp. 339–350.

83. V. Hattingh, *The use of insect growth regulators. Implications for IPM with citrus in Southern Africa as an example*, Entomophaga 41 (1996), pp. 513–518.

84. J.P. Cunha, J. Chueca, C. Garcerá, and E. Moltó, *Risk assessment of pesticide spray drift from citrus applications with air-blast sprayers in Spain*, Crop Protect. 42 (2012), pp. 116–123.

85. M. Farooq and M. Salyani, *Modeling of spray penetration and deposition on citrus tree canopies*, Trans. ASABE 47 (2004), pp. 619–627.

86. Y. Gil and C. Sinfort, *Emission of pesticides to the air during sprayer application: A bibliographic review*, Atmos. Environ. 39 (2005), pp. 5183–5193.

87. Anonymous, Admiral® Pro. Contre les cochenilles et les aleurodes. Arboriculture et cultures sous serre. Utilisable en lutte intégrée, technical document, Philagro, France, p. 2.

88. Anonymous, Pyriproxyfen (200), Available at http://www.fao.org/ag/AGP/AGPP/Pesticid/JMPR/Download/2000_eva/17Pyriprox.pdf. Accessed on January 3, 2013.

89. V. Hattingh and B.A. Tate, *Effects of field-weathered residues of insect growth regulators on some Coccinellidae (Coleoptera) of economic importance as biocontrol agents,* Bull. Entomol. Res. 85 (1995), pp. 489–493.

90. K.M. Smith, D. Smith, and A.T. Lisle, *Effect of field-weathered residues of pyriproxyfen on the predatory coccinellids* Chilocorus circumdatus *Gyllenhal and* Cryptolaemus montrouzieri *Mulsant,* Aust. J. Exp. Agri. 39 (1999), pp. 995–1000.

91. Anonymous, *Foods with Pyriproxyfen Residue,* http://www.whatsonmyfood.org/pesticide.jsp?pesticide=B24. Accessed on January 3, 2013.

92. S.E. Naranjo, J.R. Hagler, and P.C. Ellsworth, *Improved conservation of natural enemies with selective management systems for* Bemisia tabaci *(Homoptera: Aleyrodidae) in cotton.* Biocontrol Sci. Technol. 13 (2003), pp. 571–587.

93. S. Isayama and N. Tsuda, *Development of a novel formulation: Pluto® MC to control mulberry scale on tea,* Sumitomo Kagaku 2 (2008), pp. 1–12.

94. M. Eliahu, D. Blumberg, A.R. Horowitz, and I. Ishaaya, *Effect of pyriproxyfen on developing stages and embryogenesis of California red scale (CRS),* Aonidiella aurantii, Pest Manag. Sci. 63 (2007), pp. 743–746.

95. WHO, *Pyriproxyfen in Drinking-water: Use for Vector Control in Drinking-water Sources and Containers Background Document for Development of WHO Guidelines for Drinking-water Quality,* WHO/HSE/AMR/08.03/9, 2008, WHO Press, World Health Organization, Geneva, Switzerland.

96. WHO, *Pesticides and Their Application for the Control of Vectors and Pests of Public Health Importance, 6th ed. Geneva, 2006, World Health Organization, Department of Control of Neglected Tropical Diseases, Pesticides Evaluation Scheme,* Available at WHO/CDS/NTD/WHOPES/GCDPP/2006.1; http://whqlibdoc.who.int/hq/2006/WHO_CDS_NTD_WHOPES_GCDPP_2006.1_eng.pdf. Accessed on January 3, 2013.

97. WHO, *Pyriproxyfen: 4-phenoxyphenyl (RS)-2-(2-pyridyloxy)propyl ether. Geneva, World Health Organization, 2006, Specifications and Evaluations for Public Health Pesticides,* Available at http://www.who.int/whopes/quality/en/pyriproxyfen_eval_specs_WHO_jul2006.pdf, Accessed on January 3, 2013.

98. A.M.G.M. Yapabandara, *Review of the efficacy and persistence of pyriproxyfen 0.5% granules on* Anopheles sp. *and* Aedes sp. *breeding sites in Sri Lanka,* in *Proceedings of the Fifth International Conference on Urban Pests,* C.Y. Lee and W.H. Robinson, eds., Perniagaan Ph'ng, P&Y Design Network, Malaysia, 2005, pp. 319–323.

99. D.K. Lee, *Field evaluation of an insect growth regulator, pyriproxyfen, against* Aedes togoi *larvae in brackish water in South Korea,* J. Vector Ecol. 26 (2001), pp. 39–42.

100. D.H. Ross, R.G. Pennington, L.R. Cruthers, and R.L. Slone, *Efficacy of a permethrin and pyriproxyfen product for control of fleas, ticks and mosquitoes on dogs,* Canine Pract. 22 (1997), pp. 53–58.

101. D. Stanneck, K.S. Larsen, and N. Mencke, *An evaluation of the effects of pyriproxyfen on eggs and adults of the cat flea,* Ctenocephalides felis felis *(Siphonaptera: Pulicidae),* Irish Vet. J. 55 (2002), pp. 383–387.

102. E. Bouhsira, E. Lienard, P. Jacquiet, S. Warin, V. Kaltsatos, L. Baduel, and M. Franc, *Efficacy of permethrin, dinotefuran and pyriproxyfen on adult fleas, flea eggs collection, and flea egg development following transplantation of mature female fleas* (Ctenocephalides felis felis) *from cats to dogs,* Vet. Parasitol. 190 (2012), pp. 541–546.

103. M.M.G. Bitondi, I.M. Mora, Z.L.P. Simões, and V.L.C. Figueiredo, *The* Apis mellifera *pupal melanization program is affected by treatment with a juvenile hormone analogue,* J. Insect Physiol. 44 (1998), pp. 499–507.

104. M.S. Zufelato, M.M.G. Bitondi, Z.L.P. Simões, and K. Hartfelder, *The juvenile hormone analog pyriproxyfen affects ecdysteroid-dependent cuticle melanization and shifts the pupal ecdysteroid peak in the honey bee* (Apis mellifera), Arthr. Struct. Develop. 29 (2000), pp. 111–119.

105. A.E. Santos, M.M.G. Bitondi, and Z.L.P. Simões, *Hormone-dependent protein patterns in integument and cuticular pigmentation in* Apis mellifera *during pharate adult development*, J. Insect Physiol. 47 (2001), pp. 1275–1282.

106. L.Z. Pinto, M.M.G. Bitondi, and Z.L.P. Simões, *Inhibition of vitellogenin synthesis in* Apis mellifera *workers by a juvenile hormone analogue, pyriproxyfen*, J. Insect Physiol. 46 (2000), pp. 153–160.

107. A.P. Machado Baptista, G.A. Carvalho, S.M. Carvalho, C.F. Carvalho, and J.S. de Souza Bueno Filho, *Toxicidade de produtos fitossanitários utilizados em citros para* Apis mellifera, Ciência Rural, Santa Maria 39 (2009), pp. 955–961.

108. T. Utsumi, M. Miyamoto, and T. Katagi, *Ecotoxicological risk assessment of pesticides in terrestrial ecosystems*, Sumitomo Kagaku 1 (2011), pp. 1–19.

109. E.C. Yang, P.S. Wu, H.C. Chang, and Y.W. Chen, *Effect of sub-lethal dosages of insecticides on honeybee behavior and physiology*, in *Proceedings of International Seminar on Enhancement of Functional Biodiversity Relevant to Sustainable Food Production in ASPAC*, November 9–11, 2010, Tsukuba, Japan. Available at http://www.niaes.affrc. go.jp/sinfo/sympo/h22/1109/. Accessed on January 3, 2013.

110. J. Devillers, P. Aupinel, and A. Decourtye, *Réponses Individuelles et Populationnelles des Abeilles aux Perturbateurs Endocriniens Xénobiotiques*, Programme National de Recherche "Perturbateurs Endocriniens" (PNRPE)- APR 2008, Interim report, 2011.

111. P. Aupinel, D. Fortini, H. Dufour, J.N. Tasei, B. Michaud, J.F. Odoux, and M.H. Pham Delègue, *Improvement of artificial feeding in a standard* in vitro *method for rearing* Apis mellifera *larvae*, Bull. Insectol. 58 (2005), pp. 107–111.

112. P. Aupinel, D. Fortini, H. Dufour, B. Michaud, F. Marolleau, J.N. Tasei, and J.F. Odoux, *Toxicity of dimethoate and fenoxycarb to honey bee brood* (Apis mellifera), *using a new* in vitro *standardized feeding method*, Pest Manag. Sci. 63 (2007), pp. 1090–1094.

113. P. Aupinel, D. Fortini, B. Michaud, P. Mdrzycki, E. Padovani, D. Przygoda, C. Maus, J.D. Charrière, V. Kilchenmann, U. Riesberger-Galle, J.J. Vollmann, L. Jeker, M. Janke, J.F. Odoux, and J.N. Tasei, *Honey bee brood ring-test: Method for testing pesticide toxicity on honeybee brood in laboratory conditions*, Julius-Kühn-Archiv. 423 (2009), pp. 96–102.

114. K. Heylen, B. Gobin, L. Arckens, R. Huybrechts, and J. Billen, *The effects of four crop protection products on the morphology and ultrastructure of the hypopharyngeal gland of the European honeybee*, Apis mellifera, Apidologie 42 (2011), pp. 103–116.

115. J. Deseyn and J. Billen, *Age-dependent morphology and ultrastructure of the hypopharyngeal gland of* Apis mellifera *workers (Hymenoptera, Apidae)*, Apidologie 36 (2005), pp. 49–57.

116. D. Fortini, B. Michaud, and P. Aupinel, *Comparison of two methods to assess effects of insecticides on hypopharyngeal gland development of honey bee*, Poster presented at Apimondia, Montpellier, France, 2009.

117. M. Deschamps, *Impact du Pyriproxyfène, Perturbateur Endocrinien, sur l'Abeille Domestique*, Apis mellifera: *Effets d'une Exposition Chronique au Stade Larvaire*, report ACTA, 2011.

118. L. Droin, *Etude des Effets Potentiels d'une Intoxication Chronique au Stade Larvaire d'un Insecticide Régulateur de Croissance, le Pyriproxyfène, sur le Comportement des Abeilles Adultes*, Report ACTA/ISARA, 2012.

119. C.A. Mullin, M. Frazier, J.L. Frazier, S. Ashcraft, R. Simonds, D. vanEngelsdorp, and J.S. Pettis, *High levels of miticides and agrochemicals in North American apiaries: Implications for honey bee health*, PLoS ONE 5 (2010): e9754. doi:10.1371/journal. pone.0009754.

120. R.M. Johnson, M.D. Ellis, C.A. Mullin, and M. Frazier, *Pesticides and honey bee toxicity—USA*, Apidologie 41 (2010), pp. 312–331.

121. M.P. Chauzat and J.P. Faucon, *Pesticide residues in beeswax samples collected from honey bee colonies* (Apis mellifera *L.*) *in France*, Pest. Manag. Sci. 63 (2007), pp. 1100–1106.

122. M.P. Chauzat, A.C. Martel, N. Cougoule, P. Porta, J. Lachaize, S. Zeggane, M. Aubert, P. Carpentier, and J.P. Faucon, *An assessment of honeybee colony matrices*, Apis mellifera *(Hymenoptera: Apidae) to monitor pesticide presence in continental France*, Environ. Toxicol. Chem. 30 (2011), pp. 103–111.

123. J. Devillers, J.C. Doré, C. Viel, M. Marenco, F. Poirier-Duchêne, N. Galand, and M. Subirana, *Typology of French acacia honeys based on their concentrations in metallic and nonmetallic elements*, in *Honey Bees: Estimating the Environmental Impact of Chemicals*, J. Devillers and M.H. Pham-Delègue, eds., Taylor & Francis, London, UK, 2002, pp. 248–268.

124. C. Fléché, M.C. Clément, S. Zeggane, and J.P. Faucon, *Contamination des produits de la ruche et risques pour la santé humaine: Situation en France*, Rev. Sci. Tech. Off. Int. Epiz. 16 (1997), pp. 609–619.

125. Y. Le Conte, *L'eau et la colonie d'abeilles*, Abeilles Fleurs 641 (2003), pp. 29–30.

126. G.J. Devine, E.Z. Perea, G.F. Killeen, J.D. Stancil, S.J. Clark, and A.C. Morrison, *Using adult mosquitoes to transfer insecticides to* Aedes aegypti *larval habitats*, Proc. Nat. Acad. Sci. USA 106 (2009), pp. 11530–11534.

127. B. Caputo, A. Ienco, D. Cianci, M. Pombi, V. Petrarca, A. Baseggio, G.J. Devine, and A. della Torre, *The "auto-dissemination" approach: A novel concept to fight* Aedes albopictus *in urban areas*, PLoS Negl. Trop. Dis. 6 (2012): e1793. doi:10.1371/journal.pntd.0001793.

128. K.M. Doull, *The effects of different humidities on the hatching of the eggs of honeybees*, Apidologie 7 (1976), pp. 61–66.

10 Simulation of Solitary (Non-*Apis*) Bees Competing for Pollen

Jeroen Everaars and Carsten F. Dormann

CONTENTS

ABSTRACT

Bees are important pollinators, but both honey bees and wild bees are rapidly declining. One of the drivers for the decline of wild bees is land-use change, which affects the field mosaic and the fragmentation of the landscape. There is no general consensus in the literature about how wild bees respond to landscape configuration. Up to now, there are no simulation models that compare the performances of different solitary bees at the landscape scale. Body-size-related traits affect individual foraging behavior. We have therefore developed a spatially explicit individual-based simulation model, SOLBEE, with different bee traits and behavioral rules, which mimics the behavior and movement of pollen-collecting solitary bees. The model landscape is a square kilometer in size and consists of many patches with foraging habitat (varying amount) separated by inhospitable matrix (varying fragmentation levels). The foraging habitat has patch attributes such as flower density and minimum patch size. The bees differ in size (and in derived foraging traits such as velocity), nesting preference, and nest location. The system is further characterized by a timeframe of a single foraging day (with time steps of 1 s) in which bees forage and compete with each other for pollen. Bee numbers (and density within the landscape) are scaled with present foraging resources and body size. During the day, these central-place foragers displace pollen from flowers (distributed over a mosaic of patches) to their nest. The main goal is to compare how bees perform in terms of fitness (brood cells) and pollination services (number of flowers visited, foraging habitat visitation, and foraging distance) within a foraging day. Parameterization of the model input is mainly based on literature review, and the model's rule behavior was improved with a pattern-oriented approach.

In several examples, we present four simulation experiments to investigate parameter effects and parameter sensitivity (Section 10.4). We show that the model produces realistic foraging behavior progressively during the foraging day and that the responses overlap well with values from the literature. This can be considered as a validation with the exception that model bees were found to be somewhat more efficient than real bees, yielding higher numbers of brood cells. A global sensitivity analysis of the parameters in their biological range revealed that the amount of pollen per flower (*i.e.*, flower size) most influenced the number of brood cells and foraging habitat visitation. Body size was the dominant parameter for the number of visited flowers and the mean flown distance. The remaining two simulation experiments showed that the model can be considered robust against small changes (10% change in parameter value). The parameters that affected the responses here most were handling time per flower from the bee-related parameters and the amount of pollen per flower and pollen limitation (dispensing mechanism of the flower) from the landscape-related parameters. The simulation experiments yielded basic understanding of the model. Time constraints are more important for solitary bees than foraging resources, as they affect wood-nesting bees, which nest at field edges, more than soil-nesting bees.

KEYWORDS

Individual-based model (IBM), Solitary bees, Pollen foraging, Resource competition, Body size

10.1   INTRODUCTION

The honey bee (*Apis mellifera*) is the single-most important pollinator in the world [1]. However, the world counts about 20,000 wild bee species that are good pollinators as well [2]. Solitary bees are the largest subgroup of wild bees with about 14,000 species worldwide [3]. They live solitary and have a wide range of different food preferences, nest preferences, and behavioral traits (Figure 10.1). They differ considerably from eusocial and semisocial bees in foraging behavior (see, *e.g.*, [4–6]). Communication between eusocial bees can lead to near-optimal foraging [7,8], while solitary bees do not communicate about foraging patches. Solitary bees forage alone and therefore deal with limited knowledge of resource locations and their quality. While eusocial bees spend a large part of their time foraging for nectar, solitary bees focus on the collection of pollen for their offspring [9]. Evidently, "bee," "honey bee," "flower visitor," and "pollinator" cannot be used as synonyms [10].

Pollination is essential for many wild plants and human food crops. The rapid worldwide decline of honey bees and wild bees [11,12] has alarmed conservationists and politicians to take action to reduce pollinator losses [13]. The 2010 target to reduce the rate of biodiversity loss has not been met, despite increasing political effort to protect biodiversity [14]. The contribution of honey bees to crop pollination may have been overestimated [15,16], and protecting natural habitat near crop fields seems to be a key solution in providing natural pollination services (see, *e.g.*, [16–19]). In agricultural areas, habitat management for maintaining wild pollinator populations in the landscape can be a cost-effective way of securing crop pollination, especially when honey bees are decreasing [20].

(a)   (b)

(c)   (d)

FIGURE 10.1 Solitary bees are a large group of pollinators with different traits including body size and nesting preference. They share the same foraging resources. (a) A small soil-nesting solitary bee. (b) An intermediate-sized wood-nesting solitary bee. (c) A small solitary bee foraging for nectar and pollen on *Campanula patula*. (d) An intermediate-sized solitary bee foraging for nectar and pollen on *Campanula patula*. (Photo by J. Everaars.)

Agriculture is probably the largest threat for wild bees. Increased land use for agriculture leads to habitat fragmentation and loss of bee habitats and seems to be the most important driver of wild pollinator losses [21,22]. There is a consensus on how honey bee colony losses can be reduced [23,24], but defining how agriculturally dominated landscapes can be optimized for wild bees is more complex [18]. Many authors underline the need for a better mechanistic understanding of the effect of landscape configuration on wild bees in order to estimate the importance of habitat loss and fragmentation, to estimate the effects of land use on pollination services, and to support management decisions [12,25–29]. Body size and nesting preference affect the response to landscape features and fragmentation, but, up to now, no systematic exploration exists. Body size (comprising a wide range in solitary bees) affects traits such as the bee's velocity or capacity for carrying pollen (Figure 10.2). These traits influence the response of bees to landscape structures [30]. Models can help understanding how bees perform in patchy landscapes.

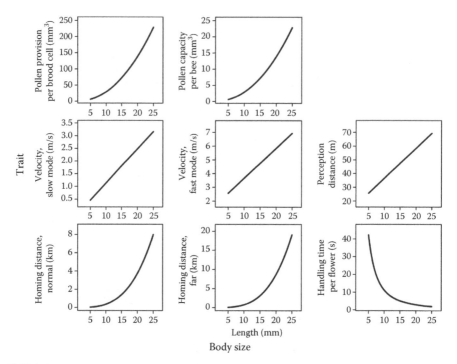

FIGURE 10.2 Effect of body size on different traits that affect the foraging behavior of solitary bees. Equations and references are given in Table 10.3.

Existing models with pollinators are of limited value for understanding the interaction between solitary bees and landscape structure. Data-based approaches include only little biological detail (see, *e.g.*, [31,32]). A few recent models included more biological realism. One model considered solitary foragers with limited memory rather than communicating eusocial bees [33]. Another study estimated field-to-field gene flow of crops by moving bumblebees based on experimental data [34]. A relatively detailed model did include local resource competition in space, bee size, and spatial configuration of patches, but focused on eusocial honey bees only [35].

No model has yet considered active pollen foraging as the main driver of pollinator movement. The existing models additionally lack several of the following components: comparison between different pollinators, partial habitats for nesting and foraging, differentiation between individuals, local competition, and size of the pollinator. Solitary bees are mainly foraging for pollen to provide their brood cells. Pollen differs from nectar in the sense that it does not replenish during the day. Solitary bees fly out alone from the nest, forage at the closest suitable flower patch, and return when they collect enough pollen and/or nectar. This typical behavior of solitary bees, which differs from eusocial bees, was the reason to develop a new model.

We developed a model with biological detail to study the interaction of wild bees with the landscape. We decided to use an individual-based, spatially explicit simulation model to simulate naive solitary bees that forage for pollen in the landscape. Time could be more constraining than energy for wild bees [36], and we therefore focused on differences in daily performance. The individual-based model (IBM, or agent-based model, ABM) is an established method for investigating animals in space and time, including solitary bees [37]. This kind of model allows us to include multiple traits (related to size and nesting preference, modeled as functional types) that determine movement and decision rules and lead to realistic depletion of pollen in space and time. Minimalistic approaches with only home-range descriptions cannot account for local competition processes [38]. IBMs with a high level of realism are often used in ecology and perceived as a welcome addition to the more theoretical approaches with a limited number of parameters [39]. In this chapter, we ask what foraging rules are required to simulate foraging behavior with body-size-related traits. At the test stage, we ask whether the model produces realistic foraging patterns in time and how well the response variables overlap with real systems. We also ask which parameters are most influential within biological parameter ranges and how sensitive parameters are against small changes.

10.2 SIMULATION MODEL FOR POLLEN-FORAGING SOLITARY BEES IN A SPATIALLY EXPLICIT LANDSCAPE

We describe here the model SOLBEE, a spatially explicit individual-based simulation model that mimics solitary bees in agriculturally dominated landscapes. Solitary bees forage for pollen individually without communication about food locations (in contrast to honey bees and bumblebees, which try to optimize foraging tasks by communicating). Our goal was to develop a mechanistic description of the pollinator–plant interaction at the landscape scale with relevant parameters. We implemented several body-size-related features that determine flight and pollen-collection behavior of the bees. The behavioral rules are implemented in an if–else format that is intuitive to follow. The landscape, with patches of flowering plants providing pollen to the bees, has several features from coarse-grained structures to fine-scaled pollen release per flower. We follow the standardized ODD (Overview, Design concepts, Details) protocol [40,41] to describe our IBM. It provides overview, general concepts, and case-specific model details in such a way that a model becomes more reproducible.

10.2.1 Purpose of the Model

This rule- and individual-based model aims at understanding how wild bees with different life-history traits interact with landscape structure and providing insight into the bee's perspective of a landscape. It is designed to be generic and applicable to multiple solitary bee species, which are defined by parameters. Many of the bee's traits, and therefore behavior as well, are related to body size. Some traits improve performance with body size (such as velocity, handling time, and resource recognition distance), while others moderate their performance (such as required pollen for a single brood cell). We expect that bees of different size perform differently when they collect pollen in a spatially explicit environment. This may also be influenced by the spatial allocation of the resources. We compare fitness differences by simulating the behavior and time allocation of solitary bees that determine the amount of pollen brought to the nest.

10.2.2 Entities, State Variables, and Scales

The model comprises a spatially explicit grid-based landscape with a given resource distribution and a community of pollen-collecting bees. Table 10.1 shows an overview of the system properties and their values.

10.2.2.1 State Variables of the Landscape (Environment and Spatial Units)

The landscape is described at two levels: at the coarse landscape level and the more detailed vegetation level (habitat units). We use for the first level a landscape generator, which enables the separation of habitat loss and fragmentation as different processes and allows for a wider range of foraging habitat availability and fragmentation than do images obtained from real land-use maps. Habitat availability (or loss) and fragmentation are focal variables for our simulations. The landscape is divided into habitat units that are either suitable or unsuitable for foraging. Agriculturally dominated landscapes are a mosaic of sharply contrasting habitat types [28]. Suitable habitat units (second level) are described by flower density, amount of pollen per flower, and the proportion of pollen that is available per pollinator visit (regulation of flower depletion). These different vegetation-level parameters serve the fine detail that is needed at the individual bee level (pollen uptake, flying between flowers) and represent homogeneous vegetation (fixed values). Auxiliary parameters describe the average distance between flowers and the initial amount of pollen for each habitat unit. We track the remaining pollen volume and the number of bee visits per grid cell during simulation.

10.2.2.2 State Variable of the Bees (the Individuals)

We consider six different types of solitary bees according to three body sizes (body length of 6, 12, and 24 mm) and two nesting preferences. Each simulation deals with a bee population of one type. Several bee traits are directly calculated from body size with allometric scaling rules. These include pollen capacity, flight velocity, flower handling time, perception distance, length of flight units ("step-lengths"), and return distance (Tables 10.1 and 10.3). Several of the bee traits are unrelated to body size such as memory size, tortuosity of the flight path, patch-leaving thresholds, and the amount of time spent at the nest for nonforaging activities. In Section 10.3.1, we give an overview of the main parameters with biological ranges and simulated values, accompanied by explanation and references.

We implemented above-ground (wood and cavity) nesting and subterranean (soil) nesting as two distinct nesting preferences. Wood-nesting bees nest in our model at the border of foraging habitat with unsuitable habitat (*i.e.*, field edges). Such habitat edges are often suitable for nesting in natural landscapes [30,42], since they provide shrubs and trees at nesting substrates. Soil-nesting bees nest in the soil at bare spots in the vegetation and nest in our model in aggregations everywhere in the foraging habitat. The nesting preference in our model has consequences for the spatial distribution of bees. Each bee has a fixed start location (nest) to which it must return (central-place foraging).

Six different bee types were selected to represent a wide range of bee genera in nature: soil-nesting small bees (6 mm): *Dufourea*, some bees from *Halictus* and

TABLE 10.1

List of Input Variables and Parameters Used in the Model with Abbreviations, Definitions, and Biological Parameter Ranges

Parameter	Short	Unit	Definition	Biological Range
Landscape:				
Landscape stochastic factor*	seed		Initial number for pseudo-random number generator	
Landscape element size	esize	m	Length of the most detailed landscape element (grain size of coarse grid)	From infinitely small to infinitely large, issue of fractality of landscapes
Foraging habitat availability*	am		Proportion of the landscape that is suitable foraging habitat	
Landscape fragmentation*	fr		Amount of fragmentation, reverse of "terrain smoothness" (Hurst exponent), synonym with habitat fragmentation	
Flower density*	fd	m^{-2}	Number of flowers	1–11,000
Pollen per flower*	ppf	mm^3	Pollen volume that one flower has available during 1 day	0.05–22
Pollen availability	plimit		Proportion of pollen of a flower that is extractable per pollinator visit	<0.1–1.0, different per species, and relatively unknown
Landscape quality*	bdc		Maximum number of brood cells that can be built from the available pollen in the landscape per individual	Unknown
Bee:				
Body length*	size	mm	Body length of a bee	4–26
Nesting preference*	nest		Category of nesting preference	Wood, soil
Flower memory*	fmem		Minimal number of most recently visited flowers that can be memorized	1 to unknown
Habitat cell memory*	cmem		Number of most recently visited habitat units that can be memorized	Unknown
Flight path tortuosity*	CRW		Density parameter of the wrapped Cauchy distribution that determines the relative amount of small turning angles during flight	Unknown
Lower patch leaving threshold	l_plt		Value of relative habitat cell quality below which a bee must leave the habitat cell	Unknown

(continued)

TABLE 10.1 (continued)

List of Input Variables and Parameters Used in the Model with Abbreviations, Definitions, and Biological Parameter Ranges

Parameter	Short	Unit	Definition	Biological Range
Bee:				
Upper patch leaving threshold	u_plt		Value of relative habitat cell quality above which a bee must stay in the habitat cell	Unknown
Time at the nest*	ntime	s	Time spent at the nest for nonforaging activities	30–800
Flytime	tt	s	Total time of activity during a foraging day	4–16 h
Bee (dependent):				
Pollen per brood cell	ppb	mm^3	Pollen volume that is needed to build one brood cell	Scaling relationship (good)
Pollen capacity per bee	pcap	mm^3	Maximum amount of pollen that can be carried per foraging bout per bee	Scaling relationship (acceptable)
Velocity medium/low	vmed	m/s	Flight velocity for flying in suitable habitat	Scaling relationship (acceptable)
Velocity high	vhi	m/s	Flight velocity for flying in unsuitable habitat	Scaling relationship (acceptable)
Handling time per flower	ht	s	Time needed to remove pollen from flower	Scaling relationship (acceptable)
Perception distance*	sightm	m	Distance radius at which bees can recognize habitat cells with flowers	Scaling relationship (unknown)
Length of flight units*	fflightm	m	Mean length of a flight unit of which a flight path is built	Scaling relationship (unknown)
General return distance	r50	km	The distance for which the probability of returning is 50%	Scaling relationship (acceptable)
Far return distance	r90	km	The distance for which the probability of returning is 90%	Scaling relationship (acceptable)
Ignorance	ig	km	Probability of (non)ignoring flower location within sight or at the present location, the inverse of habitat cell memory	Unknown

Note: Parameters with asterisk are explored in the global perturbation analysis (simulation experiment 2).

Lasioglossum; soil-nesting medium-sized bees (12 mm): most bees from *Andrena*, *Anthophora*; soil-nesting large bees (24 mm): *Centris*, *Eulaema*, *Oxaea*, *Habropoda*, *Xenoglossa*; wood-nesting small bees (6 mm): *Ceratina*, *Chelostoma*, *Heriades*, *Hyleaus*; wood-nesting medium-sized bees (12 mm): several *Osmia*, *Megachile*; wood-nesting large bees (24 mm): *Xylocopa*.

Different state variables are recorded for each foraging trip (used synonymously with foraging bout) for each bee: pollen collected during the trip, distance from the nest, number of unsuccessful and successful flower visits per trip, most recently visited locations, quality of last visited location, flight direction, start time of the foraging trip, and "future time" to schedule the next behavioral element and current spatial location. Data of different foraging trips is collected at the nest; here we record (per bee) the number of returns to the nest, pollen delivered to the nest, number of flowers probed, maximum distance from the nest, and time spent per behavioral module. Other auxiliary variables keep track of the total distance flown and trip duration. At the end of the simulation, the mean amount of pollen collected per bee is converted into the (size-related) number of brood cells. We do not use population-level variables.

We link the landscape and the bee population with a parameter that describes the overall landscape quality for bees, which is used for calculating the total number of bees in the landscape (section initializations).

10.2.2.3 Scales

The simulated landscapes have a spatial extent of 1 km, because solitary wild bees respond to landscape structures on scales up to 1,000 m [43]. We used a grain size of 50 m that mimics raster-based land-use maps and is used in landscape-scale studies with solitary bees [44,45]. Lower grain sizes result in very fragmented, grainy landscapes that do not realistically represent agriculturally dominated landscapes. However, bees perceive landscape structures in more detail [29] especially nesting substrate in strips with shrubs and trees [46] and on small tree islands [47]. Bees show differences in behavior at a fine scale (see, *e.g.*, [48,49]). We therefore subdivided the landscape maps into 5×5 m cells. Our model landscapes thus consist of 200 by 200 grid cells. From the perspective of the bee, the landscape contains foraging habitat and matrix. The foraging habitat is split up into edge (5 m wide strips, which wood-nesting and soil-nesting bees use for nesting and foraging) and interior (used for foraging by both bee types and for nesting by soil-nesting bees only).

Our model landscape has reflecting rather than absorbing boundaries, because we deal with central-place foragers that have to return to their nests. Consequences of this implementation are discussed under "Details" (Section 10.2.7). We use 14,400 time steps of 1 s, which equals a foraging period of 4 h. A behavioral unit of a bee lasts from 1 to several seconds. The decision of the exact time horizon is based on a tuning with the time spent at the nest, which is discussed in simulation experiment 1 (Section 10.4).

10.2.3 PROCESS OVERVIEW AND SCHEDULING

After the initialization phase, in which a landscape is generated, a bee community is defined (bee traits regulated by allometric scaling, see Section 10.2.5).

A nest site is assigned to each individual, upon which a foraging day starts. The virtual bees exhibit five types of behaviors. The behavior of one individual is strictly sequential in time and lasts at least for 1 s (discrete time steps). When an individual performs a behavioral type, state variables are changed. When a behavioral type lasts longer than 1 s, a waiting time is set for that individual. Individuals are processed in random order at each time step (Figure 10.3a). Waiting times imply asynchronous updating of individuals. When a foraging day is completed, values are averaged per bee and written to an output file.

These behavioral modules are performed in a loop until a certain foraging time (*flytime*) is completed after which all activities are stopped instantaneously (fair comparison between bees) and values are averaged per bee and written to an output file. During behaviors 1, 3, and 4, bees leave visitation marks in the landscape, which are used as a measure of pollination potential in the later analysis. Time penalties for each bee are given in Table 10.2.

The five behaviors are as follows (Figure 10.3b):

1. **Forage Flowers** (Forage Flowers within a Landscape Grid Cell): A bee flies from flower to flower directly (spatially implicit), encounters stochastically a full or empty flower (based on present resources), collects pollen, and decides about leaving the grid cell. Behavioral states can change into **fly around** (3) (current resources very low) and **neighboring cell** (2) (current resources low), or **fly back** (4) (enough pollen collected).

2. **Neighboring Cell** (Fly to a Suitable Neighboring Landscape Grid Cell): The bee considers the eight neighboring landscape grid cells for foraging. It accepts one of them randomly when it contains flowers and if it has not been visited recently. The behavioral state changes back to **forage flowers** (1) after moving there. On rejection of all eight cells, the behavioral state changes into **fly back** (4) (far from the nest) or into **fly around** (3).

3. **Fly Around** (Fly Around and Look for Unknown Foraging Areas): The bee performs a correlated random walk (changes direction and moves one step). The behavioral state changes to **forage flowers** (1) if either the new landscape grid cell is suitable (contains flowers and has not been visited recently) or if a suitable landscape grid cell has come within sight (move second step) or it can change into **fly back** (4) (too far from the nest). With a certain probability, suitable cells are ignored, leading to a more realistic foraging behavior (better matrix crossing and better foraging of interior habitat; see Section 10.2.8).

4. **Fly Back** (Fly Back to Nest): The bee performs a directed random walk (correlated random walk in the direction of the nest). Eventually, the state changes into **nest reached** (5).

5. **Nest Reached**: The bee delivers the pollen to the nest and spends time on nonforaging activities (while other bees still deplete the landscape). Afterward the behavioral state is set to **forage flowers** (1).

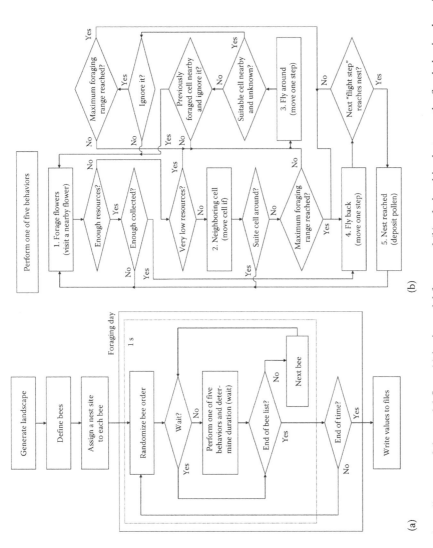

FIGURE 10.3 Flowchart diagrams of the model flow: (a) basic model frame and (b) relationships between the five behavioral modules. The relative resource quality is based on comparison of currently encountered and remembered resource quality (ratio of full and empty flowers).

TABLE 10.2
Overview of Time Penalties

Behavior	Time Penalties for	Value (Minimal 1 s)
1. Forage flowers	Poor habitat grid cell:	
	• Assessing patch quality	1 s
	Good habitat grid cell:	
	• Flying to a flower	Based on medium/low velocity (size)
	• Full flower: removing pollen from a flower	Based on handling time (size)
	• Empty flower: assessing flower	1 s
2. Neighboring cell	• Accepting or denying a surrounding cell	1 s
	• Flying to a surrounding cell	Based on medium/low velocity (size)
3. Fly around	• Distance flown per flight unit	Based on high velocity (size)
4. Fly back	• Distance flown per flight unit	Based on high velocity (size)
5. Nest reached	• Pollen deposition and other nonforaging activities	30 s (parameter *time at the nest*)

Note: More details on the selection of values can be found in Section 10.3.2.

10.2.4 DESIGN CONCEPTS

10.2.4.1 Basic Principles

We use several known foraging principles such as area-restricted search for resource items [8] and a patch "giving up time" [50]. Area-restricted search means that the animal has a limited knowledge and memory of the environment and assesses the local environment with a fixed-sized search window and inevitably chooses close resource points [8]. The decision of an individual to leave a patch is based on actually probed flowers and the estimated resource level, rather than by time spent in a patch. Resource-based "departure rules" may be more realistic than time-based ones [8], and we used resource ratios and thresholds to determine departure [51]. In a spatially explicit IBM, this means that bees react to the rapidly changing resource levels caused by other foraging individuals. We use the concept of "near–far search" as an optimization that is relevant in large patchy environments [52]: bees generally forage on nearby flowers, but switch to foraging resources farther away when nearby resource levels get low. Such behavior has been shown for solitary bees [44,53]. Not-considered concepts that are known for nectar-foraging honey bees or bumblebees include "majoring and minoring" [54], the "matching law," and the "ideal free distribution" [33], because they are not likely to apply to oligolectic (foraging on one plant species) pollen-foraging solitary bees. We do not explore or compare the used principles; they only serve to achieve realistic foraging behavior.

10.2.4.2 Emergence

Complex behavior of bees emerges from five behavioral rules, posing differences on individuals during simulation (*i.e.*, different behaviors coexist). Differences in nest positions (and bee densities) cause differences in local competition for foraging resources. The amount of pollen collected (or number of brood cells), the mean distance flown from the nest (or foraging distance), and the number of flowers visited are not coded into the model. Also a spatial "visitation map" emerges from behavioral rules and landscape structure and is used to calculate the percentage of the foraging habitat that has been visited by a bee community.

10.2.4.3   Adaptation

The individuals adapt to changing local resource conditions. After visiting and memorizing the status of a predefined minimal number of flowers (*flower memory*), the habitat cell quality (ratio of full to empty flowers encountered) is compared with the quality of the last visited habitat grid cell. According to the relative quality and an upper and lower patch-leaving threshold, a bee stays (relative quality better than upper patch-leaving threshold) or leaves.

10.2.4.4   Objectives

We measure the performance of an individual by how much pollen it can collect within a fixed time span. The costs consist of time penalties for each behavioral rule (Table 10.2). The amount of pollen collected at the nest is eventually converted to a number of brood cells according to body size. For solitary bees, efficiency (and hence fitness) can be formulated as "potential number of offspring produced from the pollen per unit time on the flower" [55], represented in the model by the number of brood cells.

10.2.4.5   Learning

The behavioral rules are static without a learning component (the same conditions in later foraging trips lead to the same decisions), because solitary bees are considered primitive foragers and display less efficient behavior than honey bees [56]. The memory of the bees is used for memorizing locations and quality and changes constantly (limited memory), but application of this memory is not optimized by learning. The sequence of behavioral modes and the outcome of decisions change with time as a result of local depletion of flowers and patches.

10.2.4.6   Prediction

The individuals cannot predict any future condition, except for the fact that they memorize recently visited habitat cells and avoid them because they predict that food conditions are low there.

10.2.4.7   Sensing

During foraging, bees assess neighboring landscape grid cells for suitable foraging habitat (resource availability) by vision and remember if they have visited them recently. When flying through unsuitable matrix, they assess the environment with a square-shaped search window. Furthermore, they sense the distance to the nest and can fly back anytime when they are too distant. The bees do not sense each other and therefore experience local competition only indirectly.

10.2.4.8   Interaction

Interaction between individuals exists indirectly through competition for resources. Local competition is expected to be highest around the nest, where resources are depleted first. Since the nest location is determined by landscape structure and nesting preference, the magnitude of interaction differs between landscapes, bee types, and within a simulation in space and time.

10.2.4.9   Stochasticity

We use stochasticity for different goals:

1. To generate landscapes: The fractal algorithm is partly based on stochastically generated noise.
2. To randomize the sequence and decisions of bees: We randomize the sequence of individuals at each time step, assign a random direction at the start of the foraging trip, choose randomly from suitable habitat cells around, and choose randomly when there are more "nearest locations" within sight.
3. To produce natural variability in behavior and nesting: Natural variability is a major element of the correlated random walk (wrapped Cauchy distribution for turning angles and a normal distribution for flight unit lengths) and different leaving thresholds. Nest sites are selected randomly. A landscape grid cell is drawn randomly until one is suitable for nesting according to different nesting preferences. Furthermore, we use specified frequencies (or a specified fraction) for ignoring suitable habitat grid cells, for determining the binary full or empty status of a flower, for clumping of nests (soil-nesting bees), and for the probability distribution of maximum return distances. In specific cases, we use stochasticity for rounding to integers (when the frequency distribution between two is not uniform).

10.2.4.10   Collectives

We do not use collectives in our model. Each bee forages solitarily without direct interaction with other bees.

10.2.4.11   Observation

We record bee performance for each individual as total pollen collected, realized and mean foraging range (distance flown from the nest), number of returns to the nest, and mean trip duration. For each behavioral module, we record the amount of time spent in it. As output variables, we use the arithmetic mean over all individuals and the standard deviation. At the landscape level, we record the total number of flower visits and visitation marks (grid cells visited, including those without flowers). From these visitation marks, we calculate the percentage of seminatural habitat that was visited. We record the number of flower visits per bee, as indicator for how the pollen vector behaves (but not as an indicator of the performance of the bee, because here we do not distinguish between full and empty flowers).

10.2.5 INITIALIZATIONS

10.2.5.1 Landscapes and Habitat

We use "noise" from a random number generator (*landscape stochastic factor*) and a Hurst exponent for fractional Brownian motion (*landscape fragmentation*) and a threshold (*foraging habitat availability*) to generate a landscape [57]. We generate landscapes symmetrically and wrapped [58]. Symmetry prevents the entire foraging habitat to be in one corner, and wrapped boundaries give the landscape the appearance to be part of a larger landscape.

Suitable habitat cells are assigned an initial pollen volume based on *flower density* and *pollen per flower*. We do not assess a range of landscape metrics for the landscapes, because most landscape metrics are highly correlated in artificial landscapes [59]. Instead, we calculate the proportion of available nest habitat (landscape grid cells suitable for nesting) according to the nesting preference of the bee.

10.2.5.2 Allometric Scaling

The bee traits relating to body size were calculated by allometric rules based on the initial body length (Table 10.3). In order to use allometric relationships from literature, we first converted body length to intertegular span—the shortest linear distance between the wing tegulae [60]—and to dry body mass (data taken from [61]). Several traits are directly based on the literature (*pollen per brood cell, general return distance, far return distance*), while others were calculated from data from multiple studies (*velocity medium/low, velocity high, handling time per flower*). We use these parameters subsequently to calculate *pollen capacity per bee, perception distance* (and perception area), and *length of flight units* (Table 10.3).

A more complex allometric trait is the maximum distance per foraging trip, based on homing distance. Homing distance can be considered as the maximum distance with knowledge of the environment, which differs between individuals and in different directions. Allometric scaling of homing distance of bees is known for typical homing distance (*r50*) and far or maximum homing distance (*r90*) from the literature [62]. We connect these two parameters by a saturation curve (Table 10.4) after calculating a shift parameter, which represents a certain minimal knowledge (see Section 10.3.2 for more details). Within this minimal distance from the nest, bees return only with a full pollen load.

10.2.5.3 Individuals

We assume our bee community to be in equilibrium with the amount of foraging resources. In natural communities, bee density is often related to flower cover [30,63] or flower diversity [64,65]. We used pollen volume as resource parameter [61] and scaled the number of individuals to the total amount of pollen present in the landscape. More specifically, the total number of individuals is calculated by dividing the pollen volume present in the landscape by the pollen volume available per individual (Table 10.4). The latter is calculated with the help of the parameter *bdc* (*bee density control* or *landscape quality for bees*) as the "potential offspring (brood cells) per individual." We set this value at 30, which means that each bee can potentially build

TABLE 10.3
Allometric Rules That Were Used in the Model

x	y	Formula	Data Source
Body length (mm)	Dry body mass (mg)	$y = 0.0398 \cdot x^{2.589}$	Data taken from [61]
Dry body mass (mg)	Intertegular span (mm)	$y = 0.77 \cdot x^{0.405}$	[60]
Dry body mass (mg)	Pollen per brood cell (mm³)	$y = 10^{(0.433+0.868 \cdot \log_{10} x)}$	[61]
Pollen per brood cell (mm³)	Pollen capacity per bee (mm³)	$y = x/10$	10 foraging trips, Section 10.3.1
Body length (mm)	Velocity medium/low (m/s)	$y = -0.214+0.135 \cdot x$	Section 10.3.1
Body length (mm)	Velocity high (m/s)	$y = 1.48+0.218 \cdot x$	Section 10.3.1
Body length (mm)	Handling time per flower(s)	$y = 919.62 \cdot x^{-1.914}$	Section 10.3.1
Velocity high (m/s)	Perception distance (m)	$y = 10 \cdot x$	Distance of 10 s flight forward
Perception distance (m)	Perception area (m²)	$y = (2 \cdot x)^2$	Search window around the bee
Perception distance (m)	Length of flight units (m)	$y = x/2$	Perception distance flown in two flight units
Intertegular span (mm)	General return distance (km)	$y=10^{(-1.643+3.242 \cdot \log_{10} x)}$	[62]
Intertegular span (mm)	Far return distance (km)	$y=10^{(-1.363+3.366 \cdot \log_{10} x)}$	[62]

Note: All bee traits (y) are directly or indirectly related to body size (x) and in the model all are determined by the variable *body length*. Parameters are explained in Table 10.1.

TABLE 10.4
Body-Size-Based Calculation of Individuals and Maximum Flight Distance on a Foraging Trip

	Multiple x	Calculation of	Formula	Notes
A	Foraging habitat availability (am), landscape area, flower density (fd), pollen per flower (ppf), landscape quality (bdc), pollen per brood cell (ppb)	Individuals	$(am \cdot \text{total area} \cdot fd \cdot ppf)/$ $(bdc \cdot ppb)$	Pollen volume available to the entire bee community (numerator) and the pollen volume available to one bee (denominator) are calculated in cm^3. The total area is 1 km^2
B	General return distance (r50), far return distance (r90), distance from the nest	Probability (of reaching a distance from the nest without knowledge of the environment, homing ability)	$(\text{Distance} - shift)/$ $(Km + \text{distance} - shift)$	$Shift = 1.125 \cdot r50$ $- 0.125 \cdot r90$ $Km = r50 + shift$
C		Maximum distance from nest (allowed to fly per foraging trip)	$(U \cdot Km/$ $1 - U) + shift$	Inverse of distance probability; U is drawn from a uniform distribution

Note: Row labels (A–C) are used in the text.

30 brood cells from the amount of pollen in the landscape independent of its size (equal performance potential for all bees). A large value means fewer bees, but also a higher landscape quality per bee. The value chosen (see Section 10.3.1 for a short review) results in realistic bee densities in the landscape with high bee numbers for small bees and lower numbers for large bees and an increase with increasing foraging habitat. A disadvantage of this approach is that model runtime varies several orders of magnitude for different parameter settings (*body length* and *foraging habitat availability* as well as *flower density* and *pollen per flower*).

Each simulation uses one bee type (according to size and nesting preference), and the initially calculated number of individuals remains constant during simulation (one foraging day without population dynamics). Each individual has several initializations. The most important is the nest assignment, which is, in the model, always near foraging resources [66]. Wood-nesting bees accept a (randomly chosen) location to nest when it is at the edge of the foraging habitat, and soil-nesting bees accept a location all over the foraging habitat but preferably in the vicinity of another bee nest in the soil (see Section 10.3.1 for details). Nest distribution depends thus directly on landscape configuration. Each bee was additionally initialized with a random direction (used in the behavior **fly around**) and a maximum distance (knowledge of the environment).

10.2.6 Input Data

The model does not use time-varying input data [41].

10.2.7 Details

Here we describe several implementation details and discuss the considerations leading to implementation decisions. Data-based selection of foraging rules and parameter values follows in separate Sections 10.3.1 and 10.3.2.

10.2.7.1 Landscapes and Habitat and Boundary Conditions

We use "noise" from a random number generator (*landscape stochastic factor*) and a Hurst exponent for fractional Brownian motion (*landscape fragmentation*) and a threshold (*foraging habitat availability*) to generate a landscape with the midpoint displacement algorithm [57]. In short, this algorithm creates a three-dimensional surface, based on four initial corner points. With "noise" from a random number generator (*landscape stochastic factor*) and a Hurst exponent for fractional Brownian motion, the additional grid points are calculated. We did not use the standard C random number generator, but an improved algorithm [67]. The three-dimensional surface is converted to a two-dimensional suitable–unsuitable map by the "valley flooding" method (in which "water" represents the suitable habitat). This method had slightly more realistic-looking landscapes than the "mountain-cut" method. We use a modified algorithm that allows for symmetrical and wrapped landscapes [58]. Symmetry prevents the entire suitable habitat to be in one corner, and wrapped boundaries give the landscape the appearance to be part of a larger landscape.

We implemented reflecting boundaries for bees foraging near the landscape border as follows. If the new coordinates for a move are calculated (with a change in x, a change in y, and the new coordinate pair), we check whether one of the coordinates lies outside the grid. If so, we multiply the change in x, y, or both with -1. This means that a line through the old coordinate pair, parallel to the border, functions as a mirror. We also prevented nesting at the border of the landscape, to prevent increased local competition by the reflecting boundaries. We scale this additional rule with body size: the minimal distance from the border where it is allowed to nest equals body length as given in meters (*i.e.*, times 1,000 for body length in mm), and we rounded it to landscape grid cells. We have chosen this method after thorough experimentation with alternatives. The selected method has the best average bee performance with the least unwanted effects, but a little higher standard deviation (larger difference between individual bees) for bee performance. The field was always large enough to have most of the bees not nesting near the landscape boundary anyway.

10.2.7.2 Patch Leaving Rules and Flower Encountering

The probability of leaving a grid cell is based on the relative quality of the habitat grid cell compared with a previously visited one. This is calculated by dividing current habitat quality by the remembered habitat quality. Habitat cell quality is the fraction of successful flower visits to the total flower visits (based on a minimal number of flowers, *fmem*). If the upper patch threshold (*u_plt*) is 1.0 and the current cell has at least the same quality as the last (value of 1.0 or higher), the probability of leaving is 0.0. If the relative cell quality comes below a certain threshold (*l_plt*), the probability of leaving is 1.0 and the bee flies up in search for a better location (behavior **fly around**). A third leaving threshold is determined stochastically (for each decision) from a uniform distribution between the upper and lower patch threshold (*i.e.*, it differs per moment). If the relative cell quality is below this third threshold, the bee will search for resources in neighboring grid cells (behavior **neighboring cell**). For the upper patch-leaving threshold, we have chosen a value of 1.0, which means that a bee only stays for certain when the current habitat grid cell is at least as good as the previous one. The lower patch threshold should not be lower than 0.5, because otherwise bees would stay in a grid cell with less than half of the quality of the previous grid cell.

For determining the success of a flower visit, we first calculated the proportion of full flowers in the habitat grid cell. We did not track the pollen volume of each flower in the landscape, but the pollen remaining per grid cell. When we divide this first over the pollen volume per flower and then over the number of flowers per grid cell, we have an estimate for the proportion of full flowers in that grid cell. Since we assume homogeneous vegetation, all habitat grid cells have the same amount of flowers based on *flower density*. For each landscape grid cell with foraging habitat, we track the amount of pollen left, as approximation for the number of full flowers. We use the proportion of full flowers as probability for encountering a full flower. If a full flower is encountered, the bee takes up pollen according to its uptake capacity, and the amount of pollen in that grid cell decreases by the same amount. The bee receives a time penalty for collecting the pollen. For both full and empty flower encounters, a bee receives a time penalty for flying to the flower (based on distance and velocity).

10.2.7.3 Determination of the Maximum Distance Allowed to Fly on a Foraging Trip (Distance of Certain Return)

We used the relatively well-studied homing distance to set a maximum foraging distance per foraging trip. First, we calculated the typical (*r50*) and far (*r90*) homing distance (distance beyond which 50% and 90% of the bees were not able to return at release, respectively) for a bee of a certain size (Table 10.3) based on a known allometric relationship [62]. We use a hyperbolic Michaelis–Menten function to describe a homing probability curve. We could solve the equation with two parameters (*r50* and *r90*) for a bee of a certain size. Since a Michaelis–Menten function normally goes through the origin and we preferred to have a range of x values (close to the nest) for which a bee would be able to return (minimal knowledge of the environment), we shifted the curve, based on the two known points (*r50* and *r90*) on this curve (Table 10.4, row B, calculation of *shift*). We did this by solving the altered equation for homing probability (Table 10.4, row B) where 50% and 90% homing probability share the same *shift* value, resulting in the equation for *shift*, which alters the saturation constant *Km* accordingly (Table 10.4, row B). The return or homing probability can then be calculated as a function of distance from the nest (Table 10.4, row B).

The response homing probability can also be interpreted as the probability that an individual has reached the distance beyond which it has no knowledge of the environment (will not be able to return home to the nest).

We know that bees fly exclusively within their homing range, which differs between individuals and within different directions. We decided to include this in the model as distance beyond which a bee is not allowed to fly and was calculated for each individual, on each foraging bout. We inversed the calculation (Table 10.4, row C), giving a stochastically determined distance per foraging trip beyond which foraging is not allowed. This reproduces a realistic set of distances from the nest where the bee should maximally return, never close to the nest and with a similar shape as for real foraging data [68]. An asymmetrical S-curve would probably be a good alternative, but we could not fit it with two known points.

10.2.8 PATTERN-ORIENTED MODELING

We used a pattern-oriented modeling (POM) approach [69,70] to parameterize the behavioral rules of the model. The visitation pattern (grid cells where bees left one or more visitation marks) showed that bees with area-restricted search and a limited memory easily fly back to old patches and grid cells and never cross the matrix (Figure 10.4b). We considered the behavior "too locally optimized." Foraging resources appear "in sight," when flying parallel to the vegetation, and such a grid cell is visited when it is not memorized (10 last visited grid cells are in memory). We added a parameter (*ignorance*) that lets the bee ignore decisions with a certain probability (details in Section 10.3.1). This probability assumes an innate preference of solitary bees to occasionally take the risk of crossing the matrix for finding

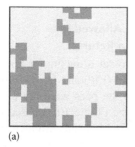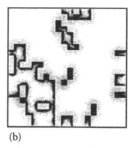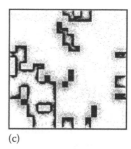

(a)   (b)   (c)

FIGURE 10.4 An example landscape with two visitation responses for bee communities with slightly different foraging behavior. (a) An example landscape with fragmented (meadow-like) fields. The darker shade of gray represents the foraging habitat in an unsuitable matrix. The smallest patch measures 0.25 ha (50 by 50 m, *landscape element size*). In this example, around 1,400 individuals (large bees) nest in woody structures at the border of the vegetation. Nests are represented by 1,400 small white dots at edge structures (partly sharing the same grid cell and hard to represent at this spatial scale, see also Figure 10.7). (b) Visitation by a community of large bees, in the first 8,000 s of simulation. The darkest shades of gray represent the highest total visitation and white no visitation (excluding nest sites). There is only flight activity around the border where the bees nest, where they find sufficient foraging resources. The grayscale is optimized to visualize single "visitation marks" as well. (c) Visitation of a bee community similar as in (b) but with introduced ignorance. Single flight paths can be recognized and the bees cross occasionally the matrix and the interior of fields with foraging habitat.

far resource patches with higher resource abundance instead of optimizing the short-term profit. We decided that a value of 0.1 (ignoring 10% of the optimal decisions) results in a more plausible foraging pattern (Figure 10.4c). Bees now sufficiently cross the matrix and search for better patches outside the local patch. They also visit more often the interior of a flower field.

10.3   LITERATURE REVIEW

This section concerns the ecology of solitary bees. We review the relevant literature that we used to develop the model and to select realistic parameter values. We aim to identify the complete biological range of the parameters and to identify the most common values. We also explain in more detail how we selected our foraging rules for the model with literature support. Finally, we explore the literature for the response variable of the model that we later use to validate the model.

10.3.1   Natural Parameter Ranges, Extremes, and Common Values

We discuss the parameters from Table 10.1 with a literature review. The results from this review, the biological range of each parameter, is included in Table 10.1.

10.3.1.1   Body Length

Body lengths of wild bees normally vary in most parts of the world between 4 and 28 mm [2]. In the model, we focused on 6, 12, and 24 mm body length for small, intermediate-sized, and large bees. We chose to model small bees of 6 mm because this covers a wide range of common bees from different genera: *Ceratina, Chelostoma, Heriades* and *Hyleaus* (wood and cavity nesting), *Dufourea*, and several bees from the genera *Halictus* and *Lasioglossum* (soil-nesting). Most bumblebees, which are wild bees but not solitary bees, range from 12 to 18 mm (10–25 for more extreme species). However, we chose a smaller group of solitary bees to represent the outer range of body sizes (24 mm). Wood-nesting bees from the genus *Xylocopa* measure 20–28 mm. Soil-nesting bees that are similarly large can be found in the genera *Centris, Eulaema, Oxaea, Habropoda,* and *Xenoglossa*. The main reason for including medium-sized bees (instead of only the two extremes) is that they are the best-studied group of solitary bees with the most reliable available data. Well-studied, intermediate-sized bees (10–14 mm) are found in the genera *Osmia* and *Megachile*. Soil-nesting, intermediate-sized bees are found in the genera *Andrena* and *Anthophora*.

10.3.1.2   Landscape Element Size

Foraging and nesting resources for bees can probably be considered to have a fractal patch size distribution in the landscape. Even in large unsuitable regions, some smaller patches of several square meters and even single isolated flowers (down to a scale of square centimeters) can function as foraging habitats. Despite this, we used a coarse grain size (50 m²) for foraging resources for several reasons. Geographic information system (GIS)-based landscape maps often use a detail of 50 m² and we also used this as grain size for foraging resources, and artificially generated landscapes with a grain size of 30 m were a good imitation of raster-based GIS maps in Hargis et al. [59]. Landscape elements with foraging resources are usually larger than those with nesting resources. A grain size (or element size) of 50 m as landscape detail (or somewhat smaller 0.2 ha blocks) has been used in several wild bee studies [44,45,71]. Furthermore, agriculturally dominated landscapes often have a patchy structure with large homogeneous blocks, rather than a fractal structure. The use of a grain size of 50 m in the landscape generator provides useful maps with foraging resources. Since nesting resources for bees are finer scaled (such as linear elements with scrubs or bare spots in the vegetation), we defined nesting resources with another method.

10.3.1.3 Flower Density, Pollen per Flower, and Pollen Availability

Flower density is a highly variable parameter. For solitary bees, the unit for visitation is the flower or flower head and not the plant. Flower densities are much higher than plant densities. For a homogeneous vegetation, flower density ranges from hundreds (*Caltha palustris*) to millions (*Thalictrum aquilegiifolium*, Ranunculaceae) per square meter [72]. High flower densities are often based on florets and not on flower heads, for example, from Centaurea species [73]. For nectar collection, these are probably realistic visiting units for bees. Active pollen collection by solitary bees is nevertheless a process on the flower(-head) level. The highest density we could find for single flower units (not composite flowers) was for *Brassica napus* (oilseed rape) and counted to more than 11,000 m^{-2} [74]. Lower densities can have any value. For clover, the flower(-head) density was counted at 50 inflorescences/m^2 [48] but Waddington reported a range of 50–250 [75]. Large flowers such as sunflowers are an exception and have a density below 10 m^{-2}. Other low flower densities reported are often a result of downscaling flower densities from heterogeneous units on a larger (hectare size) scale.

The *amount of pollen per flower* varies considerably between species. Studies used different measurement units for pollen production and assessed them with various methods and only few compared different plant species with a wide range of pollen production. The first way to measure pollen production is by number of grains, which ranges from 1,100 to 150,000 [76]. A second method is by weighing the amount of pollen, which varies from 0.35 to 6.11 mg for Ranunculaceae [72]. The third method is by calculating the pollen volume for each flower, which ranges from 0.05 to 21.7 mm^3 [61]. For both pollen number and pollen weight, lower values can be found [77,78], but we did not find higher values. Calculating the pollen volume [61] is a relatively new approach without comparable studies, but the authors put special effort in covering a wide taxonomical range. We used pollen volume, because the most qualitative information of pollen transport by bees is measured in volume [61] and therefore in the same unit.

We use an additional parameter for *pollen availability* to regulate pollen removal. Many plants release pollen in portions in order to reduce the pollen that is taken in one visit and help the plant in optimizing pollination. There are many different syndromes for regulation of pollen release, which are widely distributed in the plant kingdom [79], and there are no general rules for gradual release and depletion of pollen. Our model flower does not mimic a certain flower type. We use a pollen availability of 0.3 per visit, a very general value. This maximum proportion of available pollen (initially present) at one moment is a plant trait and independent of bee size and allows for a minimum of at least three to four bee visits to the flower before being empty. Percentages of pollen removal at a first visit (bumblebees) vary from 20% to 80% [80,81]. Depletion rates vary from almost all pollen from all plants after 2 h [82] to 60% of the pollen per flower after 3 h [83]. Some pollen will never be removed, about 1%; see [83]. We also use decision thresholds in the model (explained later) that prevent bees to spend too much time on flowers that provide little (remaining) pollen.

We tried to balance these three different flower parameters in the model. The flower density and pollen production per flower regulate the amount of pollen per square meter. Field data (see, *e.g.*, [72]) does not reveal a direct relationship between flower density and pollen production per flower, but still indicate a weak trade-off between flower number and pollen per flower (highest flower densities show a low pollen production per flower). We used 50 flowers/m^2 as flower density with a rather oligolectic bee in mind that only visits flowers of a certain species that has a more scattered distribution. In contrast, polylectic bees often visit many different flowers that have a high combined density.

There is, to our knowledge, no evidence that the size of the flower (and hence pollen production) is correlated with the size of the flower visitor; single plant species are visited by a diverse bee fauna, for example, [84,85].

Therefore, we model one flower type with a certain amount of pollen visited by all bees of all sizes. We decided to use relatively small flowers so that also the smallest bees have to visit several flowers before they have collected enough pollen. Therefore, we used a pollen volume of 0.5 mm^3. This value is also close to the mean pollen volume of 0.6426 mm^3 for one flower or flower head calculated from the data provided by Müller et al. [61]. In the analysis of the model, we explore the consequences of these parameter values, with more extreme values for both *flower density* and *pollen per flower* in simulation experiment 2 and the effect of moderate changes in *pollen availability* in simulation experiment 4.

10.3.1.4 Landscape Quality for Bees

We used the parameter *landscape quality for bees* to set a basic bee density in the landscape that also varied with body size. A higher value implies lower bee densities and more available pollen per individual. Bee densities in the literature are often given in estimated number per hectare. Bee densities differ between natural habitats and crop areas. Density of Apoidea in natural grasslands was estimated at 256–1,500 individuals per ha and at 232–4,700 ha^{-1} for alfalfa fields. Bee densities vary through the season with higher densities in spring [30]. For other habitats such as forests, woodlots, roadsides, and village parks, the estimates were lower, varying from 46 to 281 individuals per ha [86].

The parameter values for flower density and pollen per flower (Table 10.1), in combination with a value of 30 for *landscape quality for bees*, yielded between 204 and 87,209 bee individuals per km^2. This span is caused by differences in *body length* and *foraging habitat availability* in the different landscapes. A low landscape quality value of 10 in combination with high resource conditions increased the number of individuals to 92,819 and 2,093,017 for large and small bees, respectively. Scaled to 1 ha (divided by 100), it becomes clear that these are still realistic densities. For seminatural habitats, a value of 30 yields more realistic densities (2–900) than a value of 10 (900–21,000) and is in balance with our moderate resource densities. Note that for extremely high values for *flower density* and *pollen per flower* (simulation experiment 2) extremely high individual numbers can occur in the model.

10.3.1.5 Pollen per Brood Cell and Pollen Capacity per Bee

The pollen volume that a bee requires for a brood cell clearly relates to body size [61] and was not further reviewed by us. The pollen volume transported by a bee per foraging trip is unknown, but it may scale in the same way with body size as the complete brood cell volume. Therefore, we assumed an average of 10 foraging trips per brood cell to define the capacity per bee. We assume a 1:1 sex ratio to cancel out effects of sexual dimorphism [87]. It is not clear whether the number of foraging trips per brood cell can be used as a constant, since sexual dimorphism and other factors cause variability. The number of collection trips per brood cell varies from 3 to 15 trips between species [88,89]. A range of other species, with a large span in body size, need 8–12 trips per brood cell [88,90,91], which suggests that 10 is an acceptable choice.

We made one other assumption with regard to the pollen capacity of bees and their behavior. In rare cases, the smaller bees find flowers with more pollen than they can transport. Bees deal with this in different ways. While one species completes the pollen load during a single visit, another species (from the same genus) visits several flowers regardless of pollen availability [91]. We argue that a bee needs to visit at least two flowers to be biologically meaningful to flowers. When the pollen available per visit from a flower exceeds the transport capacity of a bee (which means that the bee would be fully loaded), we set the pollen available to half of the normal availability. This rule was only invoked under extreme parameter combinations (*e.g.*, very large flowers with abundant pollen in simulation experiment 2).

10.3.1.6 Velocity

In the literature, we found hymenopteran velocities from a wide taxonomic range, but almost none for solitary bees (Table 10.5). The listed values cover a wider range of body lengths than in solitary bees and were used to provide an allometric rule for high velocity and medium velocity.

10.3.1.7 Flower Handling Time

Handling times for exclusive pollen collection are rare in the literature, and most studies report handling times for mixed foraging tasks or for nectar handling alone. For the allometric scaling of pollen handling time with body length, we used three studies (Table 10.3). The first [102] gives values for five solitary bees and shows that handling time decreases with body length (Table 10.6). Solitary bees alone had a very narrow range of body sizes and we therefore included a honey bee and a bumblebee [103,104]. Data from another study with much longer handling times shows a similar negative relationship [4].

TABLE 10.5
Flight Velocities of *Hymenoptera*

Species	Hymenoptera	Bee	Solitary Bee	Body Length (mm)	Velocity Medium	Velocity High
Apis mellifera	Y	y	n	12	0.58 [92] 1.05 [92] 0.9 [93] 3.3 [94]	5.1 [94] 7 [93]
Bombus pascuorum	Y	y	n	13.5	0.37 [92] 0.5 [92]	
Bombus terrestris	Y	y	n	17.5	3 [95]	5 [95]
Elisabethiella baijnathi	Y	n	n	1.4	0.25 [96]	0.37 [97]
Osmia lignaria	Y	y	y	11.5		5 [98]
Sceliphron spirifex	Y	n	n	22		6 [99]
Vespa crabro	Y	n	n	25		5.9 [100]
Vespa mandarinia	Y	n	n	50		12.5 [101]

Note: These values were used for calculating the relationship between body size and medium velocity and high velocity. Velocities are given in meters per second. For each value, references are given in brackets.

TABLE 10.6

Flower Handling Times for Pollen Collection Based on Three Studies

Species	Hymenoptera	Bee	Solitary Bee	Body Length (mm)	Handling Time for Pollen Collection (s)
Hoplitis anthocopoides	y	y	y	10.0	5.70 [102]
Megachile relativa	y	y	y	10.0	10.40 [102]
Osmia coerulescens	y	y	y	9.5	15.70 [102]
Hoplitis producta	y	y	y	6.6	29.30 [102]
Ceratina calcarata	y	y	y	6.2	31.70 [102]
Bombus terrestris	y	y	n	26.0	4.60 [103]
Apis mellifera	y	y	n	12.0	8.10 [104]

10.3.1.8 Perception Distance and Length of Flight Units

There is no direct data on the perception distance of bees, but a body of literature exists on the visual abilities of bees showing that larger bees have better visual abilities. Eye size (facet or ommatidium diameter) of wild bees scales isometrically with thorax length [105]. Bees estimate their flight distance with the help of their velocity. Insects experience image motion (near objects appear to move faster than distant ones), mainly in a horizontal direction [106]. Honey bees estimate distance by holding a constant image velocity [107]. Because of the close link between velocity and perception, we defined the perception distance as the distance that can be flown in 10 s (view of 10 s forward). Since bees fly at several meters per second, the perception distance is several tens of meters. This may seem rather limited but wild bees generally fly 1–3 m above the ground [108] rather than higher up in the air with a wide view of the landscape.

We use flight units to model flight paths of bees, since flight paths are rather erratic than straight. Such flight paths suggest that a bee regularly evaluates (and adjusts) the flight direction. We defined a flight unit length as half of the perception distance and it was hence related to body size as well. In our implementation of movement (correlated random walk), we use slight deviations from this mean flight unit length in order to create natural-looking flight paths. We did not include any odometry in the model, because odometry only plays a minor role during foraging [109].

10.3.1.9 Flower Memory, Habitat Cell Memory, and Ignorance

Bees have different kinds of memory (*e.g.*, [110]). We use *flower memory* and *habitat cell memory* in our model, since they play an important role in decision behavior. Bees need to probe and memorize several flowers to estimate patch quality. We defined patch quality as the ratio of full to empty flowers, which reflects the probability that the next flower is full. Data from solitary bees show that the decision to stay or leave (a patch) depends on the reward of at least one to three visited flowers [111]. We used a minimum of three flowers for *flower memory*. One flower (full or empty in the model) does not suffice for a patch quality estimate. Bees improve their estimates of patch quality by visiting more (than three) flowers. Since this kind of working memory lasts from seconds to minutes [112], we did not set a maximum of flowers that a bee can remember. Within minutes, the bee either found enough pollen or left the habitat grid cell. Apart from the fixed minimal number of flowers that must be probed (*flower memory*), we implemented further flower memory during foraging within a habitat grid cell as two numbers: the number of full and the number of empty flowers. On entering a new habitat grid cell, only the final ratio of full to empty flowers is remembered and acts as the expectation of resource quality to compare the new experience (new series of probing flowers) with.

Bees remember the most recently visited grid cells with flowers (5 by 5 m units within the larger-scale foraging habitat). It is not exactly known how bees remember this kind of information, but they have a spatial memory of their route and the landscape with landmarks [113]. We defined this memory (*habitat cell memory*) as 10 grid cell locations. We cleared this memory for each new foraging trip, leading to new exploration of flower habitat grid cells (including near the nest) and new foraging directions.

We introduced an "ignore parameter" to induce matrix crossing and "far resource exploration." Wild bees do not prefer to cross the matrix [46,114]. However, absence of matrix crossing is also unrealistic since some individuals do take that risk (about 10%) [46]. Therefore, in the model, 10% of the local optimal decisions (going to a close habitat cell) are ignored. We decided to scale the parameter with *habitat cell memory* to reduce the number of independent parameters. They are inversely related (probability of 1 divided by the *habitat cell memory*), which means that for a bee with very little memory of its environment it is better to take more risks.

10.3.1.10 Flight Path Tortuosity

The turning angle distribution determines the tortuosity (erraticness) of the flight path. A higher value for *flight path tortuosity* results in more small turning angles and straighter flight paths. Turning angle distributions differ considerably between studies [115–119]. We use the correlated random walk with turning angles only to cross the matrix (area with no flowers); at the foraging habitat level, we use nearest neighbor selection rules (which indirectly lead to a very erratic flight path). For the matrix-crossing flight, we used a high value for *flight path tortuosity* (0.9), resulting in relative straight flight paths as known for areas with almost no foraging resources [118].

10.3.1.11 Time at the Nest and Foraging Time

The time that bees spent at their nest between foraging trips varies from 0.5 to 30 min [88]. A time below 5 min has been reported most often [91,120–122]. Time spent at the nest can be separated in a short time at the nest for pollen deposition (short stay) and longer time for egg laying and cell closure after a completed brood cell provision [121]. We do not know of an effect of body size (*e.g.*, data collected in [88]) and nesting preference, but this may be a knowledge gap in the literature. We use *time at the nest* as pollen deposition time and assume that larger species remove pollen more efficiently (same amount of time for depositing more pollen) and use 0.5 min as value unbiased by other activities. Instant pollen deposition at the nest (time zero) would lead to unrealistically high levels of local competition, and a high value would lead to a very low competition between bees, because they spend most time at the nest.

We simulate one foraging day to measure the bee's foraging performance, assuming that days do not differ. This also keeps simulation time acceptable compared to multiple foraging days. Foraging time (*flytime*) is often described in the literature as species activity time lasting 8–12 h [120–124], but can be considered to be half of that value (4–6 h) for the peak activity time [120,121]. Foraging time (time of day used for foraging) also relates to body size and climate. Large species can generate more heat, which is needed to start earlier [125]. However, they also forage late in the day, but have low foraging activity at the middle of the day [126]. Hence, duration of peak activity may be considered equal between small and large species. Animals often restrict their food search to the optimal time as response to the periodic nature of temporal distributions of resources [127]. We therefore consider four to eight peak foraging hours as realistic values for foraging time. The combination of *time at the nest* and *flytime* determines how much pollen has been collected by the end of the day, and possible qualitative differences of different combinations are explored in simulation experiment 1 (Section 10.4).

10.3.2 DATA SUPPORT FOR FORAGING RULES

10.3.2.1 Five Behavioral Modules

Resource competition in time with time penalties for different tasks is an important concept for modeling foraging behavior of wild bees. Bumblebees spend their time performing different behaviors: they travel between nest and foraging patches, fly between flowers within a patch, handle flowers and remove nectar and pollen, and search for other rewarding patches [36]. For solitary bees, these tasks are the same. We modified these tasks to five different behavioral modules in the model:

1. **Forage flowers** (fly to flowers, forage flowers within a landscape grid cell, fly between flowers, handle flowers, and remove pollen)
2. **Neighboring cell** (fly to a suitable neighboring landscape grid cell)
3. **Fly around** (fly around and look for unknown foraging areas)
4. **Fly back** (fly back to nest)
5. **Nest reached** (deposit collected pollen)

The finding that some bees perform an exclusively nectar-feeding trip at the end of the day [121] was not included.

There are several indications that bees first deplete flowers in the closest patches around the nest. Solitary bees deplete all flowers under high bee densities [91] and some individuals keep small foraging areas [128]. Bumblebees forage differently compared to solitary bees by following traplines (foraging paths of considerable length that guide farther and farther away from the nest), but solitary bees forage more primitively within a few hundred meters of their nest [55]. Therefore, after behavioral module 5 (nest reached), bees start foraging again from the nest location. Pollen may also still be available, since the flowers did release only a proportion of their pollen (*pollen availability*) in one time. The higher expected patch quality (memorized from farther away at the preceding foraging trip) makes that the bee flies beyond the habitat cells near the nest with low resource levels.

10.3.2.2 Use of Memory

According to the current knowledge, five memory types can be distinguished for (eusocial) bees [110]:

1. Memory of link between olfactory stimulus and reward, which lasts only seconds (flower constancy choice)
2. Short-term memory of, for example, the last visited patches, which lasts up to several minutes
3. Mid-term memory of the last location of rewarding flowers when returning to nest, memory of sequential landmarks and time of day
4. Long-term memory of flower cues
5. Long-term memory of nest location (nest recognition cues)

We represent these levels of memorization in our model as follows:

1. Bumblebees use this memory for maintaining flower constancy, but the solitary bees in our model forage on one kind of flower. Nevertheless, the model bees memorize the numbers of empty and full flowers that they have encountered and remember the current patch quality. The minimal number of flowers needed to estimate a ratio between them is defined in *flower memory*.
2. The bees in our model remember the last visited habitat grid cells. The size of this memory (number of memorized locations) is defined in *habitat cell memory*.
3. Honey bees return to rewarding patches [110], but we assumed that solitary bees do not for several reasons. We are not aware of any study showing that solitary bees directly return to the previous rewarding patch; it is even doubted for some honey bee subspecies [129]. Bees that have low associative learning capabilities, behave more like individuals, and fly shorter distances from the nest are likely to visit different locations on different bouts [129]. Instead, we assumed that the bees remember the quality of the last visited habitat grid cell (ratio of full and empty flowers encountered) for comparison with a newer location. We also indirectly assumed a memory of landmarks, resulting in a nonstraight flight path back to the nest (directed correlated walk).
4. The solitary bees in our model forage on one kind of flower. Hence, recognition of the right flower (and flower cues) is assumed, not modeled.
5. The most basic memory of the bees in our model was their nest location to which they always return.

10.3.2.3   Within-Patch (Grid Cell) Foraging Behavior
(Behavior 1: Forage Flowers)

We reduced the biological detail within a habitat grid cell and made very simplified assumptions about foraging behavior. Remember that we do not use optimized foraging behavior (shown for nectar-foraging bumblebees) for the pollen-collecting solitary bees in the model. Further, we were interested in motivations of bees for moving between grid cells (landscape level), not in detailed movement within a grid cell. We implemented a "visit nearest flower" rule and resource evaluation rules (determining flower and patch quality). We considered the following facts for our implementation:

- Pollen-collecting solitary bees do not use scent marks as much as nectar-foraging bumblebees do [130]. Flowers normally have abundant pollen (in contrast to nectar), and marking flowers is less profitable and too costly. Pollen is also mostly easily accessible than nectar. In the model, we do not use any kind of marking.
- Solitary bees are able to assess nectar reward by scent [131], but pollen recognition by scent has not been shown so far. We do not use a perception rule on this scale. Our model bee flies to a flower and receives a small time penalty for assessing pollen availability.
- Flight distance of bees flying between flowers is related to plant density [55]. We modeled the distance to the next flower as the mean distance between flowers, based on flower density.
- When bees forage on flowers, they use color contrast of multiple receptor types instead of the green contrast receptor type alone [132–134]. In order to locate a visible flower in the third dimension, bees need to adapt their velocity [106] and trade-off speed for accuracy [135]. We used therefore a lower velocity for within-patch (within-cell and between-cell) movements.

10.3.2.4   Patch Selection (Behavior 2: Neighboring Cell)

Honey bees cross vegetation structures very easily, but solitary bees were found to follow continuous vegetation structures [124]. We therefore implemented a module where only neighboring habitat cells are visited. The currently occupied cell is excluded from choice possibilities as well as recently visited habitat cells and nonforaging habitat cells. From the remaining cells, one is chosen randomly. This leads to a more or less random movement between habitat cells and a uniform distribution of turning angles with a larger-scale tendency to fly away from the nest. This movement behavior is not in contradiction with turning angle distributions found for bumblebees in rewarding flower patches [118,136].

10.3.2.5   Nest Clumping for Soil-Nesting Bees

For a more realistic distribution of nests for soil-nesting bees, we distributed them somewhat clumped rather than uniformly. Soil-nesting bees prefer moderate [91] to very dense aggregations [137,138]. Not all locations within the vegetation are suitable to nest, and clumped nest sites are also better encounter sites for mating [55]. We used moderate clumping in our model (and did not further analyze this), since we were interested in the potential difference between soil- and wood-nesting bees and not in the effect of nest clumping (it does not change the contrast of the two nesting preferences per se).

10.3.3 EXPLORATION OF FIELD DATA FOR RESPONSE VARIABLES

We discuss here response variables of the simulation model: number of brood cells, distance flown (mean and far distance from the nest), trip duration, and number of flowers visited (per day and per trip). We discuss values for small, midsize, and large bees separately to get an indication of the range of values within each group.

10.3.3.1 Number of Brood Cells

The number of brood cells that bees build in 1 day spans a large range between species and within species. Bees from the genus *Andrena* (covering probably the whole range of small to large bees) build 0.4–2 brood cells a day [88]. The small bee *Calliopsis persimilis* (<8 mm) even builds up to 6 a day [121]. Intermediate-sized bees are relatively well studied. *Osmia cornuta* builds half a cell a day [139], *Andrena humilis* builds 1.37 brood cells per day [88], *Osmia lignaria* can build 2 cells a day [44], and *Diadasia rinconis* builds 2–3 cells a day [91]. However, *Osmia bicornis* may build up to 7.1 cells a day in optimal conditions [122]. A large bee, *Creightonella frontalis*, was found to build 1 cell a day [140] and another one, *Dieunomia triangulifera*, 0.2–3 brood cells per day [141]. This suggests in summary that small bees are able to build more brood cells in 1 day than large bees.

10.3.3.2 Foraging Trip Duration

Trip duration of different solitary bees (body length 7–12 mm) lasts 6–28 min and is positively correlated with body length [142]. Another review found that trip duration for different species varies from 3 to 170 min [88]. Small bees show a particularly large range of trip durations. *Perdita opuntiae* needs about 4 min per trip, *P. texana* 6 min, and other *Perdita* species 15–30 min [91]. *Calliopsis persimilis* has foraging trips of 8 min [121], *Chelostoma florisomne* has pollen collection trips of about 20 min [143], and trips for *Lasioglossum figueresi* last 7–46 min [144]. For intermediate-sized bees, the range is less extreme. *Osmia cornuta* for pollen collection in apple orchards needs 4.7 min per trip in one experiment [120] and in another experiment 5–6 min [122] and in almond orchards around 12 min [87]. For *O. cornuta* and *O. rufa*, it is known that they have series of longer (8–17 min) and shorter trips (1–4 min) [122]. For *Andrena humilis* foraging trips range from 2 to 35 min [88]. Large solitary bees have longer foraging trip durations. *Anthophora acervorum* needs 14 min per trip in apple orchards [120], *Xylocopa pubescens* and *X. sulcatipes* need on average 20–30 min per trip (which was equal for pollen and nectar trips) but could last up to 78 min [145], *Creightonella frontalis* takes about 17 min for pollen collection [140], and for *Dieunomia triangulifera*, the trip duration varies from 20 to 190 min [141].

10.3.3.3 Flowers Visited and Flowers Visited per Trip

Small bees visit fewer flowers than large bees [85], and they visit only a few flowers per foraging trip [146]. It was estimated that *Osmia cornuta* (an intermediate-sized bee) visits about 50 flowers per trip and about 4,500–5,600 flowers a day.

Others estimate 60–90 flower visits per trip for *O. cornuta* [87], which would mean about 10,000 flowers a day. Large bees visit most flowers. *Anthophora acervorum* is estimated to visit about 250 flowers per trip and about 8,800 flowers a day [120], *Creightonella frontalis* 29 (full) flowers per trip and about 180–300 (full) flowers a day [140], *Xylocopa pubescens* and *X. sulcatipes* 15–20 (full range 4–65) flowers per trip [145], and bumblebees 20–150 flowers per trip [146].

10.3.3.4 Mean Distance from Nest and Realized Foraging Range

Small bees forage near the nest. *Perdita texana* has most foraging activity within 20 m from the nest with the longer flights around 40 m from the nest [91]. Also, Calabuig found that most solitary bees (among many small ones) forage mainly within the first 27 m [63]. An experiment with fluorescent dye showed that most pollen was transferred over short (<15 m) distances [147]. The mean activity distance for an intermediate-sized bee, *Andrena hattorfiana*, was 46–76 m (two sites), and the maximum distance was 83–130 m. Rare long dispersal movements (which can be assumed to come close to the homing distance) were up to 900 m [46]. In resource-rich environments, *Osmia cornuta* forages near the nest within 100–200 m [148]. Bumblebees (large bees) are in contrast to small solitary bees and are also frequently flying farther (100–200 m) from the edge [63]. For other bumblebees, it was found that 40% foraged within 100 m around the nest and 62.5% within 200 m [149]. The mean maximum foraging distance per colony varied from 460 to 710 m, and all bumblebees foraged within 800 m. The mean foraging distance varied from 87 to 447 [149]. When food is abundant close to the nest, as it is, for example, in a clover field, bumblebees have their main foraging activity within 18 m from the nest [150].

10.4 EXAMPLES OF SIMULATION EXPERIMENTS

We performed four simulation experiments in order to gain a basic understanding of the model and to understand the influence of the different parameters. These experiments include the basic calibration and validation of the model as well as exploration of global perturbations within the parameter space and quantifying local sensitivity of the parameter values. These simulation experiments can be considered as examples of how one can analyze a multiparameter model such as our individual-based simulation model. At the same time, these simulation experiments can yield a basic understanding of the model system. In our case, they give insights into how solitary bees experience the landscape. All parameters are listed in Table 10.7, and for each simulation experiment the used values are given.

Statistical analyses on simulated outputs were performed with R [151]. Statistical analysis of simulation results is relatively unusual since the majority of the model is deterministic, which means that all parameters have a significant influence. However, the model has many rules that interact so that we could not predict *a priori* which of the parameters would have the most effect on the response variables, and statistical tools can help in ranking the impact of different parameters. We look at different response variables: number of brood cells, number of flower visits, mean distance from the nest, and foraging habitat visitation. In simulation experiment 1, we additionally look at the mean trip duration and the maximum foraging distance.

TABLE 10.7

Simulated Parameter Values for Different Simulation Experiments

		Simulation Experiment										
		1			**2**		**3**			**4**		
		Exploration of Time Series			Sensitivity I: Global Perturbations		Sensitivity IIa: Bee Parameter Robustness			Sensitivity IIb: Landscape Analysis		
Parameter	Short	Value 1	Value 2	Value 3	Value 1	Value 2	Value 1	Value 2	Value 3	Value 1	Value 2	Value 3
Basic simulation definition												
Landscape stochastic factor	seed	821			821	188	821			−10%	100	+10%
Flytime	tt	14,400	28,800	57,600	14,400	28,800	14,400	28,800		14,400	28,800	
Landscape												
Landscape element size	esize	50			50		50			−10%	50	+10%
Foraging habitat availability	am	0.05	0.95		0.05	0.95	0.5			−0.1	0.5	+0.1
Landscape fragmentation	fr	0.5			0.05	0.95	0.5			−0.1	0.5	+0.1
Flower density	fd	50			10	100	50			−10%	50	+10%
Pollen per flower	ppf	0.5			0.1	2	0.5			−10%	0.5	+10%
Pollen availability	plimit	0.3			0.3		0.3			−10%	0.3	+10%
Bee												
Landscape quality for bees	bdc	30			10	50	30			−10%	30	+10%
Body length	size	6	12	24	6	24	12			12		
Nesting preference	nest	"wood"	"soil"		"wood"	"soil"	"wood"	"soil"		"wood"	"soil"	
Flight path tortuosity	CRW	0.9			0.1	0.9	0.1	0.8	+0.1	0.8		

(continued)

TABLE 10.7 (continued)

Simulated Parameter Values for Different Simulation Experiments

		Simulation Experiment										
		1 Exploration of Time Series			2 Sensitivity I: Global Perturbations		3 Sensitivity IIa: Bee Parameter Robustness			4 Sensitivity IIb: Landscape Analysis		
Parameter	Short	Value 1	Value 2	Value 3	Value 1	Value 2	Value 1	Value 2	Value 3	Value 1	Value 2	Value 3
Bee												
Lower patch leaving threshold	L_plt	*0.5*			*0.5*		0.1	0.5	+0.1	*0.5*		
Upper patch leaving threshold	u_plt	*1*			*1*		−0.1	0.9	+0.1	*0.9*		
Habitat cell memory	cmem	*10*			2	30	−10%	10	+10%	*10*		
Flower memory	fmem	*3*			3	10	−10%	3	+10%	*3*		
Pollen capacity per bee	pcap						−10%	Size related	+10%			
Velocity high	vhi						−10%	Size related	+10%			
Velocity medium	vmed						−10%	Size related	+10%			
Handling time per flower	ht						−10%	Size related	+10%			
General return distance	r50						−10%	Size related	+10%			
Perception distance*	sightm	*10*			10	20	−10%	Size related	+10%			
Length of flight units*	flightm	*2*			2	1	−10%	Size related	+10%			
Ignorance	ig						−10%	Memory related	+10%			
Time at the nest	ntime	**6**	**60**	**600**	1	600	*30*			*30*		

Note: Values in bold show parameters that got special attention in the analysis and values in italics were not varied in the specific simulation experiment, but are given for comparision. See Table 10.1 for parameter explanation. For parameters with asterisk (*sightm* and *flightm*), we varied in simulation experiments 1 and 2 their regulator parameter (x seconds and x units, respectively) and in simulation experiment 3 their calculated value.

10.4.1 Foraging for Pollen: Competition Processes in Time

In this first simulation, we performed a systematic exploration of the model. We assessed how the model behaves over time, and how well output parameters overlap with literature data. There are two parameters in the model that determine the global time budget for bees: foraging time (*flytime*) and the time spent at the nest for pollen deposition (*time at the nest*). Since field data cover a wide range of values (Section 10.3.1), we investigated whether different combinations have only quantitative or also qualitative effects. We simulated three time-budget scenarios: a foraging day of 4 h with 6 s at the nest after each foraging bout, a foraging day of 8 h with 60 s at the nest, and a foraging day of 16 h with 600 s at the nest. We used unequal intervals for *time at the nest* for two reasons. First, it suits the biological values best (Section 10.3.1) and second, we expect that with longer foraging days the relative effect of time at the nest decreases (due to longer foraging trips and farther foraging distances). We explored a system that is unsaturated with bees (where there is more pollen than what bees collect in one day, defined by *bdc*). We varied *body length* (6, 12, 24 mm) and explored the results for the three bee sizes separately. To reduce complexity we omitted bee sizes from the description of the results when they did not provide additional insights. We therefore describe time series for brood cells and flower visits for small and for large bees, and the other time series for large bees only. We simulated four combinations of *nesting preference* (two types) and *foraging habitat availability* (0.05 and 0.95). This gives 12 time series per bee size (Table 10.7). Within each time-budget scenario, we ranked the four time series (their values at the end of the simulated time) to trace qualitative differences.

10.4.1.1 Number of Brood Cells

Within the foraging time (*flytime*), the number of completed brood cells increased. The pattern for different time-budget scenarios was similar between small bees (Figure 10.5a) and large bees (Figure 10.5b) as well as for intermediate-sized bees (not displayed). The *flytime* determined the absolute number of brood cells that could be built in 1 day, and the *time at the nest* determined how fast (change in slope) a certain number of brood cells is built. The relative distance between different values for the *time at the nest* (6–60–600) is reflected in the relative distance between the curves. These two time-related parameters did affect the time budgets and remaining time for pollen collection as expected. The ranking of the performance (brood cells) of the bee types (ranking of the four response curves) was equal for all three time-budget scenarios.

The number of brood cells increased linearly for soil-nesting bees in all cases, which means that they were not resource limited by competition. Wood-nesting bees performed badly when they faced a high *foraging habitat availability*, which may seem counterintuitive. The increase in brood cells additionally leveled off for them, as well as for wood-nesting bees in landscapes with low foraging habitat (but weakly). These striking differences may be related to the different distribution of nests between soil- and wood-nesting bees, which induced other time constraints (Figure 10.7). Small bees were able to build more brood cells than large bees, and no bee was able to collect pollen for 30 brood cells (as defined in *bdc*).

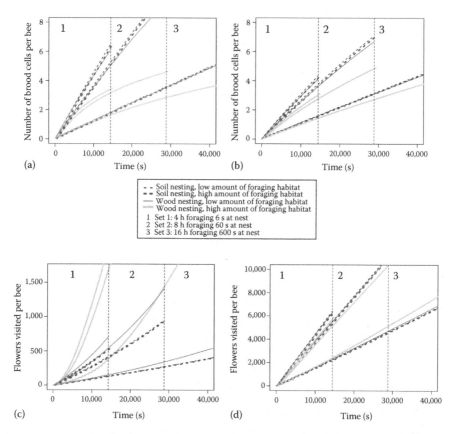

FIGURE 10.5 Development of different response variables during simulation time. Each graph contains 12 time series, of which 4 have the simulation time (*flytime*) ending at 14,400 s (first dashed vertical line), another 4 at 28,800 (second dashed vertical line), and the last 4 outside the graph on the right (at 57,600). Each subset of four represents a different scenario according to *nesting preference* and *foraging habitat availability*. The curves of each of the four scenarios are directly comparable with the curves in the other two time-budget scenarios (set 1–3). Note that we cut off the top of the y-axis to visualize the lower values so that some of the curves reach the vertical dashed line (simulation end) outside the plot. We shifted the lines for "high *foraging habitat availability*" with 500 s to make them distinguishable from other lines. (a) Brood cells built by small bees, (b) brood cells built by large bees, (c) flowers visited by small bees, and (d) flowers visited by large bees.

10.4.1.2 Number of Flower Visits

The number of flowers visited per bee (full and empty flowers) increased exponentially for all bees (Figure 10.5c and d). Hence, bees probed more flowers per time unit hinting to more empty flowers later in the day. Large bees (Figure 10.5d) visited many more flowers than small bees (Figure 10.5c). For most bees, the increase in flower visits was almost linear (Figure 10.5c and d), while for some small bees there was a strong exponential increase (Figure 10.5c).

We suspect that large bees and soil-nesting small bees (Figure 10.5c and d) had better access to full flowers than wood-nesting small bees (Figure 10.5c). Wood-nesting small bees visited more flowers per time unit later in the day and they built fewer additional brood cells per time unit (Figure 10.5a). Hence, they faced more empty flowers, leading to a higher flower visitation rate. Their nest position at the border of the habitat and their higher numbers and lower flight capability may have limited their access to less depleted areas.

The ranking of the response for the different bee types was the same again under the three time-budget scenarios, except for large bees with *ntime* 600. At this point, we cannot give a sufficient explanation for that. Time budgets for flower visits were apparently not the same time constraints that determined the pattern for the number of brood cells (Figure 10.5a and b).

10.4.1.3   Normal Foraging Activity and Far Foraging Activity

Far foraging distances (Figure 10.6b) were in most cases about twice as large as mean foraging distances (Figure 10.6a). Foraging experience (distance) increased almost linearly for soil-nesting bees and did not reach high values indicating that they found foraging resources close to the nest. Wood-nesting bees flew farther than soil-nesting bees, especially when they were in a landscape with high *foraging habitat availability* (Figure 10.6a and b). Wood-nesting bees were forced to nest in edge structures, while soil-nesting bees were more or less evenly distributed over the foraging habitat. Since in landscapes with high *foraging habitat availability* there were more bees and less edge (Figure 10.7), depletion around the nest was more likely for wood-nesting bees and resulted in farther foraging from the nest.

The expansion rate of bees decreased with time (Figure 10.5a and b). We implemented a foraging rule that prohibited flying too far from the nest (Section 10.2.7) and that could have caused this. We used parameters *r50* (typical homing distance) and *r90* (the far homing distance) to rule out exceeding these distances for 50% and 90% of the flight trips, respectively. The value for large bees was 6,760 m for *r50* and 16,017 m for *r90*. Large bees never reached these distances (Figure 10.6a and b), and this hard-coded foraging rule cannot have been the cause for the decrease in expansion rate. All patterns in Figure 10.6a and b are in agreement with Figure 10.5a and b and may indicate that longer flights directly result in fewer brood cells.

There were no qualitative differences (same ranking) between the three time-budget scenarios. Far foraging distances (Figure 10.6b) were based on rarer expansion events, and the increase was therefore less smooth and more stochastic.

10.4.1.4   Expansion over the Foraging Habitat

The model tracked at the landscape level how much of the foraging habitat was visited by the bee community (Figure 10.6c, large bees). Soil-nesting bees covered within 3,000 s the complete vegetation (100%) in landscapes with low foraging habitat availability and 90% of the vegetation in landscapes with high foraging habitat availability. This means that they had good access to all foraging resources without depleting it within a foraging day (brood cells increase with the same rate throughout the day, Figure 10.5a). Wood-nesting bees that nested at the border of the foraging

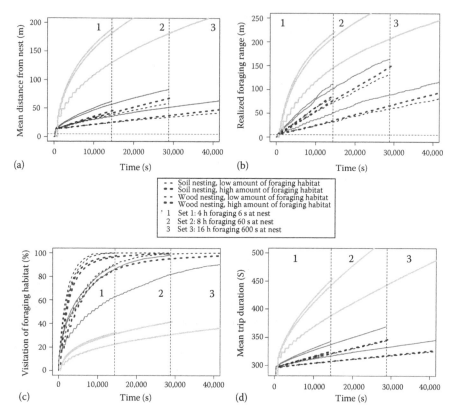

FIGURE 10.6 Progress during simulation time of expansion-related parameters for large bees. Each graph contains 12 time series, as in Figure 10.5 (see for details). (a) Mean distance from the nest, (b) maximum realized foraging range (averaged over the bee community), (c) visitation of foraging habitat, and (d) mean trip duration for large bees. Plots (a) and (b) start actually at the minimum distance of 5 m (size of one grid cell) indicated with a dashed line. The mean foraging distance seems (but is not) higher than the realized foraging range in the first few minutes, which is attributable to slightly different assessment methods in the model. Plot (d) starts at the "emerged" minimal trip duration.

habitat did not cover the whole foraging habitat. In a landscape with low *foraging habitat availability*, the bee community reached 85% within a foraging day, but in a landscape with high *foraging habitat availability*, the bees never covered more than 40% of the foraging habitat in these simulations. The ranking remained the same for all three time-budget scenarios.

10.4.1.5 Trip Duration

The mean trip duration increased during the day for all situations (Figure 10.6d), but was generally low (below 10 min). Soil-nesting bees had the shortest trips and wood-nesting bees had the longest trips, especially those in landscapes with a high foraging habitat availability, in agreement with longer foraging distances (Figure 10.6a). The ranking of the four responses did not change for the three time-budget scenarios.

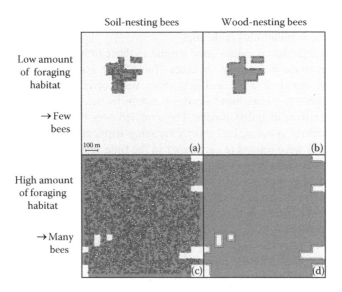

FIGURE 10.7 Four simulation examples after the initialization phase. The panels represent the four combinations of *nesting preference* and *foraging habitat availability* from simulation experiment 1. These landscape realizations have a certain amount of foraging habitat (darker gray areas in a matrix without foraging resources), an intermediate fragmentation level, and a population of large bees. A bee's nest is represented as a white dot with a dark border. High nest densities appear as merged circles (white stripes with a dark border). (a) Population of soil-nesting bees with a low *foraging habitat availability*. (b) Population of wood-nesting bees in a landscape with a low *foraging habitat availability*. (c) Population of soil-nesting bees in a landscape with a high *foraging habitat availability*. (d) Population of wood-nesting bees in a landscape with a high *foraging habitat availability*.

10.4.1.6 Time Budget Scenarios

The three different time-budget scenarios for *flytime* and *time at the nest* had only quantitative effects (with the number of flowers visited by large bees as the only exception) and did not change the relative difference between the different bee types in different landscapes. Hence, we consider it acceptable to use a single time-budget scenario for systematic investigation of other parameters. Further, we found clear differences for bees of different size and for soil-nesting and wood-nesting bees, which are worth exploring in detail.

10.4.1.7 How Well Does the Model Represent Foraging Solitary Bees in Real Systems?

We used the results (mean values at the end of a foraging day, Figures 10.5 and 10.6) from this first simulation experiment (with uncalibrated, but realistic parameter input) to compare with data from real systems. We collected data from the literature with a focus on pollen-collecting solitary bees for several response variables: brood cells per day, foraging trip duration, flowers visited (per day and per trip), mean foraging distance from the nest, and far foraging distances within a day (Section 10.3.3). We present here an overview of the value ranges (grouped for small, midsize, and large bees as solution for the sparsely available data) and compare them with the ranges from the simulations (Table 10.8).

The model reproduced values quite similar to those of natural systems with over-lapping ranges in the majority of the cases (Table 10.8). For all responses, the direction of change between small and large bees was correct, and also the intervals between the three bee sizes were similar between modeled values and real values. However, there were also differences. The modeled bees built more brood cells per day than in natural systems, had shorter foraging trips, and flew shorter distances. These responses were related to each other in the time series (Section 10.4.4.1). The number of flower visits in the model overlapped well with literature values for inter-mediate-sized bees, but not so well for small and large bees. The literature data for flower visits was sparse and lacked specification of whether or not unsuccessful flower visits were included (they were included in the model), which could explain the underestimation.

The modeled response variables yielded realistic values for small, intermediate-sized, and large bees, and the modeled ranges generally overlapped well with real systems. The model bees were however more efficient than in most natural systems. A (too) high number of brood cells may have arisen from less realistic model assump-tions such as the exclusion of nectar-foraging trips. Also collecting mud and other nest material can take considerable time [140] that is not available for collecting pol-len. Some solitary bees only build one cell per day regardless of pollen availability and remaining time [91] and are rather egg-limited [152]. Other assumptions were that bees collect pollen for 1 brood cell within 10 foraging trips (used to define the loading capacity per trip), and bees always had a full load when they return (except when flying too far), which is often not the case [153]. In the model, bees always chose the most efficient place to nest: near foraging resources, either in the soil between the vegetation or at the border of the field. In agricultural landscapes, where much of the data originates from, distances between nest and foraging resources may be larger. Hence, it makes sense that the data shows less efficient bee performance as well.

We think that the model yields realistic output and that the overlap with literature data is satisfactory, also considering the fact that we used detailed, literature-based input values without fitting any parameter. This simulation experiment has also shown that the model shows realistic patterns in time. For example, an increasing foraging trip duration during the day is a well-known time effect for pollen-foraging bees that face local pollen depletion [88,140,141].

10.4.2 WHAT DO EXTREME PARAMETER VALUES IMPLY FOR SOLITARY BEES? A GLOBAL SENSITIVITY ANALYSIS

The goal of this second simulation experiment is to quantify parameter effects within their estimated biological range (Section 10.3.1) and to identify the most important ones. We analyzed the relative effect on brood cells for 15 parameters (*flower density [fd], landscape quality for bees [bdc], pollen per flower [ppf], scal-ing parameter for perception distance [sightm], scaling parameter for length of flight units [flightm], flight path tortuosity [CRW], flower memory [fmem], habitat cell memory [cmem], time at the nest [ntime], and landscape stochastic factor [seed]*

TABLE 10.8

Value Ranges from Response Variables of the Model Compared with Ranges from the Literature

Bee Size	Literature Value Ranges			Modeled Value Ranges		
	Small	Medium	Large	Small	Medium	Large
Brood cells per day	2.0–6.0	0.5–3.0 (7.1)	0.2–3.0	3.0–10.5	3.5–9.5	3.0–7.5
Foraging trip duration (min)	(3) 4–46	1–35	14–190	3.5–8.0	4–8	5–9
Flowers visited per day	Fewer	4,500–10,000	180–9,000	500–5,000	1,500–6,000	6,000–11,500
Flowers visited per trip	Few	50–90	4–150	8–95	30–105	150–220
Mean distance from nest (m)	(15) 20–27	40–100	18–447	15–120	20–180	40–250
Realized foraging range (m)	40	80–200	(100) 460–710	25–170	45–225	80–285

Note: Modeled ranges are rounded to the widest range. Between brackets are uncertain numbers. Note that the literature ranges are rarely based on systematic experiments, but rather on single observations and rough estimates. See Section 10.3.3 for references.

and included the focal parameters *body length [size], nesting preference [nest], foraging habitat availability [am],* and *fragmentation [fr]*). We used two values for each parameter representing the biological range (Section 10.3.1). The values for *flower density, pollen per flower,* and *landscape quality for bees* were restricted by simulation resources (parameter combinations that resulted in several millions of bees required too long running times) and not simulated for the most extreme values, but the chosen values cover an acceptable extent of biological range (Table 10.7 and Section 10.3.1). We generated a set of parameters for each possible combination (2^{14} combinations) and selected 2,000 of these sets randomly for simulation. We repeated the analysis for a longer foraging time (*flytime*). We present here the most important parameter effects and parameter interactions based on linear models for the different response variables.

10.4.2.1   Brood Cells

The number of brood cells was most affected by the *pollen per flower, time at the nest,* and *landscape quality for bees*. More total pollen (*pollen per flower*) and more pollen available per bee (*landscape quality for bees*) resulted in more brood cells. A longer *time at the nest* reduced the time available for collecting pollen and reduced the number of brood cells. Several interactions between parameters were important: a longer *time at the nest* reduced the effect of pollen per flower, a high amount of *pollen per flower* reduced the effect of body size to almost none, and a long *time at the nest* as well as a low amount of *pollen per flower* reduced the effect of bee density (*landscape quality for bees*) to almost none. For a longer foraging time (*flytime*), the order of importance of the discussed parameters and interactions did not change. In exceptional cases (in 5 of the 2,000 simulations), the landscape got depleted for long foraging times, and in these cases, bees built the amount of brood cells as given by *bdc* (but not exceeding this value).

10.4.2.2   Flowers Visited

The number of visited flowers per bee was affected by *body length, time at the nest,* and *pollen per flower*. Larger bees visited more flowers, and flower visitation was also higher for a high amount of *pollen per flower*. A longer *time at the nest* decreased the number of flower visits. However, parameter interactions could change the effect. For a short *time at the nest*, more *pollen per flower* resulted in more flower visits, while for a long *time at the nest* more *pollen per flower* resulted in fewer flower visits. The aggregated number of flower visits was not separated for visits of full and empty flowers, which probably could explain some of the patterns. Visits of empty flowers must have increased for situations with increased competition pressure (such as an increased number of bees by a high amount of *pollen per flower* or more foraging activity by a short *time at the nest*). A longer foraging period did not change the importance order of the parameters.

10.4.2.3   Foraging Habitat Visitation

Foraging habitat visitation of the bee community was most affected by *pollen per flower, nesting preference, landscape quality for bees,* and *flower density*. Foraging habitat visitation (percentage of the foraging habitat visited) was higher for a higher amount of *pollen per flower* (also more bees flying around) and for soil-nesting bees (better distribution over the vegetation). A low bee density (high *landscape quality for bees*) decreases foraging habitat. Foraging habitat visitation by soil-nesting bees was hardly affected by *foraging habitat availability*, while wood-nesting bees had a lower foraging habitat visitation for a higher *foraging habitat availability* (parameter interaction). Apparently, wood-nesting bees in field edges were not able to cover the complete foraging habitat when there was much of it. For a longer foraging period, the effect of *foraging habitat availability* got more importance, and *flower density* lost importance. Both increase the amount of pollen and the number of bees in the landscape.

10.4.2.4   Mean Distance Flown

The mean distance flown from the nest per bee was most affected by *nesting preference, body length, foraging habitat availability, time at the nest,* and *landscape quality for bees*. Large wood-nesting bees in landscapes with a high *foraging habitat availability* flew the farthest. A high bee density (low *landscape quality for bees*) and a short *time at the nest* increased the mean distance flown from the nest. The effect of *foraging habitat availability* was low for soil-nesting bees but strong for wood-nesting bees. This parameter interaction was similar to that for foraging habitat visitation. The importance of parameters for a longer foraging period did switch for pairs with very similar importance (*nest–size* and *bdc–ntime*).

10.4.2.5   Conclusions from the Second Simulation Experiment

The analysis showed that *pollen per flower* (*ppf*) had a disproportionate effect on the number of brood cells and interacted with *body length* (*size*). This could mean that small bees perform better on small flowers and large bees on very large flowers. However, the other performance responses did not show a strong interaction between *ppf* and *size*. The *time at the nest* (*ntime*) and *landscape quality for bees* (*bdc*) had predictable effects on the number of brood cells and showed no important interactions with other parameters. The other response variables were also affected by *ppf*, as well as by *size, nest, ntime, bdc, fd,* and *am*.

10.4.3   Which Body-Size-Related Traits Affect Foraging Behavior Most Strongly?

The effect of biologically plausible parameter ranges does not give information about parameter effects close to the chosen common values. We therefore investigate the effect of small standardized changes in model parameters in this and the next simulation experiment. In this experiment, we investigate parameters related to flight and decision behavior of the bee and those related to the body size of the bee (*vhi, vmed, ht, r50, sightm, flightm, CRW, pcap, cmem, ig, fmem, l_plt, u_plt*). We varied each parameter by ±10%. For parameters that normally range between 1 and 0, we used 0.1% increments instead of 10%. We shifted for *CRW* and *u_plt* the default value by −0.1 (technical restriction). For *general return distance* (*r50*), we shifted the probability curve to the left or to the right (moving both *Km* and *shift* by 10% of *r50*) after *Km* and *shift* were calculated. Since most of the bee parameters are calculated from body length in different steps, we first calculated the default value and then applied the deviation (none, −10%, or +10%) randomly. Note that in case of *ig*, *sightm*, and *flightm*, there were more than three levels due to dependencies with other parameters, leading to multiple applied deviations. We generated 1,000 unique parameter sets as an alternative for generating all possible combinations. We considered wood-nesting bees of intermediate size. We repeated the analysis for soil-nesting bees (altered *nesting preference*, 1,000 new parameter settings) and a longer foraging period (*flytime* 8 h, initial 1,000 sets).

The analysis proceeded in two steps. (1) We determined which parameters were most important for variation in brood cells using a multiple regression model and checked for interactions between parameters. We simplified the linear model by minimizing the Bayesian Information Criterion (BIC). (2) We calculated the arc elasticity of the input parameters, that is, the percentage change in the response variable divided over the percentage change in input variable, measured against the average of two subsequent levels. A parameter was deemed elastic when the arc elasticity was larger than one [154, pp. 72–80]. Conversely, we considered the model parameters robust when all elasticity values remained under one. We corrected for nonsimulated parameter combinations (far more than the 1,000 simulated combinations) by predicting the response variable with the regression model. We give here an overview of the relevant effects.

10.4.3.1   Importance of the Bee's Behavioral Traits in Different Situations

The regression analysis revealed different patterns for three scenarios (wood-nesting bee, soil-nesting bee, and an elongated foraging time) in response to 10% change. *Handling time per flower* caused the most variation in the number of brood cells and the number of flowers visited under all three scenarios. *Handling time per flower* may also have prevented time being available for other activities as *time at the nest* did in simulation experiment 1. For habitat visitation and distance flown, there were differences between bee types. Foraging habitat visitation was most affected by *length of flight units* and the mean distance flown by *perception distance* for wood-nesting bees. For soil-nesting bees, both responses were most affected by the two patch-leaving thresholds (*l_plt*, *u_plt*). Expansion over the landscape was apparently not governed by the same processes as the number of brood cells. There was no effect of a longer foraging period on any of the important predictors. Parameter interactions had a negligible effect, which means that the parameters in this experiment acted additively.

10.4.3.2   Parameter Robustness of the Bee's Behavioral Traits

Elasticity values were in general low (<0.5) across the response variables (Table 10.9), which means that they were quite robust against small changes in bee-related parameters. The parameters causing most variance for each of the response variables were also causing the highest sensitivity in response (elasticity value >0.5), but never exceeded one. The highest values were for *handling time per flower* (both for wood- and soil-nesting bees), and the model can be considered moderately robust against this parameter. A longer foraging period did slightly decrease elasticity values. Habitat visitation was most robust against changes in bee-related parameters (all values below 0.5).

TABLE 10.9

Parameter Elasticity for Four Response Variables (Brood Cells, Brood Cells Built per Bee; Total Flowers, Total Number of Flower Visits per Bee; Habitat Visitation, Percentage of the Foraging Habitat That Has Been Visited; Distance Flown, Mean Distance Flown from the Nest per Bee) under Three Scenarios (Wood-Nesting Bees That Forage for 4 h, Soil-Nesting Bees That Forage for 4 h, and Wood-Nesting Bees That Forage for 8 h)

	Wood Nesting—4 h Foraging				Soil Nesting—4 h Foraging				Wood Nesting—8 h Foraging			
	Brood Cells	Total Flowers	Habitat Visitation	Distance Flown	Brood Cells	Total Flowers	Habitat Visitation	Distance Flown	Brood Cells	Total Flowers	Habitat Visitation	Distance Flown
1. Handling time per flower	**0.677**	**0.907**	0.170	0.395	**0.826**	**0.915**	0.010	0.344	**0.518**	**0.856**	0.156	0.360
2. Length of flight units	0.004	0.142	0.407	**0.689**	0.005	0.005	0.002	0.281	0.025	0.249	0.365	**0.685**
3. Perception distance	0.009	0.020	0.008	0.160	0.002	0.001	0.000	0.011	0.016	0.035	0.009	0.191
4. Velocity medium/low	0.006	0.005	0.004	0.008	0.008	0.011	0.001	0.010	0.006	0.005	0.004	0.012
5. Velocity high	0.047	0.055	0.012	0.041	0.008	0.011	0.001	0.011	0.073	0.118	0.023	0.066
6. General return distance	0.001	0.008	0.005	0.010	0.002	0.003	0.001	0.015	0.001	0.011	0.006	0.011
7. Pollen capacity per bee	0.126	0.078	0.016	0.197	0.096	0.119	0.003	0.139	0.140	0.057	0.027	0.250
8. Lower patch leaving threshold	0.008	0.116	0.126	0.386	0.006	0.036	0.009	**0.531**	0.012	0.118	0.105	0.273
9. Upper patch leaving threshold	0.086	0.235	0.098	0.445	0.047	0.258	0.016	**0.715**	0.080	0.222	0.075	0.308
10. Flower memory	0.008	0.221	0.129	0.329	0.003	0.038	0.012	0.393	0.013	0.374	0.127	0.293
11. Habitat cell memory	0.001	0.012	0.011	0.017	0.004	0.005	0.001	0.035	0.002	0.022	0.009	0.012
12. Ignorance	0.004	0.025	0.128	0.125	0.004	0.004	0.001	0.075	0.001	0.045	0.111	0.136
13. Flight path tortuosity	0.000	0.001	0.080	0.109	0.002	0.001	0.000	0.003	0.002	0.002	0.076	0.135
14. Landscape element size	0.008	0.002	0.087	0.063	0.004	0.000	0.000	0.003	0.017	0.015	0.074	0.032
15. Landscape fragmentation	0.060	0.065	0.278	0.295	0.002	0.000	0.000	0.007	0.101	0.083	0.237	0.285
16. Landscape stochastic factor	0.191	0.216	1.044	**0.809**	0.002	0.007	0.005	0.028	0.344	0.304	**0.920**	**0.750**
17. Landscape quality for bees	0.124	0.153	0.232	**0.550**	0.030	0.060	0.010	0.448	0.239	0.230	0.204	**0.570**
18. Foraging habitat availability	0.126	0.158	**0.588**	0.627	0.002	0.000	0.001	0.014	0.212	0.200	**0.502**	**0.615**

(continued)

TABLE 10.9 (continued)

Parameter Elasticity for Four Response Variables (Brood Cells, Brood Cells Built per Bee; Total Flowers, Total Number of Flower Visits per Bee; Habitat Visitation, Percentage of the Foraging Habitat That Has Been Visited; Distance Flown, Mean Distance Flown from the Nest per Bee) under Three Scenarios (Wood-Nesting Bees That Forage for 4 h, Soil-Nesting Bees That Forage for 4 h, and Wood-Nesting Bees That Forage for 8 h)

	Wood Nesting—4 h Foraging				Soil Nesting—4 h Foraging				Wood Nesting—8 h Foraging			
	Brood Cells	Total Flowers	Habitat Visitation	Distance Flown	Brood Cells	Total Flowers	Habitat Visitation	Distance Flown	Brood Cells	Total Flowers	Habitat Visitation	Distance Flown
19. Flower density	0.001	0.006	0.083	0.007	0.006	0.005	0.003	0.092	0.000	0.005	0.064	0.006.
20. Pollen per flower	**0.700**	0.085	0.250	0.277	**0.813**	0.043	0.009	0.136	**0.572**	0.175	0.189	0.267
21. Pollen availability	**0.697**	0.088	0.146	0.300	**0.807**	0.035	0.005	0.236	**0.570**	0.176	0.110	0.287

Note: A higher elasticity value means a higher sensitivity of the response variable to a change in predictor variable. An elasticity value above 1.0 means that the proportion of change in the response is higher than the proportion of change in the predictor. Values above 0.5 are highlighted in **bold**. Parameters 1–7 are based on body size and 8–13 are related to decision behavior (simulation experiment 3). Parameters 14–16 describe the landscape's structure, 17–20 the amount of pollen and the number of bees in the landscape, and 21 affects the local depletion rate.

10.4.4   How Much Do Landscape-Related Parameters Affect Solitary Bees?

In this simulation experiment, we investigated the effect of small standardized changes in model parameters that altered the landscape and the vegetation (*am, fr, esize, seed, bdc, fd, ppf, plimit*). We followed the same protocol as for the sensitivity of bee parameters in simulation experiment 3.

10.4.4.1   Importance of Landscape Characteristics

The *pollen per flower* and *pollen availability* negatively affected the number of brood cells across all three scenarios (wood-nesting bee, soil-nesting bee, and a longer foraging time). Both parameters described the local vegetation properties (as well as *flower density* and *foraging habitat availability*, which only had a low impact). The other three response variables (number of flowers visited, foraging habitat visitation, and the mean distance flown) were affected by the same dominant parameter(s), which however differed for both bee types. For wood-nesting bees, *foraging habitat availability* and *landscape stochastic factor* were most important (for all three responses and for a longer foraging period). These landscape level parameters determined the spatial distribution of foraging and nesting resources. An increase in *foraging habitat availability* resulted in more flowers visited, a lower percentage of foraging habitat visited, and longer distances flown. For soil-nesting bees, the *landscape quality for bees* was important, which determined the landscape-level bee density. None of the parameter interactions had an important effect compared to the main effects.

Some of the effects seem contradictory. An increase in local resource availability (*plimit, ppf*) decreased the number of brood cells. An increase in *foraging habitat availability* increased the distance flown, but decreased the percentage of the foraging habitat visited. We suspect that local bee density around the nest (Figure 10.7) is important for these patterns and may be affected by different parameters for both bee types (wood-nesting bees responded most to *bdc*) and should be considered in future analyses. The *landscape stochastic factor*, used for adding spatial structure in the landscape generation process, had a stronger effect than *landscape fragmentation* and *landscape element size*.

10.4.4.2   Parameter Robustness of Landscape Characteristics

Elasticity for landscape parameters was also, in general, low (<0.5) across the response variables, which means that they were quite robust against small changes in landscape-related parameters. The parameters causing most variance for each of the response variables were also causing the highest sensitivity in response, but values never exceeded one. Other high values were for *pollen per flower* and *pollen availability* (both for wood- and soil-nesting bees), and the model can be considered moderately robust against these parameters. A longer foraging period did slightly decrease elasticity values. The total number of flowers visited per bee was most robust against changes in landscape-related parameters.

An exception was the very high elasticity of *landscape stochastic factor*, exceeding one. We used the set 90–100–110 for this analysis, but the set 99–100–101 would lead to the same response (due to the stochastic effect of this parameter) and would reduce the elasticity values by a factor of 10 (due to 1% change instead of 10% in the independent variable). The comparison of the *landscape stochastic factor* with other parameters in this way appears not to be meaningful. However, it does imply that the required number of replicates for robust results still needs to be tested.

10.5   UNDERSTANDING HOW SOLITARY BEES INTERACT WITH THE LANDSCAPE

10.5.1   GENERAL CONCLUSIONS

The model produced realistic output, but our simplifications led to more efficient foraging bees that do not spend time on additional activities or unexpected conditions existing in the real world. The model seems to be driven by local bee densities (affected by vegetation parameters), time budgets (affected by performance limitations and size of the bee), and the spatial distribution of pollen and nest locations, which affected the response variables differently. The model seems to be suitable to explore further ecological questions dealing with pollen-collecting solitary bees.

Different species of solitary bees clearly responded differently to landscape features. It was a remarkable result that wood-nesting bees that nest in field edges performed worse when there were more foraging resources. Wood-nesting bees needed to fly farther from the nest and were therefore under certain conditions more time limited than other bees and built fewer brood cells. Body size also had a large impact on performance (brood cells, flowers visited, foraging distance), but bees of different size did in general respond qualitatively similar to other parameter changes. From all size-related traits, *handling time per flower* has had the most effect on the bee's performance. The amount of *pollen per flower* and *landscape quality for bees* had a very strong effect as well, by affecting the number of bees in the landscape. Despite strong and some unexpected effects in the model, all parameters were robust against small changes.

The simulation experiments showed that the selection of values for some parameters deserves more attention for selection than others. The most uncertain and unknown parameters (*sightm, flightm, CRW, fmem, cmem*) did not have strong effects on the response variables and do not require further study. The strong effect of *pollen per flower* (*ppf*) suggests that the model is suitable for comparing different vegetation types such as fields of clover (little pollen) and fields of sunflowers (very much pollen). However, the model does require very long running times for large flowers (resources for millions of bees), making such comparisons unattractive. Alternatively, one can also consider meadow-like vegetation with moderately sized flowers (moderate pollen production) as a legitimate and realistic scenario for future simulations, since large flowers are not common in most landscapes. The time-budget parameters *time at the nest* and foraging time (*flytime*) affected the responses, but they did not affect the relative impact of other parameters. We suggest a moderate value for *time at the nest*, since it is biologically more plausible. A value of 4 h as foraging time is just as suitable as 8 h to measure performance within a foraging day and saves simulation time. We also consider the selected value for *bdc* (*landscape quality for bees*) as biologically more plausible than its extremes (see Section 10.3.1). The other important parameters *body length*, *nesting preference*, and *foraging habitat availability* should be incorporated in all future simulation experiments.

10.5.2 GENERAL INTEREST AND LIMITATIONS

The main strength of the model is its high level of realism. In such a mechanistic model, all essential processes are controllable with parameters, which enables investigation of each parameter independently. This can be especially valuable in ecology, since it can rule out variance present in field experiments. In ecological field experiments, it is often impossible to regulate all environmental conditions independently. All of the parameters can potentially be measured, and none of the parameters has an abstract meaning, which make the model suitable for many applications. The main response simulated (brood cells) is also very direct. It registers the direct performance of the bee in amount of pollen that can be collected in a certain amount of time, which is a relative novel approach for the ecology of bees. Most studies at the landscape scale deal with aggregated responses such as species numbers and individual numbers, while the model can study the potential offspring (brood cells) for a certain bee species in a certain landscape.

We experienced also some limitations. The many parameters in the model introduce uncertainty. We have put substantial effort in reducing these parameter uncertainties by extensive literature review (Section 10.3.1), effect exploration (simulation experiment 1), and sensitivity analysis (simulation experiments 2, 3, and 4), but this did not inform us about all uncertainties. We did, for example, test what the consequences would be if each of the bee traits would be 10% higher or lower than that estimated by the scaling relationship (simulation experiment 3), but we did not test what consequences altered scaling relations would have (possibly leading to larger deviations than 10% for small or large bees). In addition to this, we ignored that some parameters could scale with body size as well (but have not been shown yet to do so for bees or similar insects), such as for *time at the nest*, the perceived *grain size*, and the preference for a certain flower size (*pollen per flower*). The model design not only involved assumptions based on observations, but some were also made for simplifying reasons. The model assumed that small and large bees use the same proportion of their flower visiting time for pollen collection, that male and female cells are built in equal ratio (mean number of trips required per cell was applied as fixed value), and that bees choose to nest near their foraging resource. These simplifications may have caused that the model bees were more efficient than real bees (Section 10.4.1.2).

Ideally, we would like to validate the model for fixed positions in the parameter space, but the field data only allowed us to show that their ranges overlap well with the model's response variables. The response variable "brood cells" (per individual, per day) is difficult to link to field studies since most studies give the number of brood cells per trap nest (undefined number of individuals) or per lifetime (undefined number of days). Further, most experimental studies have measured only a fraction of the parameters used as input for the model, and up to now there was no complete set of environmental parameters available for the model. This is another clear disadvantage of a large number of parameters. Some of the important parameters such as flower density and activity time of the bee could be measured, but no single study at the landscape level did so far. Other parameters, such as the local bee density in the environment or the amount of pollen per flower, may be hard to measure accurately in the field. A suggestion is to roughly estimate the parameters for field studies to get a complete set of parameters. Landscape parameters can be assessed from land-use maps validated by field surveys and be completed with estimated flower densities for each land-use class. A more difficult-to-measure parameter such as pollen production per flower could be approximated by flower size. The bee's body length can be measured in the field as well and serve as proxy for behavioral foraging traits. Measuring such values in the field and using them as input for the model to simulate specific situations have a high potential for understanding observed field patterns.

In general, more field research would benefit the model. Fundamental and descriptive ecological research could strengthen theoretical relationships between body size and traits that affect time budgets. Measurement efforts to assess or at least estimate the complete set of parameters for several example landscapes would strengthen further model validation. However, due to the labor-intensive nature of parameter assessment in the field, such measurements are restricted by financial resources. This financial issue would favor applying the model in its current state, since it is quite elaborate already. In the case of future application of the model, it is important to ask whether it is acceptable that the model simulates more efficient bees than in real systems (*e.g.*, by nesting close to their foraging resource) and whether it is of minor importance for the specific research question.

10.5.3 Outlook

We found several interesting results that are worth exploring further, mainly concerning bee type and landscape structure. Soil-nesting and wood-nesting bees responded differently to landscape structure, and body size also affected performance. Species differ, and systematic simulations could help to find out how their performance differs in different landscapes and what the causes are. Both the spatial resource distribution of pollen and nest locations influence how bees behave in the model, and they determine depletion patterns. How does fragmentation of the spatial patch mosaic influence the interaction between bees and their performance? How do bee size, nesting preference, and landscape structure affect pollination services?

Finally, the model has the potential to study other topics and learn more about bee biology. What is, for example, the effect of different foraging trade-offs? How do handling time, pollen requirements, number of flower visits, and patch-size selection affect one another? Are there optimal strategies concerning different "giving up times" at the flower and the patch level? Can bees of different size optimize their foraging behavior by choosing an optimal flower size?

ACKNOWLEDGMENTS

This study was supported by the European Commission within FP 6 Integrated Project ALARM: Assessing LArge scale environmental Risks with tested Methods (GOCE-CT-2003-506675), the Helmholtz Association (VH-NG 247), and the Helmholtz Impulse and Networking Fund through HIGRADE: Helmholtz Interdisciplinary Graduate School for Environmental Research. We thank Volker Grimm for helpful comments on this chapter.

REFERENCES

1. M.A. Aizen, L.A. Garibaldi, S.A. Cunningham, and A.-M. Klein, *How much does agriculture depend on pollinators? Lessons from long-term trends in crop production*, Ann. Bot. 103 (2009), pp. 1579–1588.

2. C.D. Michener, *The Bees of the World*, Johns Hopkins University Press, Baltimore, MD, 2000.

3. J.L. Neff, *Components of nest provisioning behavior in solitary bees (Hymenoptera: Apoidea)*, Apidologie 39 (2008), pp. 30–45.

4. M. Chagnon, J. Gingras, and D. Deoliveira, *Complementary aspects of strawberry pollination by honey and indigenous bees (Hymenoptera)*, J. Econ. Entomol. 86 (1993), pp. 416–420.

5. A.-M. Klein, I. Steffan-Dewenter, and T. Tscharntke, *Fruit set of highland coffee increases with the diversity of pollinating bees*, Proceed. Royal Soc. Lond. B: Biol. Sci. 270 (2003), pp. 955–961.

6. M.A. Aizen and P. Feinsinger, *Habitat fragmentation, native insect pollinators, and feral honey-bees in Argentine Chaco Serrano*, Ecol. Appl. 4 (1994), pp. 378–392.

7. E. Andrieu, A. Dornier, S. Rouifed, B. Schatz, and P.-O. Cheptou, *The town Crepis and the country Crepis: How does fragmentation affect a plant-pollinator interaction?* Acta Oecol. 35 (2009), pp. 1–7.

8. G.H. Pyke, *Animal movements: An optimal foraging approach*, in *The Ecology of Animal Movement*, J.R. Swingland and P.J. Greenwood, eds., Oxford University Press, Oxford, UK, 1983, pp. 7–31.

9. E.G. Linsley, *The ecology of solitary bees*, Hilgardia 27 (1958), pp. 543–599.

10. C. Westerkamp, *Honeybees are poor pollinators—Why?* Plant Syst. Evol. 177 (1991), pp. 71–75.

11. N.R.C. Committee on the Status of Pollinators in North America, *Status of Pollinators in North America*, The National Academies Press, Washington, DC, 2007.

12. I.H. Williams, *Insect pollination and crop production: A European perspective*, in *Pollinating Bees—The Conservation Link between Agriculture and Nature*, P.G. Kevan and V.L. Imperatriz-Fonseca, eds., Ministry of Environment, Brasília, Brazil, 2002, pp. 59–65.

13. I. Steffan-Dewenter, S.G. Potts, and L. Packer, *Pollinator diversity and crop pollination services are at risk*, Trends Ecol. Evol. 20 (2005), pp. 651–652.

14. S.H.M. Butchart, M. Walpole, B. Collen, A. van Strien, J.P.W. Scharlemann, R.E.A. Almond, J.E.M. Baillie, B. Bomhard, C. Brown, J. Bruno, K.E. Carpenter, G.M. Carr, J. Chanson, A.M. Chenery, J. Csirke, N.C. Davidson, F. Dentener, M. Foster, A. Galli, J.N. Galloway, P. Genovesi, R.D. Gregory, M. Hockings, V. Kapos, J.F. Lamarque, F. Leverington, J. Loh, M.A. McGeoch, L. McRae, A. Minasyan, M.H. Morcillo, T.E.E. Oldfield, D. Pauly, S. Quader, C. Revenga, J.R. Sauer, B. Skolnik, D. Spear, D. Stanwell-Smith, S.N. Stuart, A. Symes, M. Tierney, T.D. Tyrrell, J.C. Vie, and R. Watson, *Global biodiversity: Indicators of recent declines*, Science 328 (2010), pp. 1164–1168.

15. T.D. Breeze, A.P. Bailey, K.G. Balcombe, and S.G. Potts, *Pollination services in the UK: How important are honeybees?* Agric. Ecosyst. Environ. 142 (2011), pp. 137–143.

16. L.A. Garibaldi, I. Steffan-Dewenter, C. Kremen, J.M. Morales, R. Bommarco, S.A. Cunningham, L.G. Carvalheiro, N.P. Chacoff, J.H. Dudenhoffer, S.S. Greenleaf, A. Holzschuh, R. Isaacs, K. Krewenka, Y. Mandelik, M.M. Mayfield, L.A. Morandin,

S.G. Potts, T.H. Ricketts, H. Szentgyörgyi, B.F. Viana, C. Westphal, R. Winfree, and A.-M. Klein, *Stability of pollination services decreases with isolation from natural areas despite honey bee visits*, Ecol. Lett. 14 (2011), pp. 1062–1072.

17. L.G. Carvalheiro, C.L. Seymour, R. Veldtman, and S.W. Nicolson, *Pollination services decline with distance from natural habitat even in biodiversity-rich areas*, J. Appl. Ecol. 47 (2010), pp. 810–820.

18. T.H. Ricketts, J. Regetz, I. Steffan-Dewenter, S.A. Cunningham, C. Kremen, A. Bogdanski, B. Gemmill-Herren, S.S. Greenleaf, A.-M. Klein, M.M. Mayfield, L.A. Morandin, A. Ochieng', S.G. Potts, and B.F. Viana, *Landscape effects on crop pollination services: Are there general patterns?* Ecol. Lett. 11 (2008), pp. 499–515.

19. C. Kremen, N.M. Williams, R.L. Bugg, J.P. Fay, and R.W. Thorp, *The area requirements of an ecosystem service: Crop pollination by native bee communities in California*, Ecol. Lett. 7 (2004), pp. 1109–1119.

20. S.A. Corbet, I.H. Williams, and J.L. Osborne, *Bees and the pollination of crops and wild flowers in the European community*, Bee World 72 (1991), pp. 47–59.

21. S.G. Potts, J.C. Biesmeijer, C. Kremen, P. Neumann, O. Schweiger, and W.E. Kunin, *Global pollinator declines: Trends, impacts and drivers*, Trends Ecol. Evol. 25 (2010), pp. 345–353.

22. J.C. Biesmeijer, S.P.M. Roberts, M. Reemer, R. Ohlemüller, M. Edwards, T. Peeters, A.P. Schaffers, S.G. Potts, R. Kleukers, C.D. Thomas, J. Settele, and W.E. Kunin, *Parallel declines in pollinators and insect-pollinated plants in Britain and the Netherlands*, Science 313 (2006), pp. 351–354.

23. P. Neumann and N.L. Carreck, *Honey bee colony losses*, J. Apic. Res. 49 (2010), pp. 1–6.

24. D. van Engelsdorp, J.D. Evans, C. Saegerman, C. Mullin, E. Haubruge, B.K. Nguyen, M. Frazier, J. Frazier, D. Cox-Foster, Y.P. Chen, R.M. Underwood, D.R. Tarpy, and J.S. Pettis, *Colony collapse disorder: A descriptive study*, PLoS ONE 4 (2009), p. e6481.

25. M.E. Knight, J.L. Osborne, R.A. Sanderson, R.J. Hale, A.P. Martin, and D. Goulson, *Bumblebee nest density and the scale of available forage in arable landscapes*, Insect Conserv. Divers. 2 (2009), pp. 116–124.

26. C.A. Kearns, D.W. Inouye, and N.M. Waser, *Endangered mutualisms: The conservation of plant-pollinator interactions*, Annu. Rev. Ecol. Syst. 29 (1998), pp. 83–112.

27. J.H. Cane and V.J. Tepedino, *Causes and extent of declines among native North American invertebrate pollinators: Detection, evidence, and consequences*, Conserv. Ecol. 5 (2001). Available at http://www.consecol.org/vol5/iss1/art1/.

28. T. Tscharntke and R. Brandl, *Plant-insect interactions in fragmented landscapes*, Annu. Rev. Entomol. 49 (2004), pp. 405–430.

29. E. Lonsdorf, C. Kremen, T.H. Ricketts, R. Winfree, N.M. Williams, and S.S. Greenleaf, *Modelling pollination services across agricultural landscapes*, Ann. Bot. 103 (2009), pp. 1589–1600.

30. J. Banaszak, *Effect of habitat heterogeneity on the diversity and density of pollinating insects*, in *Interchanges of Insects between Agricultural and Surrounding Landscapes*, B.S. Ekbom, M.E. Irwin, and Y. Robert, eds., Kluwer Academic, Dordrecht, the Netherlands, 2000, pp. 123–140.

31. L.D. Harder and W.G. Wilson, *Theoretical consequences of heterogeneous transport conditions for pollen dispersal by animals*, Ecology 79 (1998), pp. 2789–2807.

32. R.B. Aronson and T.J. Givnish, *Optimal central-place foragers—A comparison with null hypotheses*, Ecology 64 (1983), pp. 395–399.

33. F. Thuijsman, B. Peleg, M. Amitai, and A. Shmida, *Automata, matching and foraging behavior of bees*, J. Theor. Biol. 175 (1995), pp. 305–316.

34. J.E. Cresswell, *A mechanistic model of pollinator-mediated gene flow in agricultural safflower*, Basic Appl. Ecol. 11 (2010), pp. 415–421.

35. R. Dukas and L. Edelstein-Keshet, *The spatial distribution of colonial food provisioners*, J. Theor. Biol. 190 (1998), pp. 121–134.

36. C. Westphal, I. Steffan-Dewenter, and T. Tscharntke, *Foraging trip duration of bumblebees in relation to landscape-wide resource availability*, Ecol. Entomol. 31 (2006), pp. 389–394.

37. K. Ulbrich and K. Seidelmann, *Modeling population dynamics of solitary bees in relation to habitat quality*, Web Ecol. 2 (2001), pp. 57–64.

38. D. Austin, W.D. Bowen, and J.I. McMillan, *Intraspecific variation in movement patterns: Modeling individual behaviour in a large marine predator*, Oikos 105 (2004), pp. 15–30.

39. V. Grimm and S.F. Railsback, *Individual-Based Modeling and Ecology*, Princeton Series in Theoretical and Computational Biology, Princeton University Press, Princeton, NJ, 2005.

40. V. Grimm, U. Berger, F. Bastiansen, S. Eliassen, V. Ginot, J. Giske, J. Goss-Custard, T. Grand, S.K. Heinz, G. Huse, A. Huth, J.U. Jepsen, C. Jørgensen, W.M. Mooij, B. Müller, G. Pe'er, C. Piou, S.F. Railsback, A.M. Robbins, M.M. Robbins, E. Rossmanith, N. Rüger, E. Strand, S. Souissi, R.A. Stillman, R. Vabø, U. Visser, and D.L. DeAngelis, *A standard protocol for describing individual-based and agent-based models*, Ecol. Model. 198 (2006), pp. 115–126.

41. V. Grimm, U. Berger, D.L. DeAngelis, J.G. Polhill, J. Giske, and S.F. Railsback, *The ODD protocol A review and first update*, Ecol. Model. 221 (2010), pp. 2760–2768.

42. R.M. Ewers and R.K. Didham, *Confounding factors in the detection of species responses to habitat fragmentation*, Biol. Rev. 81 (2006), pp. 117–142.

43. I. Steffan-Dewenter, U. Münzenberg, and T. Tscharntke, *Pollination, seed set and seed predation on a landscape scale*, Proc. Royal Soc. Lond. Ser. B: Biol. Sci. 268 (2001), pp. 1685–1690.

44. N.M. Williams and V.J. Tepedino, *Consistent mixing of near and distant resources in foraging bouts by the solitary mason bee* Osmia lignaria, Behav. Ecol. 14 (2003), pp. 141–149.

45. V. Monsevičius, *Fauna of wild bees in Lithuania and trends of its changes*, in *Changes in Fauna of Wild Bees in Europe*, J. Banaszak, ed., Pedagogical University, Bydgoszcz, Poland, 1995, pp. 27–39.

46. M. Franzén, M. Larsson, and S.G. Nilsson, *Small local population sizes and high habitat patch fidelity in a specialised solitary bee*, J. Insect. Conserv. 13 (2009), pp. 89–95.

47. D.R. Artz and K.D. Waddington, *The effects of neighbouring tree islands on pollinator density and diversity, and on pollination of a wet prairie species*, Asclepias lanceolata *(Apocynaceae)*, J. Ecol. 94 (2006), pp. 597–608.

48. T. Diekötter, K.J. Haynes, D. Mazeffa, and T.O. Crist, *Direct and indirect effects of habitat area and matrix composition on species interactions among flower-visiting insects*, Oikos 116 (2007), pp. 1588–1598.

49. J. Joshi, P. Stoll, H.P. Rusterholz, B. Schmid, C. Dolt, and B. Baur, *Small-scale experimental habitat fragmentation reduces colonization rates in species-rich grasslands*, Oecologia 148 (2006), pp. 144–152.

50. E.L. Charnov, *Optimal foraging, marginal value theorem*, Theor. Popul. Biol. 9 (1976), pp. 129–136.

51. A. Basset, M. Fedele, and D.L. DeAngelis, *Optimal exploitation of spatially distributed trophic resources and population stability*, Ecol. Model. 151 (2002), pp. 245–260.

52. U. Motro and A. Shmida, *Near-far search—An evolutionarily stable foraging strategy*, J. Theor. Biol. 173 (1995), pp. 15–22.

53. M. Beil, H. Horn, and A. Schwabe, *Analysis of pollen loads in a wild bee community (Hymenoptera: Apidae)—A method for elucidating habitat use and foraging distances*, Apidologie 39 (2008), pp. 456–467.

54. B. Heinrich, *Majoring and minoring by foraging bumblebees*, Bombus vagans—*Experimental analysis*, Ecology 60 (1979), pp. 245–255.

55. G.C. Eickwort and H.S. Ginsberg, *Foraging and mating-behavior in Apoidea*, Annu. Rev. Entomol. 25 (1980), pp. 421–446.

56. R. Campan and M. Lehrer, *Discrimination of closed shapes by two species of bee*, Apis mellifera *and* Megachile rotundata, J. Exp. Biol. 205 (2002), pp. 559–572.

57. D. Saupe, *Algorithms for random fractals*, in *The Sciences of Fractal Images*, H.-O. Peitgen and D. Saupe, eds., Springer-Verlag, New York, 1988, pp. 71–113.

58. K.A. With, R.H. Gardner, and M.G. Turner, *Landscape connectivity and population distributions in heterogeneous environments*, Oikos 78 (1997), pp. 151–169.

59. C.D. Hargis, J.A. Bissonette, and J.L. David, *The behavior of landscape metrics commonly used in the study of habitat fragmentation*, Landscape Ecol. 13 (1998), pp. 167–186.

60. J.H. Cane, *Estimation of bee size using intertegular span (Apoidea)*, J. Kans. Entomol. Soc. 60 (1987), pp. 145–147.

61. A. Müller, S. Diener, S. Schnyder, K. Stutz, C. Sedivy, and S. Dorn, *Quantitative pollen requirements of solitary bees: Implications for bee conservation and the evolution of bee-flower relationships*, Biol. Conserv. 130 (2006), pp. 604–615.

62. S.S. Greenleaf, N.M. Williams, R. Winfree, and C. Kremen, *Bee foraging ranges and their relationship to body size*, Oecologia 153 (2007), pp. 589–596.

63. I. Calabuig, *Solitary bees and bumblebees in a Danish agricultural landscape*, Ph.D. diss., University of Copenhagen, Copenhagen, Denmark, 2000.

64. A. Gathmann, H.J. Greiler, and T. Tscharntke, *Trap-nesting bees and wasps colonizing set-aside fields—Succession and body-size, management by cutting and sowing*, Oecologia 98 (1994), pp. 8–14.

65. T. Pawlikowski, *Struktura zgrupowań dzikich pszczołowatych (Hymenoptera, Apoidea) z obszarów rolnych o różnych typach parcelacji powierzchni uprawnej. [The structure of wild bee (Hymenoptera, Apoidea) communities from farming areas of different field sizes]*, Acta Univ. Nic. Copernici, Biol. 33 (1989), pp. 31–46.

66. P. Westrich, *Habitat requirements of central European bees and the problems of partial habitats*, in *The Conservation of Bees*, A. Matheson, S.L. Buchmann, C. O'Toole, P. Westrich, and I.H. Williams, eds., Academic Press, London, UK, 1996, pp. 1–16.

67. S. Kirkpatrick and E.P. Stoll, *A very fast shift-register sequence random number generator*, J. Comput. Phys. 40 (1981), pp. 517–526.

68. D.P. Abrol and R.P. Kapil, *On homing ability and pollination effectiveness of bees*, Mysore J. Agric. Sci. 28 (1994), pp. 249–252.

69. V. Grimm, E. Revilla, U. Berger, F. Jeltsch, W.M. Mooij, S.F. Railsback, H.H. Thulke, J. Weiner, T. Wiegand, and D.L. DeAngelis, *Pattern-oriented modeling of agent-based complex systems: Lessons from ecology*, Science 310 (2005), pp. 987–991.

70. T. Wiegand, F. Jeltsch, I. Hanski, and V. Grimm, *Using pattern-oriented modeling for revealing hidden information: A key for reconciling ecological theory and application*, Oikos 100 (2003), pp. 209–222.

71. N.M. Williams and C. Kremen, *Resource distributions among habitats determine solitary bee offspring production in a mosaic landscape*, Ecol. Appl. 17 (2007), pp. 910–921.

72. K. Szklanowska, *Pollen flows of crowfoot family (Ranunculaceae) from some natural plant communities*, in *Changes in Fauna of Wild Bees in Europe*, J. Banaszak, ed., Pedagogical University, Bydgoszcz, Poland, 1995, pp. 201–214.

73. B. Denisow, *Blooming and pollen production of several representatives of the genus* Centaurea L, J. Apicult. Sci. 50 (2006), pp. 13–20.

74. Z. Kołtowski, *The effect of pollinating insects on the yield of winter rapeseed (*Brassica napus *L. var. napus f. biennis) cultivars*, J. Apicult. Sci. 49 (2005), pp. 29–41.

75. K.D. Waddington, *Foraging patterns of Halictid bees at flowers of* Convolvulus arvensis, Psyche 83 (1976), pp. 112–119.

76. L.D. Harder, *Pollen-size comparisons among animal-pollinated angiosperms with different pollination characteristics*, Biol. J. Linn. Soc. 64 (1998), pp. 513–525.

77. R. Gallardo, E. Dominguez, and J.M. Munoz, *Pollen ovule ratio, pollen size, and breeding system in* Astragalus *(Fabaceae) subgenus Epiglottis—A pollen and seed allocation approach*, Am. J. Bot. 81 (1994), pp. 1611–1619.

78. B. Denisow and M. Bozek, *Blooming and pollen production of two* Lamium L. *species*, J. Apicult. Sci. 52 (2008), pp. 21–30.

79. C. Erbar and P. Leins, *Portioned pollen release and the syndromes of secondary pollen presentation in the Campanulales-Asterales-complex*, Flora 190 (1995), pp. 323–338.

80. L.D. Harder, *Behavioral responses by bumble bees to variation in pollen availability*, Oecologia 85 (1990), pp. 41–47.

81. L.D. Harder, *Pollen removal by bumble bees and its implications for pollen dispersal*, Ecology 71 (1990), pp. 1110–1125.

82. D.S. Willis and P.G. Kevan, *Foraging dynamics of* Peponapis pruinosa *(Hymenoptera, Anthophoridae) on pumpkin (*Cucurbita pepo*) in southern Ontario*, Can. Entomol. 127 (1995), pp. 167–175.

83. C. Schlindwein, D. Wittmann, C.F. Martins, A. Hamm, J.A. Siqueira, D. Schiffler, and I.C. Machado, *Pollination of* Campanula rapunculus L. *(Campanulaceae): How much pollen flows into pollination and into reproduction of oligolectic pollinators?* Plant Syst. Evol. 250 (2005), pp. 147–156.

84. S.S. Greenleaf and C. Kremen, *Wild bees enhance honey bees' pollination of hybrid sunflower*, Proc. Nat. Acad. Sci. USA. 103 (2006), pp. 13890–13895.

85. P. Hoehn, T. Tscharntke, J.M. Tylianakis, and I. Steffan-Dewenter, *Functional group diversity of bee pollinators increases crop yield*, Proc. Royal Soc. B: Biol. Sci. 275 (2008), pp. 2283–2291.

86. J. Banaszak, *Natural resources of wild bees in Poland and an attempt at estimation of their changes*, in *Changes in Fauna of Wild Bees in Europe*, J. Banaszak, ed., Pedagogical University, Bydgoszcz, Poland, 1995, pp. 9–25.

87. J. Bosch, *The nesting behaviour of the mason bee* Osmia cornuta *(Latr) with special reference to its pollinating potential (Hymenoptera, Megachilidae)*, Apidologie 25 (1994), pp. 84–93.

88. M. Franzén and M. Larsson, *Pollen harvesting and reproductive rates in specialized solitary bees*, Ann. Zool. Fenn. 44 (2007), pp. 405–414.

89. M. Giovanetti and E. Lasso, *Body size, loading capacity and rate of reproduction in the communal bee* Andrena agilissima *(Hymenoptera; Andrenidae)*, Apidologie 36 (2005), pp. 439–447.

90. M. Larsson and M. Franzén, *Critical resource levels of pollen for the declining bee* Andrena hattorfiana *(Hymenoptera, Andrenidae)*, Biol. Conserv. 134 (2007), pp. 405–414.

91. J.L. Neff and B.N. Danforth, *The nesting and foraging behavior of* Perdita texana *(Cresson) (Hymenoptera, Andrenidae)*, J. Kans. Entomol. Soc. 64 (1991), pp. 394–405.

92. J. Kunze and L. Chittka, *Bees and butterflies fly faster when plants feed them more nectar*, in *Goettingen Neurobiology Report 1996*, N. Elsner and H. Schnitzler, eds., Thieme Verlag, Stuttgart, Germany, 1996, pp. 109–109.

93. A. Barron and M.V. Srinivasan, *Visual regulation of ground speed and headwind compensation in freely flying honey bees (*Apis mellifera L.*)*, J. Exp. Biol. 209 (2006), pp. 978–984.

94. W. Nachtigall, U. Hanauer-Thieser, and M. Morz, *Flight of the honey bee. 7. Metabolic power versus flight speed relation*, J. Comp. Physiol. B: Biochem. System. Environ. Physiol. 165 (1995), pp. 484–489.

95. C.P. Ellington, K.E. Machin, and T.M. Casey, *Oxygen-consumption of bumblebees in forward flight*, Nature 347 (1990), pp. 472–473.

96. A.B. Ware and S.G. Compton, *Dispersal of adult female fig wasps. 2. Movements between trees*, Entomol. Exp. Appl. 73 (1994), pp. 231–238.

97. S.G. Compton, M.D.F. Ellwood, A.J. Davis, and K. Welch, *The flight heights of chalcid wasps (Hymenoptera, Chalcidoidea) in a lowland Bornean rain forest: Fig wasps are the high fliers*, Biotropica 32 (2000), pp. 515–522.

98. C. Guédot, J. Bosch, and W.P. Kemp, *Relationship between body size and homing ability in the genus* Osmia *(Hymenoptera; Megachilidae)*, Ecol. Entomol. 34 (2009), pp. 158–161.

99. W. Nachtigall, *Formation of clay globules and flight departure with the building material by the thread-waisted potter wasp* Sceliphron spirifex *(Hymenoptera: Sphecidae)*, Entomol. Gen. 25 (2001), pp. 161–170.

100. T.J. Dean, *Chapter 1: Fastest flyer*, in *University of Florida Book of Insect Record*. Available at http://entomology.ifas.ufl.edu/walker/ufbir.

101. R. Piper, *Extraordinary Animals: An Encyclopedia of Curious and Unusual Animals*, Greenwood Press, Westport, CT, 2007.

102. K. Strickler, *Specialization and foraging efficiency of solitary bees*, Ecology 60 (1979), pp. 998–1009.

103. N.E. Raine and L. Chittka, *Pollen foraging: Learning a complex motor skill by bumblebees (*Bombus terrestris*)*, Naturwissenschaften 94 (2007), pp. 459–464.

104. H.J. Young and M.L. Stanton, *Influences of floral variation on pollen removal and seed production in wild radish*, Ecology 71 (1990), pp. 536–547.

105. U. Jander and R. Jander, *Allometry and resolution of bee eyes (Apoidea)*, Arthr. Struct. Dev. 30 (2002), pp. 179–193.

106. A. Dafni and P.G. Kevan, *Hypothesis on adaptive features of the compound eye of bees—Flower-specific specializations*, Evol. Ecol. 9 (1995), pp. 236–241.

107. M.V. Srinivasan, S.W. Zhang, M. Lehrer, and T.S. Collett, *Honeybee navigation en route to the goal: Visual flight control and odometry*, J. Exp. Biol. 199 (1996), pp. 237–244.

108. J.R. Riley, D.R. Reynolds, A.D. Smith, A.S. Edwards, J.L. Osborne, I.H. Williams, and H.A. McCartney, *Compensation for wind drift by bumble-bees*, Nature 400 (1999), p. 126.

109. L. Chittka, N.M. Williams, H. Rasmussen, and J.D. Thomson, *Navigation without vision: Bumblebee orientation in complete darkness*, Proc. Royal Soc. Lond. Ser. B: Biol. Sci. 266 (1999), pp. 45–50.

110. R. Menzel, *Behavioral and neural mechanisms of learning and memory as determinants of flower constancy*, in *Cognitive Ecology of Pollination*, L. Chittka and J.D. Thomson, eds., Cambridge University Press, Cambridge, UK, 2001, pp. 21–40.

111. G. Ne'eman, O. Shavit, L. Shaltiel, and A. Shmida, *Foraging by male and female solitary bees with implications for pollination*, J. Insect. Behav. 19 (2006), pp. 383–401.

112. L. Chittka and N.E. Raine, *Recognition of flowers by pollinators*, Curr. Opin. Plant Biol. 9 (2006), pp. 428–435.

113. R. Menzel, R. Brandt, A. Gumbert, B. Komischke, and J. Kunze, *Two spatial memories for honeybee navigation*, Proc. Royal Soc. Lond. Ser. B: Biol. Sci. 267 (2000), pp. 961–968.

114. A.H. Powell and G.V.N. Powell, *Population-dynamics of male Euglossine bees in Amazonian forest fragments*, Biotropica 19 (1987), pp. 176–179.

115. P. Schmid-Hempel, *How do bees choose flight direction while foraging*, Physiol. Entomol. 10 (1985), pp. 439–442.

116. T.T. Makino, K. Ohashi, and S. Sakai, *How do floral display size and the density of surrounding flowers influence the likelihood of bumble bee revisitation to a plant?* Funct. Ecol. 21 (2007), pp. 87–95.

117. R.V. Cartar and L.A. Real, *Habitat structure and animal movement: The behaviour of bumble bees in uniform and random spatial resource distributions*, Oecologia 112 (1997), pp. 430–434.

118. B. Heinrich, *Resource heterogeneity and patterns of movement in foraging bumblebees*, Oecologia 40 (1979), pp. 235–245.

119. A.M. Reynolds, A.D. Smith, R. Menzel, U. Greggers, D.R. Reynolds, and J.R. Riley, *Displaced honey bees perform optimal scale-free search flights*, Ecology 88 (2007), pp. 1955–1961.

120. H. Teppner, *Bienen und Obstbaum-Bestäubung*, Obst Wein Garten 65 (1996), pp. 3–7.

121. B.N. Danforth, *Provisioning behavior and the estimation of investment ratios in a solitary bee*, Calliopsis *(Hypomacrotera)* persimilis *(Cockerell) (Hymenoptera: Andrenidae)*, Behav. Ecol. Sociobiol. 27 (1990), pp. 159–168.

122. J.-N. Tasei, *Le comportement de nidification chez* Osmia *(Osmia)* cornuta *Latr. et* Osmia *(Osmia)* rufa *L. (Hymenoptera Megachilidae)*, Apidologie 4 (1973), pp. 195–225.

123. E.G. Linsley, *Temporal patterns of flower visitation by solitary bees, with particular reference to the southwestern United States*, J. Kans. Entomol. Soc. 51 (1978), pp. 531–546.

124. J. Bosch and M. Blas, *Foraging behavior and pollinating efficiency of* Osmia cornuta *and* Apis mellifera *on almond (Hymenoptera, Megachilidae and Apidae)*, Appl. Entomol. Zool. 29 (1994), pp. 1–9.

125. G.N. Stone, *Patterns of evolution of warm-up rates and body temperatures in-flight in solitary bees of the genus Anthophora*, Funct. Ecol. 8 (1994), pp. 324–335.

126. P.G. Willmer and G.N. Stone, *Behavioral, ecological, and physiological determinants of the activity patterns of bees*, Adv. Study Behav. 34 (2004), pp. 347–466.

127. W.J. Bell, *Searching behavior patterns in insects*, Annu. Rev. Entomol. 35 (1990), pp. 447–467.

128. J.D. Thomson and L. Chittka, *Pollinator individuality: When does it matter?* in *Cognitive Ecology of Pollination*, L. Chittka and J.D. Thomson, eds., Cambridge University Press, Cambridge, UK, 2001, pp. 191–213.

129. M.J. Couvillon, G. DeGrandi-Hoffman, and W. Gronenberg, *Africanized honeybees are slower learners than their European counterparts*, Naturwissenschaften 97 (2010), pp. 153–160.

130. T. Yokoi and K. Fujisaki, *Recognition of scent marks in solitary bees to avoid previously visited flowers*, Ecol. Res. 24 (2009), pp. 803–809.

131. A.D. Howell and R. Alarcón, *Osmia bees (Hymenoptera: Megachilidae) can detect nectar-rewarding flowers using olfactory cues*, Anim. Behav. 74 (2007), pp. 199–205.

132. J. Spaethe, J. Tautz, and L. Chittka, *Visual constraints in foraging bumblebees: Flower size and color affect search time and flight behavior*, Proc. Nat. Acad. Sci. USA 98 (2001), pp. 3898–3903.

133. M. Giurfa and M. Lehrer, *Honeybee vision and floral displays: From detection to close-up recognition*, in *Cognitive Ecology of Pollination*, L. Chittka and J.D. Thomson, eds., Cambridge University Press, Cambridge, UK, 2001, pp. 61–82.

134. L. Chittka, J. Spaethe, A. Schmidt, and A. Hickelsberger, *Adaptation, constraint, and chance in the evolution of flower color and pollinator color vision*, in *Cognitive Ecology of Pollination*, L. Chittka and J.D. Thomson, eds., Cambridge University Press, Cambridge, UK, 2001, pp. 106–126.

135. L. Chittka, A.G. Dyer, F. Bock, and A. Dornhaus, *Psychophysics—Bees trade off foraging speed for accuracy*, Nature 424 (2003), p. 388.

136. M. Goverde, K. Schweizer, B. Baur, and A. Erhardt, *Small-scale habitat fragmentation effects on pollinator behaviour: Experimental evidence from the bumblebee* Bombus veteranus *on calcareous grasslands*, Biol. Conserv. 104 (2002), pp. 293–299.

137. H.E. Julier and T.H. Roulston, *Wild bee abundance and pollination service in culti-
     vated pumpkins: Farm management, nesting behavior and landscape effects*, J. Econ.
     Entomol. 102 (2009), pp. 563–573.

138. J.H. Cane, *Soils of ground-nesting bees (Hymenoptera, Apoidea)—Texture, moisture,
     cell depth and climate*, J. Kans. Entomol. Soc. 64 (1991), pp. 406–413.

139. J. Bosch and N. Vicens, *Body size as an estimator of production costs in a solitary bee*,
     Ecol. Entomol. 27 (2002), pp. 129–137.

140. P.G. Willmer and G.N. Stone, *Incidence of entomophilous pollination of lowland coffee
     (*Coffea canephora*)—The role of leaf cutter bees in Papua New Guinea*, Entomol. Exp.
     Appl. 50 (1989), pp. 113–124.

141. R.L. Minckley, W.T. Wcislo, D. Yanega, and S.L. Buchmann, *Behavior and phenology
     of a specialist bee (*Dieunomia*) and sunflower (*Helianthus*) pollen availability*, Ecology
     75 (1994), pp. 1406–1419.

142. A. Gathmann and T. Tscharntke, *Foraging ranges of solitary bees*, J. Anim. Ecol. 71
     (2002), pp. 757–764.

143. M. Munster-Swendsen and I. Calabuig, *Interaction between the solitary bee* Chelostoma
     florisomne *and its nest parasite* Sapyga clavicornis*—Empty cells reduce the impact of
     parasites*, Ecol. Entomol. 25 (2000), pp. 63–70.

144. W.T. Wcislo, A. Wille, and E. Orozco, *Nesting biology of tropical solitary and social
     sweat bees*, Lasioglossum *(Dialictus) figueresi Wcislo and* Lasioglossum *(D.) aeneiven-
     tre (Friese)*, Insect. Soc. 40 (1993), pp. 21–40.

145. P.G. Willmer, *The role of insect water-balance in pollination ecology—*Xylocopa *and
     Calotropis*, Oecologia 76 (1988), pp. 430–438.

146. A. Sih and M.S. Baltus, *Patch size, pollinator behavior, and pollinator limitation in
     catnip*, Ecology 68 (1987), pp. 1679–1690.

147. F. Van Rossum, *Pollen dispersal and genetic variation in an early-successional forest
     herb in a peri-urban forest*, Plant Biol. 11 (2009), pp. 725–737.

148. N. Vicens and J. Bosch, *Nest site orientation and relocation of populations of the
     orchard pollinator* Osmia cornuta *(Hymenoptera: Megachilidae)*, Environ. Entomol. 29
     (2000), pp. 69–75.

149. S. Wolf and R.F.A. Moritz, *Foraging distance in* Bombus terrestris *L. (Hymenoptera:
     Apidae)*, Apidologie 39 (2008), pp. 419–427.

150. A.D. Brian, *The foraging of bumble bees. Part I. Foraging behaviour*, Bee World 35
     (1954), pp. 61–67.

151. R Development Core Team, *R: A language and environment for statistical computing*,
     2.9.0. R Foundation for Statistical Computing, eds., Vienna, Austria, 2009. Software
     available at http://www.r-project.org.

152. C. O'Toole and A. Raw, *Bees of the World*, Facts on File, New York, 1991.

153. T.L. Pitts-Singer and J. Bosch, *Nest establishment, pollination efficiency, and repro-
     ductive success of* Megachile rotundata *(Hymenoptera: Megachilidae) in relation to
     resource availability in field enclosures*, Environ. Entomol. 39 (2010), pp. 149–158.

154. T. Bradley and P. Patton, *Essential Mathematics for Economics and Business*, 2nd edn.,
     John Wiley & Sons, Chichester, UK, 2002.

11 Estimating the Potential Range Expansion and Environmental Impact of the Invasive Bee-Hawking Hornet, *Vespa velutina nigrithorax*

Claire Villemant, Franck Muller,
Quentin Rome, Adrien Perrard,
Morgane Barbet-Massin, and Frédéric Jiguet

CONTENTS

ABSTRACT

Vespa velutina nigrithorax, an Asian bee-hawking hornet, was unintentionally introduced in southwestern France before 2004 and is currently spreading widely across the country. By modeling the climatic suitability of the yellow-legged hornet at a global scale using various niche models, we estimated the potential invasion risk of this invasive species across the world, with a focus on Europe. We used eight different modeling techniques within an ensemble forecast framework to show that the invasion success in southwestern France could have been predicted using data from the native Asian range of the species. We further used data from both the native and invaded ranges (including a recently established population in Korea) to better predict the potential invasion range across all continents. Results are discussed in terms of the interest of ecological niche modeling for invasion biology, realized niche of the invasive wasp, potential threats to native entomofauna, and economic impacts of this new predator. Particular attention is paid to beekeeping activities that are nowadays already threatened by a wide panel of adversary factors. However, as far as we know, the true impact of the alien hornet on colony losses and on honey production has not yet been evaluated in France or in its area of origin. Regions at risk hold the highest densities of beehives in Europe, which could suffer from the potential predation of the putative invading hornet of both honey bees and wild pollinators. Furthermore, the impacts of *V. velutina* on pollinators must be quickly investigated, as they might well have more durable consequences than the already publicly known nuisance to apiculture.

KEYWORDS

Beekeeping, Invasive species, Native range, Niche modeling, Social Hymenoptera, Vespidae

11.1   INTRODUCTION

Honey bees are essential pollinators of crops [1,2]. In Europe, important pollinator declines have been reported, which jeopardize pollination services in agricultural ecosystems and have great economic impacts [3–6]. The recent accidental introduction of the yellow-legged Asian bee-hawking hornet, *Vespa velutina*, into Europe represents a new threat to beekeeping activities. The hornet was unintentionally introduced in southwestern France before 2004 and since then has rapidly spread, covering about half of the country by 2011. Its arrival in northern Spain was also reported in 2010 and in Portugal and Belgium in 2011 [7–9]. Being a bee-hawking predator, this species has been noticed by the French population since 2007. Considered as a dangerous stinging insect and a major threat to domestic bees, its expansion could challenge the economic viability of beekeeping and affect pollination services. Therefore, modeling the potential invasion extent of the invasive hornet appeared necessary to predict regions at risk and, hence, to help with planning future surveys of invasion range expansion and possible dedicated protection measures. The invasion progression currently observed matched with the predictions assessed using climatic suitability models [9]. We detail here the history of invasion and the biological cycle of the invasive wasp, its realized niche, and its potential range expansion in the world. We discuss the potential economic impacts of this new predator to beekeeping activities, which are nowadays already threatened by a wide panel of adversary factors [4,5,10–15].

11.2 HISTORY AND ORIGIN OF THE INVASION

The establishment of *Vespa velutina* in France represents the first successful invasion of an exotic social wasp into Europe [16]. Twenty other species have been introduced in various countries around the world, 12 of which were established in the Americas [17]. The potential for vespid wasps to gain assisted passage from humans is high as fertilized queens of many species seek sheltered locations to undergo diapause. Such shelters are often found in human goods, which may then be transported to new locations all around the world [17].

The yellow-legged hornet was recorded for the first time in Aquitaine in 2005 [18] but locally collected data suggested that hibernating founder queens could have been imported from China before 2004 through the horticultural trade [19]. This hornet has also established itself in Korea in the 2000s, where its spread remains very limited [20,21] compared to France. A key difference, however, is that Korea has six other hornet species that may compete with the alien species, whereas France has only one, the native European hornet *Vespa crabro* [9]. Moreover, one of the Korean hornets (*V. mandarinia*) is not only a fierce predator of social wasps and bees but also the most dominant when competing for food with other hornet species [22]. *Vespa crabro* is larger but has less populous colonies than *Vespa velutina*, and its nesting sites are different. While it has a similar prey spectrum, it does not mainly focus its attacks on social Hymenoptera as *Vespa velutina* does [23,24] and is considered in Europe as a mild predator of honey bees [25].

Asia is the center of evolutionary diversification of hornets [22]; 20 of the 22 known *Vespa* species are naturally restricted to Asia and Oceania and two others expanded westwards: the Oriental hornet *Vespa orientalis* reached the Mediterranean basin, where it is also known as a fear predator of honey bees [26,27], while the European hornet occupied the whole of Eurasia. So far, *Vespa crabro* is the only hornet species acclimatized in the United States after it was released in the nineteenth century to control forest caterpillar outbreaks [28]. Since then, it does not seem to have caused any particular damage [17].

Vespa velutina, which comprises 12 color variants in its natural area of distribution, is distributed from north Afghanistan to Bhutan, eastern China, Indochina, and Indonesia [29,30]. The French invasive lineage that belongs, as the Korean ones [20], to the black brown variant *nigrithorax* (Figure 11.1a), is restricted to

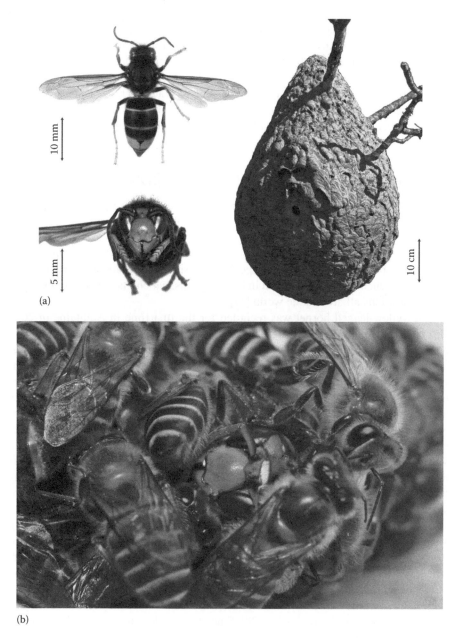

(a)

(b)

FIGURE 11.1 **(See color insert.)** (a) Adult worker and nest of *Vespa velutina nigrithorax*. (b) Heat-balling behavior of *Apis cerana* on *Vespa velutina auraria* (Nepal).

the temperate parts of this area: from West Bengal (India) and Bhutan to north-eastern China [19]. Molecular comparisons between native and invasive populations of *Vespa velutina* are in course to test the hypotheses of a Chinese origin and of the potential pathway modalities of its introduction in France. The analysis of the microsatellite allelic frequencies has evidenced a strong consanguinity among the invasive population, which may indicate a single introduction of one or more queens [31].

11.3   LIFE CYCLE

Apart from reported damages on hives, little was known about *Vespa velutina* in its native range [32–34]. Studies performed since 2007 in France greatly improved the knowledge on the biology of this species [7,9,24,35,36]. Like many social wasps, *Vespa velutina* produces annual colonies, initiated by a single queen. Each fertilized foundress builds a primary nest in spring (February–April), earlier than that of *Vespa crabro* (April–May). She takes care of its brood until the first batch of adult workers emerges and takes over the work of enlarging the nest and feeding larvae, leaving the queen to focus on egg-laying activity. Thus, the colony initiated by a single individual develops by producing up to 15,000 individuals through the season (from April to November). In autumn, a nest may include up to 2,000 workers that rear more than 1,000 future queens and drones. The sexual offspring leave the nest in late autumn, generally after the death of the mother queen, and take part in mating flights. After reproductive swarming, the colony, composed only of remaining workers, males, and brood, decline to finally pass on before midwinter. Thus, a colony never survives more than 1 year, and the only survivors are the future foundresses that seek winter shelters to hibernate. This efficient life cycle initiated by only one individual makes social insects, such as hornets, redoubtable invaders [17].

The primary nest is generally founded in a sheltered place (hive, hut, hole in a wall, roof edge, bramble, etc.); however, as for many other hornet species [22], when the environment becomes adverse or the primary site too narrow for the growing nest, the colony relocates after building a secondary nest in a more open and higher location—mainly tree tops [9,36]. Initially spheroid, the nest often becomes ovoid at the end of the season, reaching on average 40–60 cm in diameter, while the largest nests may attain about 1 m in height and 80 cm in diameter. *Vespa velutina* nests are easy to distinguish from nests of all other European social wasps by their narrow entrance always open at the lateral side of the envelope (Figure 11.1). After the death of the colony in winter, the empty nest may remain more or less undamaged in place during several weeks or months.

11.4   PREDATION BEHAVIOR

Like other hornet species, *Vespa velutina* is a generalist predator that attacks a wide range of arthropod preys. It may also tear out flesh pellets from vertebrate dead bodies, as well as from fish and shrimp in open markets [9,36]. The hornet generally attacks its prey in flight and immediately hangs on to a support, keeps the prey's thorax, which contains the nutritious flight muscles, and discards the rest. The flesh pellet is then brought to the nest to feed larvae with proteins, while adults only consume sweet liquids and an energetic protein-rich liquid regurgitated by larvae [22]. Workers transport sweet liquids (sap exuding from trunks, honeydew, nectar, flesh of ripe fruit, etc.) in their crop to feed their fellows remaining in the nest by trophallaxis, including the queen and future foundresses.

11.4.1 Prey Spectrum

Observations made in French agricultural and natural areas showed that honey bees and social wasps may represent more than two-thirds of the preys captured by a colony, while flies (mainly hover flies, flesh and carrion flies) account for most of the rest of its diet. Other arthropods only represent a small fraction of preys, but they contrariwise belong to a wide variety of insects and spiders [35,36]. Our observations in Asian countries where the yellow-legged hornet is common showed that this predator focuses its attacks on feeding sources that provide insects more or less continuously along the day, like beehives and yellow jacket nests, carcasses and herds flown by flies, or honey flowers sought by pollinators [37]. In France, for example, umbels are such honey flowers that provide, in large number, pollinator preys (notably hover flies) to hornets. The predation impact of *Vespa velutina* on local entomofauna rapidly increases during summer with the size of the colony, and it reaches its maximum in October when workers feed the sexual brood [38]. The invasive hornet impact on the diversity and biomass of the invertebrate fauna in the invaded region is still under study [38].

11.4.2 Predation on Honey Bees

In the French territory, *Vespa velutina* multiplication was likely favored by the dense and wide occurrence of the European honey bee, which is one of its main preys. However, while it represents generally more than 70% of its diet in urbanized areas with poorly diversified entomofauna, it is much less captured (about 30%) in agricultural or natural and forestry sites where many other preys are available [24,35,36].

In Asia, many hornet species hunt honey bees but only several of them focus their attacks on hives. Apart from the giant hornet *Vespa mandarinia*, which attacks hives in groups, other species including *Vespa velutina* are bee-hawking hornets that predate bees individually, waiting for returning foraging bees in front of hives [22] or catching them when foraging in the field [39]. *Vespa velutina* is considered as a fierce enemy of honey bee colonies in Kashmir, India, Nepal [40], and China [41], as well as in Korea [21], countries where several other hornet species may hover at the same time in front of hives. Nevertheless, no real quantification of the impact of these hornets is available in the literature. Ken et al. [42] reported that 20%–30% of a colony of the Eastern honey bee *Apis cerana* might succumb to the predation of *Vespa velutina* in China, while losses appeared even greater for introduced European honey bees. In the presence of bee-hawking hornets, the cessation of forager activity and the consecutive stop of honey storage in the beehive are the main factors that impact the winter survival of honey bee colonies [43].

11.4.3   HONEY BEE DEFENSE STRATEGY

Having faced their hornet predators for long, native Asian honey bees have developed efficient strategies to defend their colonies. For example, when attacked by *Vespa mandarinia* or the yellow hornet *Vespa simillima*, the Eastern bee *Apis cerana* and the dwarf bee *Apis florea* form a ball of workers around the intruder and kill it by heat stroke [41,44] (Figure 11.1b). Similarly, *Apis mellifera cypria* in Cyprus forms a ball to kill the Oriental hornet, although the underlying killing mechanism (asphyxia) is different [26]. Observations made in the field in France showed that *Apis mellifera mellifera* is able to withstand the attack of this new predator [27,43]. However, several decades after its introduction in Asia, the European honey bee became able to display the same balling and killing behavior as *Apis cerana* although with less efficiency [41,45].

11.5   CURRENT INVASION RANGE EXPANSION

Figure 11.2 shows the invasion progress of the alien hornet from 2004 to 2011. Since 2006, the monitoring of *Vespa velutina* presence in France is made by individual public warning through an online biodiversity database held by the MNHN [46]. The web page dedicated to *Vespa velutina* provides general information on the invasive hornet and the species with which it can be confused. Articles, fact sheets, and a slideshow are also downloadable [46]. Even if data are incomplete, notably for

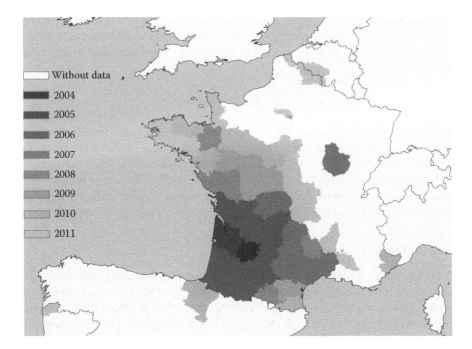

FIGURE 11.2   **(See color insert.)** Annual distribution of *Vespa velutina* in Europe.

the first 2 years of observation (in addition to the still uncompleted 2010 and 2011 recordings), it is clear that the yellow-legged hornet rapidly spread out during this short period with a range expansion at around 100 km per year [7]. A few nests have also been recorded more than 200 km away from the invasion front [7], suggesting accidental human transport or migration of foundresses as reported for *Vespa crabro* by Mulhauser and Vernier [47]. First experiments showed that *Vespa velutina* foundresses can fly about 30 km in one day [36].

The first two nests were reported by a bonsai producer in Lot-et-Garonne (southwestern French counties) in 2004 [19]. Then the number of recorded nests gradually increased to reach 1,637 in 32 counties in 2009 [7,9]. The precise number of nests recorded in 2010 and 2011 is still unknown due to ongoing verification processes of public observations. In fact, misidentifications of adults and nests, mainly with representatives of other vespid species, lead to almost 30% of wrong records. The potential extent map obtained from modeling (see thereafter) would have been strongly overvalued with these incorrect records [35].

Nevertheless in 2011, already acquired records showed that the presence of the yellow-legged hornet already covered 50 French counties, corresponding to 270,000 km^2 of invaded area [8]. Moreover, the invasive hornet has now reached neighboring countries: Spain in 2010 [48,49] and Portugal and Belgium in 2011 [8,50].

The habitat of more than 4,000 nests correctly georeferenced between 2007 and 2009 was determined from the CORINE Land Cover database [51]. A nest distribution analysis showed that about 49% were located in urban or periurban areas, 43% in farmlands, 7% in woodlands, and 1% in wetlands. However, we must also consider that public warning is not homogeneous and that record frequency also depends on the opening of the environment, the nests being more likely visible and easily located in urbanized and open territories than in forests and closed nonurbanized environments [35,36].

11.6 INVASION RISK MODELING

Impacts on domestic bees and native insect communities could locally challenge beekeeping economics and affect pollination services. Thus, modeling the potential invasion extent of the yellow-legged hornet around the world appeared necessary to predict regions at risk and, hence, to help with planning future surveys of potentially invaded areas and possible dedicated control measures, a prerequisite for replacing the reactive nature of current solutions with a proactive, predictive approach.

11.6.1 MODELS

To predict the potential invasion risk by *Vespa velutina*, we used ecological niche modeling to infer suitable distribution ranges. Combining presence data of the taxon and bioclimatic variables allows modeling the species niche, while further projecting this niche geographically provides worldwide predictions of distribution suitability. For introduced invasive species, the presence data should include locations from both native and invaded ranges, and niche modeling tools are best multiplied than combined in an ensemble forecast framework. We used eight climatic variables for the niche modeling, extracted from the BIOCLIM database as five arc-minute grids (http://www.worldclim.org/ [52]). We considered the annual mean temperature, the temperature seasonality, the maximum temperature of the warmest month, the minimum temperature of the coldest month, the annual precipitation, the precipitation of the wettest month, the precipitation of the driest month, and the precipitation seasonality (coefficient of variation). Climatic variables are supposed to be the main contributors to species niche delimitation at large scales [53], and these variables have previously been used for insect niche modeling [54]. Temperature and precipitation seasonality are computed as the standard deviations of the monthly values. The use of climatic variables only assumes that current range limits are mainly driven by climate, which is a reasonable assumption at such a continental scale.

Climatic suitability was modeled by running eight different niche-based modeling techniques using the BIOMOD platform [55]. These models are (1) generalized linear model (GLM); (2) generalized additive model (GAM); (3) classification tree analysis (CTA), a classification method running a 50-fold cross-validation to select the best trade-off between the number of leaves of the tree and the explained deviance; (4) artificial neural networks (ANNs), a machine learning method, with the mean of three runs used to provide predictions and projections, as each simulation gives slightly different results; (5) multivariate adaptive regression splines (MARS); (6) mixture discriminant analysis (MDA), a classification method that uses MARS function for the regression part of the model; (7) generalized boosting model (GBM), a machine learning method that combines a boosting algorithm and a regression tree algorithm to construct an "ensemble" of trees; and (8) random forest (RF), a machine learning method that is a combination of tree predictors such that each tree depends on the values of a random vector sampled independently and with the same distribution for all trees in the forest. In order to evaluate the predictive performance of a species distribution model, we used a random subset of 70% of the data to calibrate every model and the remaining 30% for the evaluation. Models were evaluated using a receiver operating characteristic (ROC) curve and the area under the curve (AUC; [56]). We replicated the data splitting five times and calculated the average AUC of the cross-validations, which gives a more robust estimate of the predictive performance of each model. The final calibration of every model for making predictions uses 100% of the data available.

Presence data concerned only the variant *Vespa velutina nigrithorax*, with the French and Korean localities considered as nonnative. Two biases occurred in the presence data and necessitated a dedicated treatment: (1) records from the invaded range were by far more numerous ($n \sim 1,700$) than records from the native range ($n = 69$) and (2) the species range is still expanding in France, with more presence records close to the introduction center; so we randomly draw two localities per administrative county to avoid potential overweighting of historical invaded sites. At the time of the modeling exercise (end of 2010), the hornet occurred in 39 French counties and the random selection of at most 2 records per county resulted in a subset of 69 records—the same sample size as available for the native range. The final consensus map obtained from the models included outputs from 10 random subsets of invaded locations.

11.6.2 POTENTIAL RANGE EXPANSION

The consensus map obtained from the models showed that *Vespa velutina nigrithorax* could successfully invade many other parts of the world since the scenario of introduction through international trade that occurred in France could well be repeated [9].

Results emphasized that the area of invasion is globally discriminated from most of the Asian area of origin by its higher levels of precipitations during the driest month of the year—this represents an extension of the realized niche of the native grounds. Many countries of Western Europe exhibited a high probability of being invaded with a higher risk along the Atlantic and northern Mediterranean coasts. Coastal areas of the Balkan Peninsula, Turkey, and Near East also appeared suitable and could potentially be colonized later. All these European areas could well be reached by the still expanding French introduced population. Reduced invasion risks only concern dryer European southern regions (Figure 11.3). If accidentally introduced, the invasive hornet could also acclimatize in many other parts of the world (Figure 11.4), notably regions already invaded by *Vespula germanica*, with which it shares a similar climatic niche [9]. This yellow jacket is widely distributed in Eurasia and was (as the European *Vespula vulgaris*) unintentionally but successfully introduced in many regions of the world during the past decades; as well as being widespread, these two wasps have become significant pests in most countries they have invaded [17].

FIGURE 11.3   (See color insert.) Predicted potential invasion risk of *V. v. nigrithorax* in Europe, based on ensemble forecast models using eight climatic data from WorldClim. The suitability probability is increasing from dark blue to red. (From Villemant, C. et al., *Biol. Conserv.*, 144, 2142, 2011.)

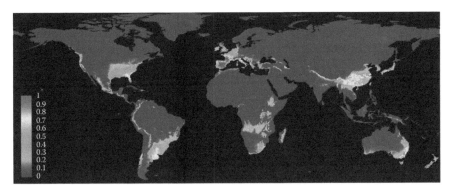

FIGURE 11.4 **(See color insert.)** Predicted potential invasion risk of *V. v. nigrithorax* in the world, based on ensemble forecast models using eight climatic data from WorldClim. The suitability probability is increasing from dark blue to red. (From Villemant, C. et al., *Biol. Conserv.*, 144, 2142, 2011.)

11.6.3 INVASION RISKS AND BEEKEEPING ACTIVITY IN EUROPE

When comparing the map of invasion risks (Figure 11.3) with the density of managed honey bee colonies in Europe and Turkey (Figure 11.5a [57,58]), we can notice that many European countries, and notably the Balkan area, combine a high level of beekeeping activity with a high risk of *Vespa velutina* acclimatization. Though ranking fourth in Europe with more than one million registered hives [59], France does not show as a whole a high level of beehive density due to the extent of its agricultural lands. However, when considering the previous ratio at the regional level (Figure 11.5b [59,60]), it appears that—in addition to the whole southwest of France—the Rhone valley, Corsica, and Alsace gather the greatest managed beehive

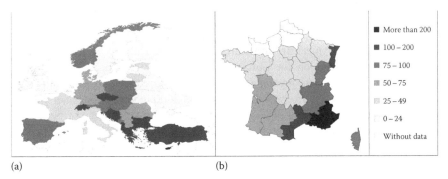

(a)   (b)

FIGURE 11.5 Ratio between the number of registered (or estimated) managed hives and the surface of agricultural land (beehive number per 10 km²) in (a) Europe and Turkey and (b) the French regions. Data (2007 for land surface, 2008 for beehives) from FAO agricultural resources [59]. For countries from which official data are lacking in FAOSTAT, estimated data are from the Apiservices database [60]. Beehive numbers in France (2004) are from the Gem-Oniflhor (2005) report [58] and land surface data from the 2000's French agricultural survey [57].

numbers with the highest risks of *Vespa velutina* arrival. Moreover in Spain, which counts the greatest number of beehives (more than 2.5 million in 2008) in Europe, recently underwent the arrival of the bee-hawking hornet in the Basque country, colonizing from the French invasive population. In 2011, the invasive hornet also reached the north of Portugal. In our models, both Basque country and northern Portugal faced a high risk of invasion. On the other hand, the potential expansion of *Vespa velutina* could be constrained by the presence of a congeneric competitor [61] in countries like Albania, Greece, and Turkey, where it would face *Vespa orientalis*, another active bee-hawking hornet [26,27]. Indeed, it seems that the success of the invasion in France, explained by the ongoing colonization of new realized climatic niches, might well have been favored by a release from congeneric competition [62].

11.7   POTENTIAL CONSEQUENCES OF THE INVASION

Due to the size and the increasing number of its colonies, as well as the duration of its period of activity, the yellow-legged hornet is able to remove a considerable biomass of arthropods, so that vulnerable prey species may become threatened [36], as observed in southern countries invaded by European yellow jackets. In New Zealand, for example, the amount of preys killed by *Vespula vulgaris* and *V. germanica* has been considered to be similar to that consumed by the entire local insectivorous bird fauna [63]. However, while consuming a large variety of preys, the ecological and economic consequences of the spread of the invasive hornet would be particularly strong through its impact on honey bees and other pollinators. Such a potential impact must however be carefully assessed by comparison of faunas of the same areas before and after invasion [64]. Being able to predict the risk of acclimatization of *Vespa velutina* in a given region will thus greatly help with planning where to survey prey populations in a before/after invasion design in order to provide insights into a wide variety of impacts associated with such invasions.

11.7.1   IMPACT ON BEEKEEPING

For several decades, American and European managed honey bees have been in decline, though information for Europe remains patchy and localized [3,5,11]. In France, colony numbers are greater today when compared to 1961 populations, but have constantly decreased after reaching a peak in 2000 [65]. A wide panel of adversary factors, such as pests and diseases, pesticides, and loss of forage or beekeeping practices, are involved in this phenomenon throughout Europe [4,12–15]. The invasion of the yellow-legged hornet adds to this long list a new protagonist that will undoubtedly focus its attack on weakened bee colonies. Its damages on honey bee colonies are well established in the native range [39,45] and could be amplified in the invaded ranges. Even if the predation pressure may not be detrimental, at least bee behavior is modified; attacked hives spend more resources to face hornet attacks than to store reserves and produce honey. This could challenge the survival of the hive, especially during the winter, but also (as a direct or indirect consequence) the activity of honey production itself. Beekeepers could abandon this activity if their hives do not produce enough honey to ensure reasonable incomes. This would represent a direct loss in free pollination services provided by the honey producers. This loss in services is associated with an economic cost, which can be estimated as the costs of funding the maintenance of the hives or of artificially pollinating crops. However, as far as we know, the true impact of the alien hornet on colony losses and honey production has not yet been evaluated in France or in its area of origin.

11.7.2 THREAT TO LOCAL ENTOMOFAUNA

If actively preyed, locally rare species may be endangered by the presence of the invasive hornet. Ongoing studies on the species prey spectrum comprise barcode referencing of the preys [38]. Indeed, visually determining the preys is very often challenging, as only insect thorax pieces are brought back to the nest. Consequently, barcoding is of great interest as a tool for better describing the prey spectrum. In addition, comparing these results to barcoded reference collections will allow an easy detection of rare or endangered species preyed by *Vespa velutina*.

In Europe, populations of social wasps should notably be impacted by *Vespa velutina*, as yellow jackets may represent a third of its prey spectrum. Moreover, the European hornet, whose colonies are three times smaller than those of *Vespa velutina* [66], may be threatened when competing for food with the huge colonies of its invasive congener. Further studies, however, are required to assess the prey overlap between the two species [36].

On the other hand, thoughtless reactions of the French public facing *Vespa velutina* invasion appeared to be more deleterious to the entomofauna than the pest problem itself. Uncontrolled mass trappings performed every year in France, in and outside the invaded area, by beekeepers and general public kill a huge number of nontarget insects [67,68], while uncontrolled nest destructions have a side effect by their greater impact on social vespid populations than in the past.

11.7.3 THREAT TO POLLINATION SERVICES

Pollinators play a key role in most wild plant communities and agroecosystems [2,5,11,69]. A recent review by Potts et al. [6] reminds us that the value of insect pollination to European agriculture is estimated to be worth ~€22 billion per year [2] with 84% of European crop varieties being dependent, at least in part, on insect pollinators [70]; wildflowers being also highly dependent on insects for their reproduction, with an estimated 78%–94% of flowering species relying on biotic pollination [71]. Furthermore, the persistence of a plant community can be affected by a loss of diversity of its pollinating fauna. Thus, the functional diversity of plant–pollinator networks may be critical for the functioning of ecosystems and should be carefully protected [72,73]. By predating various taxonomic groups of pollinators, such as Apidae and Syrphidae especially, *Vespa velutina* is already interfering with native plant–pollinator networks. These impacts must soon be investigated, as they might well have more durable consequences than the already publicly known consequences on apiculture.

Furthermore, hoverflies are not only efficient pollinators but many of them are also natural predators of aphids in their larval state and thus play an important role in reducing these agricultural pests [74].

11.7.4 Impact on Humans

Bees and social wasps have a well-known and painful sting that can occasionally cause a life-threatening allergic reaction, but death from such envenomation remains a rare event. Multiple stings that generally result from an accidental nest disturbance are even rare. Moreover, with supportive care, most nonallergic victims should be able to survive attacks from hundreds of wasps [75]. In France, there is concern from the general public when the enormous nests of *Vespa velutina* are discovered after leaf fall, often hanging from tree crowns at tens of meters above the ground [7]. As the species is common in urbanized areas, there is potential for an increase in the likelihood of humans being stung and a consequent increasing risk of allergic reactions [36]. However, the high location of most of the nests limits the risk of colony disturbance, and so far the rate of Hymenoptera stings appears not to have increased in regions colonized by the yellow-legged hornet [76].

11.8   CONTROL

When they happen to shelter in goods that get shipped, hibernating wasp foundresses have indeed a relatively low probability of being detected with standard protocols used for custom inspection [77]. On the other hand, as quoted by Thomas [78] for invasive yellow jackets in New Zealand, mass destruction of hornet founder queens in spring seems to have virtually no effect on nest density in the following summer months [17,36,63,78,79], while side effects on nontarget species may be important [68]. The best control measure is to kill off a colony by spraying permethrin inside the nest after dark, when foraging activities cease. The nest is then removed and burned. However, nests are often difficult to locate before leaf fall, when sexual progeny is already produced [36]. While it is impossible to stop or slow the natural spread of the invasive hornet, the unique possibility is to use specific baited mass traps to protect hives from very fierce hornet attacks. The development of such specific traps is still under investigation [80], while it is too late now to consider any efficient global eradication methods for this bee-hawking hornet. Efforts have now to be made to slow down the range expansion and to mitigate the impacts on domestic and wild pollinators, with priorities defined from the modeled invasion risk across Europe.

11.9   CONCLUSION

Social wasps represent a particularly destructive and successful group of invasive invertebrates. Their sociality, including a number of unique mechanisms enhancing survival and reproduction, is certainly the key factor that substantially contributes to their success as invaders [81]. Their main general advantage may be the flexibility arising from having both individual and colony responses. This confers upon them a remarkable efficiency to compete and exploit food sources, buffering against environmental changes. They develop populous colonies with an effective predator defense and produce numerous foundresses that provide large dispersal capacity [81,82].

Given the potential economic and biological impacts of the invasive yellow-legged hornet, ongoing developments are focusing on better understanding the invasion dynamics by developing modeling approaches to predict the rate of range expansion, which is necessary for a temporal risk assessment and the further implementation of effective management strategies [83,84].

Furthermore, it would be of great interest to investigate the perturbations *Vespa velutina* might engender to the complete pollinator community. Theoretical approaches could be used, modeling theoretical mutualistic networks of plants and pollinators and applying an orientated change in the structure of the plant–pollinator network (mimicking depredation by the hornet on some of the pollinator agents). Results would help to understand how these changes might modify stability properties and architecture of the whole mutualistic network.

REFERENCES

1. E.E. Southwick and L. Southwick JR., *Estimating the economic value of honey bees (Hymenoptera: Apidae) as agricultural pollinators in the United States*, J. Econ. Entomol. 85 (1992), pp. 621–633.
2. N. Gallai, J.-M. Salles, J. Settele, and B.E. Vaissière, *Economic valuation of the vulnerability of world agriculture confronted with pollinator decline*, Ecol. Econ. 68 (2009), pp. 810–821.
3. J.C. Biesmeijer, S.P.M. Roberts, M. Reemer, R. Ohlemüller, M. Edwards, T. Peeters, A.P. Schaffers, S.G. Potts, R. Kleukers, C.D. Thomas, J. Settele, and W.E. Kunin, *Parallel declines in pollinators and insect-pollinated plants in Britain and the Netherlands*, Science 313 (2006), pp. 351–354.
4. N. Carreck and P. Neumann, *Honey bee colony losses*, J. Apicult. Res. 49 (2010), pp. 1–6.
5. S. Potts, S. Roberts, R. Dean, G. Marris, M. Brown, R. Jones, P. Neumann, and J. Settele, *Declines of managed honey bees and beekeepers in Europe*, J. Apicult. Res. 49 (2010), pp. 15–22.
6. S.G. Potts, J.C. Biesmeijer, R. Bommarco, A. Felicioli, M. Fischer, P. Jokinen, D. Kleijn, A.-M. Klein, W.E. Kunin, P. Neumann, L.D. Penev, T. Petanidou, P. Rasmont, S.P.M. Roberts, H.G. Smith, P.B. Sørensen, I. Steffan-Dewenter, B.E. Vaissière, M. Vilà, and A. Vujić, *Developing European conservation and mitigation tools for pollination services: Approaches of the STEP (status and trends of European pollinators) project*, J. Apicult. Res. 50 (2011), pp. 152–164.
7. Q. Rome, F. Muller, O. Gargominy, and C. Villemant, *Bilan 2008 de l'invasion de* Vespa velutina *Lepeletier en France (Hymenoptera: Vespidae)*, B. Soc. Entomol. Fr. 114 (2009), pp. 297–302.
8. Q. Rome, F. Muller, and C. Villemant, *Expansion 2011 de* Vespa velutina Lepeletier *(Hymenoptera, Vespidae) en Europe*, B. Soc. Entomol. Fr. 117 (2012), p. 114.
9. C. Villemant, M. Barbet-Massin, A. Perrard, F. Muller, O. Gargominy, F. Jiguet, and Q. Rome, *Predicting the invasion risk by the alien bee-hawking yellow-legged hornet* Vespa velutina nigrithorax *across Europe and other continents with niche models*, Biol. Conserv. 144 (2011), pp. 2142–2150.
10. P. De la Rúa, R. Jaffé, R. Dall'Olio, I. Muñoz, and J. Serrano, *Biodiversity, conservation and current threats to European honeybees*, Apidologie 40 (2009), pp. 263–284.
11. S.G. Potts, J.C. Biesmeijer, C. Kremen, P. Neumann, O. Schweiger, and W.E. Kunin, *Global pollinator declines: Trends, impacts and drivers*, Trends Ecol. Evol. 25 (2010), pp. 345–353.

12. N. Bacandritsos, A. Granato, G. Budge, I. Papanastasiou, E. Roinioti, M. Caldon, C. Falcaro, A. Gallina, and F. Mutinelli, *Sudden deaths and colony population decline in Greek honey bee colonies*, J. Invertebr. Pathol. 105 (2010), pp. 335–340.

13. J.D. Ellis, J.D. Evans, and J. Pettis, *Colony losses, managed colony population decline, and colony collapse disorder in the United States*, J. Apicult. Res. 49 (2010), pp. 134–136.

14. E. Genersch, W. von der Ohe, H. Kaatz, A. Schroeder, C. Otten, R. Büchler, S. Berg, W. Ritter, W. Mühlen, S. Gisder, M. Meixner, G. Liebig, and P. Rosenkranz, *The German bee monitoring project: A long term study to understand periodically high winter losses of honey bee colonies*, Apidologie 41 (2010), pp. 332–352.

15. M. Higes, R. Martín-Hernández, A. Martínez-Salvador, E. Garrido-Bailón, A.V. González-Porto, A. Meana, J.L. Bernal, M.J. Del Nozal, and J. Bernal, *A preliminary study of the epidemiological factors related to honey bee colony loss in Spain*, Environ. Microbiol. Reports 2 (2010), pp. 243–250.

16. J.-Y. Rasplus, C. Villemant, M.R. Paiva, G. Delvare, and A. Roques, *Hymenoptera*, in *Arthropod Invasions in Europe*, A. Roques, M. Kenis, D. Lees, C. Lopez-Vaamonde, W. Rabitsch, J.-Y. Rasplus, D. Roy, eds., BioRisk 4(2) (2010), pp. 669–776.

17. J.R. Beggs, E.G. Brockerhoff, J.C. Corley, M. Kenis, M. Masciocchi, F. Muller, Q. Rome, and C. Villemant, *Ecological effects and management of invasive alien Vespidae*, BioControl 56 (2011), pp. 505–526.

18. J. Haxaire, J.-P. Bouguet, and J.-P. Tamisier, Vespa velutina *Lepeletier, 1836, une redoutable nouveauté pour la faune de France (Hym, Vespidae)*, B. Soc. Entomol. Fr. 111 (2006), p. 194.

19. C. Villemant, J. Haxaire, and J. Streito, *Premier bilan de l'invasion de* Vespa velutina *Lepeletier en France (Hymenoptera, Vespidae)*, B. Soc. Entomol. Fr. 111 (2006), p. 535.

20. J.-K. Kim, M. Choi, and T.-Y. Moon, *Occurrence of* Vespa velutina *Lepeletier from Korea, and a revised key for Korean* Vespa *species (Hymenoptera: Vespidae)*, Entomol. Res. 36 (2006), p. 112.

21. C. Jung, D.-W. Kim, H.-S. Lee, and H. Baek, *Some biological characteristics of a new honeybee pest,* Vespa velutina nigrithorax *Buysson 1905 (Hymenoptera: Vespidae)*, Korean J. Apicult. 24 (2008), pp. 61–65.

22. M. Matsuura and S. Yamane, *Biology of the Vespine Wasps*, Springer-Verlag, Berlin, Germany, 1990.

23. J.P. Spradbery, *Wasps: An Account of the Biology and Natural History of Solitary and Social Wasps with Particular Reference to Those of the British Isles*, Sidgwick & Jackson, London, UK, 1973.

24. A. Perrard, J. Haxaire, A. Rortais, and C. Villemant, *Observations on the colony activity of the Asian Hornet* Vespa velutina *Lepeletier 1836 (Hymenoptera: Vespidae: Vespinae) in France*, Ann. Soc. Entomol. Fr. 45 (2009), p. 10.

25. D. Baracchi, G. Cusseau, D. Pradella, and S. Turillazzi, *Defence reactions of* Apis mellifera ligustica *against attacks from the European hornet* Vespa crabro, Ethol. Ecol. Evol. 22 (2010), pp. 281–294.

26. A. Papachristoforou, A. Rortais, G. Zafeiridou, G. Theophilidis, L. Garnery, A. Thrasyvoulou, and G. Arnold, *Smothered to death: Hornets asphyxiated by honeybees*, Curr. Biol. 17 (2007), pp. R795–R796.

27. A. Rortais, C. Villemant, O. Gargominy, Q. Rome, J. Haxaire, A. Papachristoforou, and G. Arnold, *A new enemy of honeybees in Europe: The Asian hornet* Vespa velutina, in *Atlas of Biodiversity Risks—From Europe to the Globe, from Stories to Maps*, J. Settele, L.D. Penev, T.A. Georgiev, R. Grabaum, V. Grobelnik, V. Hammen, S. Klotz, M. Kotarac, and I. Kuehn, eds., Pensoft, Sofia, Moscow, Russia, 2010, p. 181.

28. F.R. Shaw and J. Weidhaas JR., *Distribution and habits of the giant hornet in North America*, J. Econ. Entomol. 49 (1956), p. 275.

29. J.M. Carpenter and J. Kojima, *Checklist of the species in the subfamily Vespinae (Insecta: Hymenoptera: Vespidae)*, Nat. Hist. B. Ibaraki Univ. 1 (1997), pp. 51–92.

30. L.T. Nguyen, F. Saito, J. Kojima, and J.M. Carpenter, *Vespidae of Vietnam (Insecta: Hymenoptera) 2 Taxonomic notes on Vespinae*, Zool. Sci. 23 (2006), pp. 95–104.

31. M. Arca, C. Capdevielle-Dulac, C. Nadeau, C. Villemant, G. Arnold, and J.-F. Silvain, *Genetic Characterization of the Invasive Populations of* Vespa velutina *in France*, Apimondia, Montpellier, France, 2009.

32. M. Matsuura, *Nesting habit of several species of the genus* Vespa *in Formosona*, Kontyu 41 (1973), pp. 286–293.

33. S.J. Martin, *Hornets (Hymenoptera: Vespinae) of Malaysia*, Malayan Nat. J. (1995), p. 71.

34. M. Nakamura and S. Sonthichai, *Nesting habits of some hornet species (Hymenoptera, Vespidae) in Northern Thailand*, Kasetsart J. Nat. Sci. 38 (2004), p. 196.

35. Q. Rome, A. Perrard, F. Muller, and C. Villemant, *Monitoring and control modalities of a honeybee predator, the yellow-legged hornet* Vespa velutina nigrithorax *(Hymenoptera: Vespidae)*, Aliens 31 (2011), pp. 7–15.

36. C. Villemant, F. Muller, S. Haubois, A. Perrard, E. Darrouzet, and Q. Rome, *Bilan des travaux (MNHN et IRBI) sur l'invasion en France de* Vespa velutina, *le frelon asiatique prédateur d'abeilles*, in *Journée Scientifique Apicole JSA*, J.-M. Barbançon and M. L'Hostis, eds., ONIRIS-FNOSAD, Nantes, France, 2011, pp. 3–12.

37. A. Perrard, F. Muller, Q. Rome, and C. Villemant, *Observations sur le Frelon asiatique à pattes jaunes*, Vespa velutina Lepeletier, 1836 (Hymenoptera, Vespidae), B. Soc. Entomol. Fr. 116 (2011), pp. 159–164.

38. F. Muller, Q. Rome, A. Perrard, and C. Villemant, *Potential influence of habitat type and seasonal variations on prey spectrum of* Vespa velutina, *the Asian hornet, in Europe*, Apimondia, Montpellier, France, 2009.

39. D.P. Abrol, *Ecology, behaviour and management of social wasp,* Vespa velutina *Smith (Hymenoptera: Vespidae), attacking honeybee colonies*, Korean J. Apicult. 9 (1994), p. 5.

40. N.B. Ranabhat and A.S. Tamrakar, *Study on seasonal activity of predatory wasps attacking honeybee* Apis cerana Fab. *colonies in southern belt of Kaski district, Nepal*, J. Nat. Hist. Museum 23 (2009), pp. 125–128.

41. K. Tan, S.E. Radloff, J.J. Li, H.R. Hepburn, M.X. Yang, L.J. Zhang, and P. Neumann, *Bee-hawking by the wasp,* Vespa velutina, *on the honeybees* Apis cerana *and* A. mellifera, Naturwissenschaften 94 (2007), pp. 469–72.

42. T. Ken, H.R. Hepburn, S.E. Radloff, Y. Yusheng, L. Yiqiu, Z. Danyin, and P. Neumann, *Heat-balling wasps by honeybees*, Naturwissenschaften 92 (2005), pp. 492–495.

43. M. Arca, A. Papachristoforou, N. Maher, G. Arnold, and A. Rortais, *Defensive behaviour of* Apis mellifera *against the invasive Asian hornet (*Vespa velutina*) in south south-west of France*, Apimondia Workshop, Montpellier, France, 2009.

44. M. Ono, T. Igarashi, E. Ohno, and M. Sasaki, *Unusual thermal defense by a honeybee against mass attack by hornets*, Nature 377 (1995), pp. 334–336.

45. D.P. Abrol, *Defensive behaviour of* Apis cerana f *against predatory wasps*, J. Apicult. Sci. 50 (2006), p. 39.

46. MNHN, Vespa velutina Lepeletier, 1836, in *INPN, Inventaire national du Patnimoine naturel*, Muséum national d'Histoire naturelle ed., 2003–2012. Available at http://inpn.mnhn.fr/espece/cd_nom/433589

47. B. Mulhauser and R. Vernier, *Une migration groupée de fondatrices de frelon,* Vespa crabro L *(Hymenoptera, Vespidae)*, Mitt. Schweiz. Entomol. G. 67 (1994), pp. 61–65.

48. L. Castro and S. Pagola-Carte, Vespa velutina Lepeletier (Hymenoptera: Vespidae), *recolectada en la Peninsula Ibérica*, Heteropterus, Rev. Entomol. 10 (2010), pp. 193–196.

49. S. López, M. González, and A. Goldarazena, Vespa velutina *Lepeletier, 1836 (Hymenoptera: Vespidae): First records in Iberian Peninsula*, EPPO Bull. 41 (2011), pp. 439–441.

50. J.M. Grosso-Silva and M. Maia, Vespa velutina *Lepeletier, 1836 (Hymenoptera, Vespidae), New Species for Portugal*, Arquivos Entomolóxicos 6 (2012), pp. 53–54.

51. European Union, SOeS, *CORINE land cover*, 2006. Available at http://www.statistiques. developpement-durable.gouv.fr/donnes-ligne/li/1825.html

52. R.J. Hijmans, S.E. Cameron, J.L. Parra, P.G. Jones, and A. Jarvis, *Very high resolution interpolated climate surfaces for global land areas*, Int. J. Climatol. 25 (2005), pp. 1965–1978.

53. M. Luoto, R. Virkkala, and R.K. Heikkinen, *The role of land cover in bioclimatic models depends on spatial resolution*, Global Ecol. Biogeogr. 16 (2007), pp. 34–42.

54. K.A. Medley, *Niche shifts during the global invasion of the Asian tiger mosquito,* Aedes albopictus *Skuse (Culicidae), revealed by reciprocal distribution models*, Global Ecol. Biogeogr. 19 (2010), pp. 122–133.

55. W. Thuiller, B. Lafourcade, R. Engler, and M.B. Araújo, *BIOMOD—A platform for ensemble forecasting of species distributions*, Ecography 32 (2009), pp. 369–373.

56. A.H. Fielding and J.F. Bell, *A review of methods for the assessment of prediction errors in conservation presence/absence models*, Environ. Conserv. 24 (1997), pp. 38–49.

57. Agreste, *Recensement agricole 2000*, 2001. Available at http://www.agreste.agriculture. gouv.fr/spip.php?page=de_research&id_subsique-464&

58. GEM, Oniflhor, *Audit de la Filière Miel 2004*, 2005. Available at www.itsap.asso.fr/ downloads/publications/filiere_apicole2004.pdf.

59. FAO, *FAOSTAT, Production, live animals*, 2010, 2011. Available at http://faostat.fao. org/site/573/DesktopDefault.aspx?PageID=573#ancor

60. Apiservices, *France apicole 1994 par régions*, in Apiservices Virtual Beekeeping Gallery, 2012. Available at http://www.beekeeping.com/countries/france_regions.htm

61. W.E. Snyder and E.W. Evans, *Ecological effects of invasive arthropod generalist predators*, Annu. Rev. Ecol. Evol. Syst. 37 (2006), pp. 95–122.

62. D.I. Bolnick, T. Ingram, W.E. Stutz, L.K. Snowberg, O.L. Lau, and J.S. Paull, *Ecological release from interspecific competition leads to decoupled changes in population and individual niche width*, Proc. Royal Soc. B Biol. 277 (2010), p. 1789.

63. R.J. Harris, *Diet of the Wasps* Vespula vulgaris *and* V. germanica *in honeydew beech forest of the South Island*, N. Z. J. Ecol. 18 (1991), pp. 159–170.

64. D.A. Holway, L. Lach, A.V. Suarez, N.D. Tsutsui, and T.J. Case, *The causes and consequences of ant invasions*, Annu. Rev. Ecol. Syst. 33 (2002), pp. 181–233.

65. D. Van Engelsdorp and M.D. Meixner, *A historical review of managed honey bee populations in Europe and the United States and the factors that may affect them*, J. Invertebr. Pathol. 103 (2010), pp. S80–S95.

66. M.E. Archer, *Taxonomy, distribution and nesting biology of species of the genera* Provespa *Ashmead and* Vespa *Linneaus (Hymenoptera, Vespidae)*, Entomol. Month. Mag. 144 (2008), pp. 69–101.

67. P. Dauphin and H. Thomas, *Quelques données sur le contenu des "pièges à frelons asiatiques" posés à Bordeaux (Gironde) en 2009*, B. Soc. Linn. Bordeaux 144 (2009), pp. 287–297.

68. Q. Rome, F. Muller, T. Théry, J. Andrivot, S. Haubois, E. Rosenstiehl, and C. Villemant, *Impact sur l'entomofaune des pièges à bière ou à jus de cirier utilisés dans la lutte contre le frelon asiatique*, in *Journée Scientifique Apicole JSA*, J.-M. Barbançon and M. L'Hostis, eds., ONIRIS-FNOSAD, Nantes, France, 2011, pp. 3–12.

69. A.-M. Klein, B.E. Vaissière, J.H. Cane, I. Steffan-Dewenter, S.A. Cunningham, C. Kremen, and T. Tscharntke, *Importance of pollinators in changing landscapes for world crops*, Proc. Royal Soc. B Biol. 274 (2007), pp. 303–313.

70. I. Williams, *The dependence of crop production within the European Union on pollination by honey bees*, Agricult. Zool. Rev. 6 (1994), pp. 229–257.

71. J. Ollerton, R. Winfree, and S. Tarrant, *How many flowering plants are pollinated by animals?* Oikos 120 (2011), pp. 321–326.

72. C. Fontaine, I. Dajoz, J. Meriguet, and M. Loreau, *Functional diversity of plant–pollinator interaction webs enhances the persistence of plant communities*, PLoS Biol. 4 (2005), pp. 129–135.

73. E. Thébault and C. Fontaine, *Stability of ecological communities and the architecture of mutualistic and trophic networks*, Science 329 (2010), pp. 853–856.

74. E. Branquart and J. Hemptinne, *Selectivity in the exploitation of floral resources by hoverflies (Diptera: Syrphinae)*, Ecography 23 (2000), pp. 732–742.

75. R.S. Vetter, P.K. Visscher, and S. Camazine, *Mass envenomations by honey bees and wasps*, West. J. Med. 170 (1999), pp. 223–227.

76. L. de Haro, M. Labadie, P. Chanseau, C. Cabot, I. Blanc-Brisset, and F. Penouil, *Medical consequences of the Asian black hornet (*Vespa velutina*) invasion in southwestern France*, Toxicon 55 (2010), pp. 650–652.

77. M. Kenis, M.-A. Auger-Rozenberg, R. Roques, L. Timms, C. Péré, J.W. Cock, J. Settele, S. Augustin, and C. Lopez-Vaamonde, *Ecological effects of invasive alien insects*, Biol. Invasions 11 (2009), pp. 21–45.

78. C.R. Thomas, *The European wasp (*Vespula germanica *Fab) in New Zealand*, N. Z. Dep. Sci. Ind. Res. Inform. Ser. 27 (1960), pp. 1–74.

79. R.J. Harris, *Frequency of overwintered* Vespula germanica *(Hymenoptera: Vespidae) colonies in scrubland-pasture habitat and their impact on prey*, N. Z. J. Zool. 23 (1996), pp. 11–17.

80. N. Maher and D. Thiéry, *Comparison of trap designs against the yellow-legged hornet (*Vespa velutina*)*, Apimondia Workshop, Montpellier, France, 2009.

81. R.F.A. Moritz, S. Härtel, and P. Neumann, *Global invasions of the western honeybee (*Apis mellifera*) and the consequences for biodiversity*, Ecoscience 12 (2005), pp. 289–301.

82. H. Moller, *Lessons for invasion theory from social insects*, Biol. Conserv. 78 (1996), pp. 125–142.

83. A. Hastings, K. Cuddington, K.F. Davies, C.J. Dugaw, S. Elmendorf, A. Freestone, S. Harrison, M. Holland, J. Lambrinos, U. Malvadkar, B.A. Melbourne, K. Moore, C. Taylor, and D. Thomson, *The spatial spread of invasions: New developments in theory and evidence*, Ecol. Lett. 8 (2005), pp. 91–101.

84. T. Kadoya, H.S. Ishii, R. Kikuchi, S.-I. Suda, and W.I. Washitani, *Using monitoring data gathered by volunteers to predict the potential distribution of the invasive alien bumblebee* Bombus terrestris, Biol. Conserv. 142 (2009), pp. 1011–1017.

Index

T - #0367 - 101024 - C8 - 234/156/17 - PB - 9781138374706 - Gloss Lamination